Methods of Applied Mathematics

FRANCIS B. HILDEBRAND

Professor of Mathematics, Emeritus
Massachusetts Institute of Technology

SECOND EDITION

DOVER PUBLICATIONS, INC.
New York

This Dover edition, first published in 1992, is an unabridged and slightly corrected republication of the second edition (1965) of the work first published in 1952 by Prentice-Hall, Inc., Englewood Cliffs, N.J.

Manufactured in the United States of America
Dover Publications, Inc., 31 East 2nd Street, Mineola, N.Y. 11501

Library of Congress Cataloging-in-Publication Data

Hildebrand, Francis Begnaud.
Methods of applied mathematics / Francis B. Hildebrand. — 2nd ed.
p. cm.
"This . . . is an unabridged and slightly corrected republication of the 2nd ed. (1965) of the work first pub. in 1952 by Prentice-Hall"—T.p. verso.
Includes index.
ISBN 0-486-67002-3 (pbk.)
1. Mathematics. I. Title.
QA37.2.H5 1992
510—dc20 92-64
CIP

Preface

The principal aim of this volume is to place at the disposal of the engineer or physicist the basis of an intelligent working knowledge of a number of facts and techniques relevant to some fields of mathematics which often are not treated in courses of the "Advanced Calculus" type, but which are useful in varied fields of application.

Many students in the fields of application have neither the time nor the inclination for the study of detailed treatments of each of these topics from the classical point of view. However, efficient use of facts or techniques depends strongly upon a substantial understanding of the basic underlying principles. For this reason, care has been taken throughout the text either to provide a rigorous proof, when the proof is believed to contribute to an understanding of the desired result, or to state the result as precisely as possible and indicate why it might have been *formally* anticipated.

In each chapter, the treatment consists of showing how typical problems may arise, of establishing those parts of the relevant theory which are of principal practical significance, and of developing techniques for analytical and numerical analysis and problem solving.

Whereas experience gained from a course on the Advanced Calculus level is presumed, the treatments are largely self-contained, so that the nature of this preliminary course is not of great importance.

In order to increase the usefulness of the volume as a basic or supplementary text, and as a reference volume, an attempt has been made to organize the material so that there is little essential interdependence among the chapters, and considerable flexibility exists with regard to the omission of topics within chapters. In addition, a substantial amount of supplementary material is included in annotated problems which complement numerous exercises, of varying difficulty, arranged in correspondence with successive sections of the text at the end of the chapters. Answers to all problems are either incorporated into their statement or listed at the end of the book.

The first chapter deals principally with *linear algebraic equations, quadratic and Hermitian forms*, and operations with *vectors and matrices*, with special emphasis on the concept of characteristic values. A brief summary of corresponding results in *function space* is included for comparison, and for convenient reference. Whereas a considerable amount of material is presented, particular care was taken here to arrange the demonstrations in such a way that maximum flexibility in selection of topics is present.

The first portion of the second chapter introduces the variational notation and derives the Euler equations relevant to a large class of problems in the *calculus of variations*. More than usual emphasis is placed on the significance of natural boundary conditions. Generalized coordinates, Hamilton's principle, and Lagrange's equations are treated and illustrated within the framework of this theory. The chapter concludes with a discussion of the formulation of minimal principles of more general type, and with the application of direct and semidirect methods of the calculus of variations to the exact and approximate solution of practical problems.

The concluding chapter deals with the formulation and theory of linear *integral equations*, and with exact and approximate methods for obtaining their solutions, particular emphasis being placed on the several equivalent interpretations of the relevant Green's function. Considerable supplementary material is provided here in annotated problems.

The present text is a revision of corresponding chapters of the first edition, published in 1952. It incorporates a number of changes in method of presentation and in notation, as well as some new material and additional problems and exercises. A revised and expanded version of the earlier material on difference equations and on finite difference methods is to appear separately.

Many compromises between mathematical elegance and practical significance were found to be necessary. However, it is hoped that the text will serve to ease the way of the engineer or physicist into the more advanced areas of applicable mathematics, for which his need continues to increase, without obscuring from him the existence of certain *difficulties*, sometimes implied by the phrase "It can be shown," and without failing to warn him of certain *dangers* involved in formal application of techniques beyond the limits inside which their validity has been well established.

The author is indebted to colleagues and students in various fields for help in selecting and revising the content and presentation, and particularly to Professor Albert A. Bennett for many valuable criticisms and suggestions.

FRANCIS B. HILDEBRAND

Contents

CHAPTER ONE

Matrices and Linear Equations 1

CHAPTER TWO

Calculus of Variations and Applications 119

CHAPTER THREE

Integral Equations 222

Methods of Applied Mathematics

CHAPTER ONE

Matrices and Linear Equations

1.1. Introduction. In many fields of analysis we find it necessary to deal with an *ordered set* of elements, which may be numbers or functions. In particular, we may deal with an ordinary *sequence* of the form

$$a_1, a_2, \cdots, a_n$$

or with a two-dimensional *array* such as the rectangular arrangement

$$\begin{matrix} a_{11}, & a_{12}, & \cdots, & a_{1n} \\ a_{21}, & a_{22}, & \cdots, & a_{2n} \\ \cdots & \cdots & \cdots & \cdots \\ a_{m1}, & a_{m2}, & \cdots, & a_{mn}, \end{matrix}$$

consisting of m rows and n columns.

When suitable laws of equality, addition and subtraction, and multiplication are associated with sets of such rectangular arrays, the arrays are called *matrices*, and are then designated by a special symbolism. The laws of combination are specified in such a way that the matrices so defined are of frequent usefulness in both practical and theoretical considerations.

Since matrices are perhaps most intimately associated with sets of *linear algebraic equations*, it is desirable to investigate the general nature of the solutions of such sets of equations by elementary methods, and hence to provide a basis for certain definitions and investigations which follow.

1.2. Linear equations. The Gauss-Jordan reduction. We deal first with the problem of attempting to obtain solutions of a set of m linear equations in n unknown variables $x_1, x_2, \cdots, x_n$, of the form

$$\left.\begin{array}{l} a_{11}x_1 + a_{12}x_2 + \cdots + a_{1n}x_n = c_1, \\ a_{21}x_1 + a_{22}x_2 + \cdots + a_{2n}x_n = c_2, \\ \cdots\cdots\cdots\cdots\cdots\cdots\cdots\cdots \\ a_{m1}x_1 + a_{m2}x_2 + \cdots + a_{mn}x_n = c_m \end{array}\right\}, \tag{1}$$

by direct calculation.

Under the assumption that (1) does indeed possess a solution, the *Gauss-Jordan reduction* proceeds as follows:

First Step. Suppose that $a_{11} \neq 0$. (Otherwise, renumber the equations or variables so that this is so.) Divide both sides of the first equation by a_{11}, so that the resultant equivalent equation is of the form

$$x_1 + a'_{12}x_2 + \cdots + a'_{1n}x_n = c'_1. \tag{2}$$

Multiply both sides of (2) successively by $a_{21}, a_{31}, \cdots, a_{m1}$, and subtract the respective resultant equations from the second, third, $\cdots$, mth equations of (1), to reduce (1) to the form

$$\left.\begin{array}{l} x_1 + a'_{12}x_2 + \cdots + a'_{1n}x_n = c'_1, \\ \qquad a'_{22}x_2 + \cdots + a'_{2n}x_n = c'_2, \\ \qquad \cdots\cdots\cdots\cdots\cdots\cdots \\ \qquad a'_{m2}x_2 + \cdots + a'_{mn}x_n = c'_m \end{array}\right\}. \tag{3}$$

Second Step. Suppose that $a'_{22} \neq 0$. (Otherwise, renumber the equations or variables so that this is so.) Divide both sides of the second equation in (3) by a'_{22}, so that this equation takes the form

$$x_2 + a''_{23}x_3 + \cdots + a''_{2n}x_n = c''_2, \tag{4}$$

and use this equation, as in the first step, to eliminate the coefficient of x_2 in *all other equations* in (3), so that the set of equations becomes

$$\left.\begin{array}{l} x_1 \quad + a''_{13}x_3 + \cdots + a''_{1n}x_n = c''_1, \\ \quad x_2 + a''_{23}x_3 + \cdots + a''_{2n}x_n = c''_2, \\ \qquad a''_{33}x_3 + \cdots + a''_{3n}x_n = c''_3, \\ \qquad \cdots\cdots\cdots\cdots\cdots\cdots \\ \qquad a''_{m3}x_3 + \cdots + a''_{mn}x_n = c''_m \end{array}\right\}. \tag{5}$$

Remaining Steps. Continue the above process r times until it terminates, that is, until $r = m$ *or* until the coefficients of all x's are *zero* in all equations following the rth equation. We shall speak of these $m - r$ equations as the *residual equations*, when $m > r$.

There then exist two alternatives. *First*, it may happen that, with $m > r$, one or more of the residual equations has a nonzero right-hand member, and hence is of the form $0 = c_k^{(r)}$ (where in fact $c_k^{(r)} \neq 0$). In this case, the assumption that a solution of (1) exists leads to a *contradiction*, and hence *no solution exists*. The set (1) is then said to be *inconsistent* or *incompatible*.

Otherwise, no contradiction exists, and the set (1) of m equations is reduced to a set of r equations which, after a transposition, can be written in the form

$$\left.\begin{aligned} x_1 &= \gamma_1 + \alpha_{11}x_{r+1} + \cdots + \alpha_{1,n-r}x_n, \\ x_2 &= \gamma_2 + \alpha_{21}x_{r+1} + \cdots + \alpha_{2,n-r}x_n, \\ &\cdots\cdots\cdots\cdots\cdots\cdots\cdots\cdots \\ x_r &= \gamma_r + \alpha_{r1}x_{r+1} + \cdots + \alpha_{r,n-r}x_n \end{aligned}\right\} . \tag{6}$$

where the γ's and α's are specific constants related to the coefficients in (1). Since each of the steps in the reduction of the set (1) to the set (6) is *reversible*, it follows that the two sets are *equivalent*, in the sense that each set implies the other. Hence, in this case the most general solution of (1) expresses each of the r variables $x_1, x_2, \cdots, x_r$ as a specific constant plus a specific linear combination of the remaining $n - r$ variables, each of which can be assigned *arbitrarily*.

If $r = n$, a *unique* solution is obtained. Otherwise, we say that an *$(n - r)$-parameter family* of solutions exists. The number $n - r = d$ may be called the *defect* of the system (1). We notice that if the system (1) is consistent and r is less than m, then $m - r$ of the equations (namely, those which correspond to the residual equations) are actually ignorable, and hence must be implied by the remaining r equations.

The reduction may be illustrated by considering the four simultaneous equations

$$\left.\begin{aligned} x_1 + 2x_2 - x_3 - 2x_4 &= -1, \\ 2x_1 + x_2 + x_3 - x_4 &= 4, \\ x_1 - x_2 + 2x_3 + x_4 &= 5, \\ x_1 + 3x_2 - 2x_3 - 3x_4 &= -3 \end{aligned}\right\} . \tag{7}$$

It is easily verified that after two steps in the reduction one obtains the equivalent set

$$\left.\begin{aligned} x_1 + x_3 \qquad &= 3, \\ x_2 - x_3 - x_4 &= -2, \\ 0 &= 0, \\ 0 &= 0 \end{aligned}\right\} .$$

Hence the system is of defect *two*. If we write $x_3 = C_1$ and $x_4 = C_2$, it follows that the general solution can be expressed in the form

$$x_1 = 3 - C_1, \quad x_2 = -2 + C_1 + C_2, \quad x_3 = C_1, \quad x_4 = C_2, \tag{8a}$$

where C_1 and C_2 are arbitrary constants. This two-parameter family of solutions can also be written in the symbolic form

$$\{x_1, x_2, x_3, x_4\} = \{3, -2, 0, 0\} + C_1\{-1, 1, 1, 0\} + C_2\{0, 1, 0, 1\}. \tag{8b}$$

It follows also that the third and fourth equations of (7) must be consequences of the first two equations. Indeed, the third equation is obtained by subtracting the first from the second, and the fourth by subtracting one-third of the second from five-thirds of the first.

The Gauss-Jordan reduction is useful in actually obtaining numerical solutions of sets of linear equations,* and it has been presented here also for the purpose of motivating certain definitions and terminologies which follow.

1.3. Matrices. The set of equations (1) can be visualized as representing a *linear transformation* in which the set of n numbers $\{x_1, x_2, \cdots, x_n\}$ is transformed into the set of m numbers $\{c_1, c_2, \cdots, c_m\}$.

The rectangular array of the coefficients a_{ij} specifies the transformation. Such an array is often enclosed in square brackets and denoted by a single boldface capital letter,

$$\mathbf{A} \equiv [a_{ij}] \equiv \begin{bmatrix} a_{11} & a_{12} & \cdots & a_{1n} \\ a_{21} & a_{22} & \cdots & a_{2n} \\ \cdots & \cdots & \cdots & \cdots \\ a_{m1} & a_{m2} & \cdots & a_{mn} \end{bmatrix}, \tag{9}$$

and is called an $m \times n$ *matrix* when certain laws of combination, yet to be specified, are laid down. In the symbol a_{ij}, representing a typical element, the *first* subscript (here i) denotes the row and the *second* subscript (here j) the column occupied by the element.

* In place of eliminating x_k from *all* equations except the kth, in the kth step, one may eliminate x_k only in those equations *following* the kth equation. When the process terminates, after r steps, the rth unknown is given explicitly by the rth equation. The $(r - 1)$th unknown is then determined by substitution in the $(r - 1)$th equation, and the solution is completed by working back in this way to the first equation. The method just outlined is associated with the name of *Gauss*. In order that the "round-off" errors be as small as possible, it is usually desirable that the sequence of eliminations be ordered such that the coefficient of x_k in the equation used to eliminate x_k is as large as possible in absolute value, relative to the remaining coefficients in that equation.

A modification of this method, due to Crout (Reference 3), which is particularly well adapted to the use of desk computing machines, is described in an appendix.

The sets of quantities x_i $(i = 1, 2, \cdots, n)$ and c_i $(i = 1, 2, \cdots, m)$ are conventionally represented as matrices of *one column* each. In order to emphasize the fact that a matrix consists of only one column, it is sometimes convenient to denote it by a lower-case boldface letter and to enclose it in braces, rather than brackets, and so to write

$$\mathbf{x} \equiv \{x_i\} \equiv \begin{Bmatrix} x_1 \\ x_2 \\ \cdot \\ \cdot \\ \cdot \\ x_n \end{Bmatrix}, \qquad \mathbf{c} \equiv \{c_i\} \equiv \begin{Bmatrix} c_1 \\ c_2 \\ \cdot \\ \cdot \\ \cdot \\ c_m \end{Bmatrix}. \tag{10a,b}$$

For convenience in writing, the elements of a one-column matrix are frequently arranged horizontally,

$$\mathbf{x} = \{x_1, x_2, \cdots, x_n\},$$

the use of braces then being *necessary* to indicate the transposition.

Other symbols, such as parentheses or double vertical lines, are also used to enclose matrix arrays.

If we interpret (1) as stating that the matrix **A** transforms the one-column matrix **x** into the one-column matrix **c**, it is natural to write the transformation in the form

$$\mathbf{A}\,\mathbf{x} = \mathbf{c}, \tag{11}$$

where $\mathbf{A} = [a_{ij}]$, $\mathbf{x} = \{x_i\}$, and $\mathbf{c} = \{c_i\}$.

On the other hand, the set of equations (1) can be written in the form

$$\sum_{k=1}^{n} a_{ik} x_k = c_i \qquad (i = 1, 2, \cdots, m), \tag{12a}$$

which leads to the matrix equation

$$\left\{\sum_{k=1}^{n} a_{ik} x_k\right\} = \{c_i\}. \tag{12b}$$

Hence, if (11) and (12b) are to be equivalent, we are led to the *definition*

$$\mathbf{A}\,\mathbf{x} = [a_{ik}]\{x_k\} \equiv \left\{\sum_{k=1}^{n} a_{ik} x_k\right\}. \tag{13}$$

Formally, we merely replace the *column* subscript in the general term of the *first* factor by a *dummy index k*, replace the *row* subscript in the general term of the *second* factor by the same dummy index, and sum over that index.*

The definition clearly is applicable only when the number of *columns* in the *first* factor is equal to the number of *rows* (elements) in the *second* factor. Unless this condition is satisfied, the product is undefined.

We notice that a_{ik} is the element in the ith row and kth column of $\mathbf{A}$, and that x_k is the kth element in the one-column matrix $\mathbf{x}$. Since i ranges from 1 to m in a_{ij}, the definition (13) states that the product of an $m \times n$ matrix into an $n \times 1$ matrix is an $m \times 1$ matrix (m elements in one column). The ith element in the product is obtained from the ith row of the first factor and the single column of the second factor, by multiplying together the first elements, second elements, and so forth, and adding these products together algebraically.

Thus, for example, the definition leads to the result

$$\begin{bmatrix} 1 & 0 \\ 2 & 1 \\ -1 & 2 \end{bmatrix} \begin{Bmatrix} 1 \\ 2 \end{Bmatrix} = \begin{Bmatrix} 1 \cdot 1 + 0 \cdot 2 \\ 2 \cdot 1 + 1 \cdot 2 \\ -1 \cdot 1 + 2 \cdot 2 \end{Bmatrix} = \begin{Bmatrix} 1 \\ 4 \\ 3 \end{Bmatrix}.$$

Now suppose that the n variables $x_1, \cdots, x_n$ are expressed as linear combinations of s new variables $y_1, \cdots, y_s$, that is, that a set of relations holds of the form

$$x_i = \sum_{k=1}^{s} b_{ik} y_k \qquad (i = 1, 2, \cdots, n). \tag{14}$$

If the original variables satisfy (12a), the equations satisfied by the new variables are obtained by introducing (14) into (12a). In addition to replacing i by k in (14), for this introduction, we must replace k in (14) by a *new*

* Very frequently, in the literature, use is made of the so-called *summation convention*, in which the sigma symbol is omitted in a sum such as

$$\sum_{k=1}^{n} a_{ik} x_k$$

with the understanding that the notation $a_{ik}x_k$ then is to indicate the result of *summing* the product with respect to the *repeated* index, over the range of that index. Similarly, with this convention one would write $a_{ik}b_{kl}c_{lj}$ when summations with respect to both k and l are intended. An explicit statement then must be made when the *element* a_{kk} is to be distinguished from the *sum*

$$\sum_{k=1}^{n} a_{kk}$$

or in other cases when the summation convention temporarily is to be abandoned. The summation convention will not be used in this text.

dummy index, say l, to avoid ambiguity of notation. The result of the substitution then takes the form

$$\sum_{k=1}^{n} a_{ik}\left(\sum_{l=1}^{s} b_{kl}y_l\right) = c_i \qquad (i = 1, 2, \cdots, m), \tag{15a}$$

or, since the order in which the finite sums are formed is immaterial,

$$\sum_{l=1}^{s}\left(\sum_{k=1}^{n} a_{ik}b_{kl}\right)y_l = c_i \qquad (i = 1, 2, \cdots, m). \tag{15b}$$

In matrix notation, the transformation (14) takes the form

$$\mathbf{x} = \mathbf{B}\,\mathbf{y} \tag{16}$$

and, corresponding to (15a), the introduction of (16) into (11) gives

$$\mathbf{A}(\mathbf{B}\,\mathbf{y}) = \mathbf{c}. \tag{17}$$

But if we write

$$p_{ij} = \sum_{k=1}^{n} a_{ik}b_{kj} \qquad \begin{pmatrix} i = 1, 2, \cdots, m \\ j = 1, 2, \cdots, s \end{pmatrix} \tag{18}$$

equation (15b) takes the form

$$\sum_{l=1}^{s} p_{il}y_l = c_i \qquad (i = 1, 2, \cdots, m),$$

and hence, in accordance with (12a) and (13), the matrix form of the transformation (15b) is

$$\mathbf{P}\,\mathbf{y} = \mathbf{c}, \tag{19}$$

where $\mathbf{P} = [p_{ij}]$.

Thus it follows that the result of operating on $\mathbf{y}$ by $\mathbf{B}$, and on the product by $\mathbf{A}$ [given by the left-hand member of (17)], is the same as the result of operating on $\mathbf{y}$ directly by the matrix $\mathbf{P}$. We accordingly *define* this matrix to be the product $\mathbf{A}\,\mathbf{B}$,

$$\mathbf{A}\,\mathbf{B} = [a_{ik}][b_{kj}] \equiv \left[\sum_{k=1}^{n} a_{ik}b_{kj}\right]. \tag{20}$$

The desirable relation

$$\mathbf{A}(\mathbf{B}\,\mathbf{y}) = (\mathbf{A}\,\mathbf{B})\mathbf{y}$$

then is a consequence of this definition.

Recalling that the first subscript in each case is the row index and the second the column index, we see that if the first factor of (20) has m rows and n columns, and the second n rows and s columns, the index i in the right-hand member may vary from 1 to m while the index j in that member may vary from 1 to s. Hence, the *product of an $m \times n$ matrix into an $n \times s$ matrix is an $m \times s$ matrix*. The element p_{ij} in the ith row and jth column of the product is formed by multiplying together corresponding elements of the ith *row* of the *first* factor and the jth *column* of the *second* factor, and adding the results algebraically. In particular, the definition (20) properly reduces to (13) when $s = 1$.

Thus, for example, we have

$$\begin{bmatrix} 1 & 0 & 1 \\ 1 & -2 & 1 \end{bmatrix} \begin{bmatrix} 1 & 2 & 1 \\ 1 & 0 & 1 \\ 2 & 1 & 0 \end{bmatrix}$$

$$= \begin{bmatrix} (1\cdot 1 + 0\cdot 1 + 1\cdot 2) & (1\cdot 2 + 0\cdot 0 + 1\cdot 1) & (1\cdot 1 + 0\cdot 1 + 1\cdot 0) \\ (1\cdot 1 - 2\cdot 1 + 1\cdot 2) & (1\cdot 2 - 2\cdot 0 + 1\cdot 1) & (1\cdot 1 - 2\cdot 1 + 1\cdot 0) \end{bmatrix}$$

$$= \begin{bmatrix} 3 & 3 & 1 \\ 1 & 3 & -1 \end{bmatrix}.$$

We notice that $\mathbf{A}\,\mathbf{B}$ is defined only if the number of *columns* in $\mathbf{A}$ is equal to the number of *rows* in $\mathbf{B}$. In this case, the two matrices are said to be *conformable* in the order stated.

If $\mathbf{A}$ is an $m \times n$ matrix and $\mathbf{B}$ an $n \times m$ matrix, then $\mathbf{A}$ and $\mathbf{B}$ are conformable in either order, the product $\mathbf{A}\,\mathbf{B}$ then being a *square* matrix of order m and the product $\mathbf{B}\,\mathbf{A}$ a square matrix of order n. Even in the case when $\mathbf{A}$ and $\mathbf{B}$ are square matrices of the same order the products $\mathbf{A}\,\mathbf{B}$ and $\mathbf{B}\,\mathbf{A}$ are not generally equal. For example, in the case of two square matrices of order two we have

$$\begin{bmatrix} a_{11} & a_{12} \\ a_{21} & a_{22} \end{bmatrix} \begin{bmatrix} b_{11} & b_{12} \\ b_{21} & b_{22} \end{bmatrix} = \begin{bmatrix} a_{11}b_{11} + a_{12}b_{21} & a_{11}b_{12} + a_{12}b_{22} \\ a_{21}b_{11} + a_{22}b_{21} & a_{21}b_{12} + a_{22}b_{22} \end{bmatrix},$$

and also

$$\begin{bmatrix} b_{11} & b_{12} \\ b_{21} & b_{22} \end{bmatrix} \begin{bmatrix} a_{11} & a_{12} \\ a_{21} & a_{22} \end{bmatrix} = \begin{bmatrix} a_{11}b_{11} + a_{21}b_{12} & a_{12}b_{11} + a_{22}b_{12} \\ a_{11}b_{21} + a_{21}b_{22} & a_{12}b_{21} + a_{22}b_{22} \end{bmatrix}.$$

Thus, in multiplying $\mathbf{B}$ by $\mathbf{A}$ in such cases, we must carefully distinguish *pre*multiplication ($\mathbf{A}\,\mathbf{B}$) from *post*multiplication ($\mathbf{B}\,\mathbf{A}$).

Two $m \times n$ matrices are said to be *equal* if and only if corresponding elements in the two matrices are equal.

The *sum* of two $m \times n$ matrices $[a_{ij}]$ and $[b_{ij}]$ is defined to be the matrix $[a_{ij} + b_{ij}]$. Further, the product of a number k and a matrix $[a_{ij}]$ is defined to be the matrix $[ka_{ij}]$, in which *each* element of the original matrix is multiplied by k.

From the preceding definitions, it is easily shown that, if $\mathbf{A}$, $\mathbf{B}$, and $\mathbf{C}$ are each $m \times n$ matrices, *addition* is *commutative* and *associative*:

$$\mathbf{A} + \mathbf{B} = \mathbf{B} + \mathbf{A}, \qquad \mathbf{A} + (\mathbf{B} + \mathbf{C}) = (\mathbf{A} + \mathbf{B}) + \mathbf{C}. \tag{21}$$

Also, if the relevant products are defined, *multiplication* of matrices is *associative*,

$$\mathbf{A}(\mathbf{B}\,\mathbf{C}) = (\mathbf{A}\,\mathbf{B})\mathbf{C}, \tag{22}$$

and *distributive*,

$$\mathbf{A}(\mathbf{B}+\mathbf{C}) = \mathbf{A}\,\mathbf{B} + \mathbf{A}\,\mathbf{C}, \qquad (\mathbf{B}+\mathbf{C})\mathbf{A} = \mathbf{B}\,\mathbf{A} + \mathbf{C}\,\mathbf{A}, \tag{23}$$

but, in general, *not commutative.**

It is consistent with these definitions to divide a given matrix into smaller submatrices, the process being known as the *partitioning* of a matrix. Thus, for example, we may partition a square matrix $\mathbf{A}$ of order three *symmetrically* as follows:

$$\mathbf{A} = \left[\begin{array}{cc|c} a_{11} & a_{12} & a_{13} \\ a_{21} & a_{22} & a_{23} \\ \hline a_{31} & a_{32} & a_{33} \end{array}\right] \equiv \begin{bmatrix} \mathbf{B}_{11} & \mathbf{B}_{12} \\ \mathbf{B}_{21} & \mathbf{B}_{22} \end{bmatrix}$$

where the elements of the partitioned form are the *matrices*

$$\mathbf{B}_{11} = \begin{bmatrix} a_{11} & a_{12} \\ a_{21} & a_{22} \end{bmatrix}, \qquad \mathbf{B}_{12} = \begin{bmatrix} a_{13} \\ a_{23} \end{bmatrix},$$

$$\mathbf{B}_{21} = [a_{31} \quad a_{32}], \qquad \mathbf{B}_{22} = [a_{33}].$$

If, for example, a second square matrix of order three is similarly partitioned, the submatrices can be treated as single elements and the usual laws of matrix multiplication and addition can be applied to the two matrices so partitioned, as is easily verified.

More generally, if two conformable matrices in a product are partitioned, necessary and sufficient conditions that this statement apply are that to each vertical partition line separating *columns* r and $r + 1$ in the *first* factor there correspond a horizontal partition line separating *rows* r and $r + 1$ in the *second* factor, and that no additional *horizontal* partition lines be present in the *second* factor.

In particular, if we think of the matrices $\mathbf{B}$ and $\mathbf{C}$ in the relation $\mathbf{A}\,\mathbf{B} = \mathbf{C}$ as being partitioned into columns, we rediscover the fact that each column of $\mathbf{C}$ can be obtained by premultiplying the corresponding column of $\mathbf{B}$ by the matrix $\mathbf{A}$. Similarly, by thinking of $\mathbf{A}$ and $\mathbf{C}$ as being partitioned into *rows*, we see that each row of $\mathbf{C}$ is the result of postmultiplying the corresponding row of $\mathbf{A}$ by the matrix $\mathbf{B}$.

* Except when otherwise noted, it will be supposed in this text that the *scalars* which comprise the *elements* of matrices are real or complex numbers. However, many of the results to be established are also valid (when suitably interpreted) for many other sets of permissible elements.

1.4. Determinants. Cramer's rule. In this section we review certain properties of *determinants*. Associated with any *square* matrix $[a_{ij}]$ of order n we define the *determinant* $|\mathbf{A}| = |a_{ij}|$,

$$|\mathbf{A}| = \begin{vmatrix} a_{11} & a_{12} & \cdots & a_{1n} \\ a_{21} & a_{22} & \cdots & a_{2n} \\ \cdots & \cdots & \cdots & \cdots \\ a_{n1} & a_{n2} & \cdots & a_{nn} \end{vmatrix},$$

as a *number* obtained as the sum of all possible products in each of which there appears one and only one element from each row and each column, each such product being prefixed by a plus or minus sign according to the following rule: *Let the elements involved in a given product be joined in pairs by line segments. If the total number of such segments sloping upward to the right is even, prefix a plus sign to the product. Otherwise, prefix a negative sign.**

From this definition, the following properties of determinants, which greatly simplify their actual evaluation, are easily established:

1. If all elements of any row or column of a square matrix are zeros, its determinant is zero.
2. The value of the determinant is unchanged if the rows and columns of the matrix are interchanged.
3. If two rows (or two columns) of a square matrix are interchanged, the sign of its determinant is changed.
4. If all elements of one row (or one column) of a square matrix are multiplied by a number k, the determinant is multiplied by k.
5. If corresponding elements of two rows (or two columns) are equal or in a constant ratio, then the determinant is zero.
6. If each element in one row (or one column) is expressed as the sum of two terms, then the determinant is equal to the sum of two determinants, in each of which one of the two terms is deleted in each element of that row (or column).
7. If to the elements of any row (column) are added k times the corresponding elements of any other row (column), the determinant is unchanged.†

* This statement of the rule of signs is equivalent to the statement which involves *inversions* of subscripts. It possesses the advantage of being readily applicable in actual cases when the elements are numbers (or functions) and are not provided with explicit subscripts. Also, with this statement of the rule, the proofs of the properties which follow are in general simplified.

† It can be shown that if we impose the condition that Properties 4 and 7 hold, and in addition impose the requirement that the determinant be unity when the diagonal elements are unity and all other elements are zero, then these conditions imply all other properties of determinants, and may serve as the definition of a determinant.

From these properties others may be deduced. For example, from Property 7 it follows immediately that to any row (column) of a square matrix may be added any *linear combination* of the *other* rows (columns) without changing the determinant of the matrix. By combining this result with Property 1, we deduce that *if any row (column) of a square matrix is a linear combination of the other rows (columns), then the determinant of that matrix is zero.*

If the row and column containing an element a_{ij} in a square matrix $\mathbf{A}$ are deleted, the determinant of the remaining square array is called the *minor* of a_{ij}, and is denoted here by M_{ij}. The *cofactor* of a_{ij}, denoted here by A_{ij}, is then defined by the relation

$$A_{ij} = (-1)^{i+j} M_{ij}. \tag{24}$$

Thus if the sum of the row and column indices of an element is *even*, the cofactor and the minor of that element are identical; otherwise they differ in sign.

It is a consequence of the definition of a determinant that *the cofactor of a_{ij} is the coefficient of a_{ij} in the expansion of* $|\mathbf{A}|$. This fact leads to the important *Laplace expansion formulas*:

$$|\mathbf{A}| = \sum_{k=1}^{n} a_{ik} A_{ik} \qquad \text{and} \qquad |\mathbf{A}| = \sum_{k=1}^{n} a_{kj} A_{kj}, \tag{25a,b}$$

for any relevant value of i or j. These formulas state that *the determinant of a square matrix is equal to the sum of the products of the elements of any single row or column of that matrix by their cofactors.*

If a_{ik} is replaced by a_{rk} in (25a), the result $\Sigma\, a_{rk} A_{ik}$ must accordingly be the determinant of a new matrix in which the elements of the ith row are replaced by the corresponding elements of the rth row, and hence must vanish if $r \neq i$ by virtue of Property 5. An analogous result follows if a_{kj} is replaced by a_{ks} in (25b), when $s \neq j$. Thus, in addition to (25), we have the relations

$$\sum_{k=1}^{n} a_{rk} A_{ik} = 0 \quad (r \neq i), \qquad \sum_{k=1}^{n} a_{ks} A_{kj} = 0 \quad (s \neq j). \tag{26a,b}$$

These results lead directly to *Cramer's rule* for solving a set of n linear equations in n unknown quantities, of the form

$$\sum_{k=1}^{n} a_{ik} x_k = c_i \qquad (i = 1, 2, \cdots, n), \tag{27}$$

in the case when the determinant of the matrix of coefficients is not zero,

$$|a_{ij}| \neq 0. \tag{28}$$

For if we *assume* the existence of a solution and multiply both sides of

(27) by A_{ir}, where r is any integer between 1 and n, and then sum the results with respect to i, there follows (after an interchange of order of summation)

$$\sum_{k=1}^{n}\left(\sum_{i=1}^{n} a_{ik}A_{ir}\right)x_k = \sum_{i=1}^{n} c_i A_{ir} \qquad (r = 1, 2, \cdots, n). \tag{29}$$

By virtue of (25b) and (26b), the inner sum on the left in (29) vanishes unless $k = r$ and is equal to $|\mathbf{A}|$ in that case. Hence (29) takes the form

$$|\mathbf{A}|\, x_r = \sum_{i=1}^{n} A_{ir}c_i \qquad (r = 1, 2, \cdots, n). \tag{30}$$

Thus, if the system (27) possesses a solution, then that solution also must satisfy the equation set (30). When $|\mathbf{A}| \neq 0$, the only possible solution of (27) accordingly is given by

$$x_r = \frac{1}{|\mathbf{A}|}\sum_{i=1}^{n} A_{ir}c_i \qquad (r = 1, 2, \cdots, n). \tag{31}$$

To verify that (31) does in fact satisfy (27), we introduce (31) into (27), first changing the dummy index i in (31) to another symbol, say j, to avoid ambiguity, after which the left-hand side of the ith equation of (27) becomes

$$\frac{1}{|\mathbf{A}|}\sum_{k=1}^{n} a_{ik}\sum_{j=1}^{n} A_{jk}c_j = \frac{1}{|\mathbf{A}|}\sum_{j=1}^{n}\left(\sum_{k=1}^{n} a_{ik}A_{jk}\right)c_j$$

after an interchange of order of summation. The use of (26a) and (25a) properly reduces this expression to

$$\frac{1}{|\mathbf{A}|}\left(\sum_{k=1}^{n} a_{ik}A_{ik}\right)c_i = c_i,$$

and hence establishes the validity of (31) when $|\mathbf{A}| \neq 0$.

The expansion on the right in (30) differs from the right-hand member of the expansion

$$|\mathbf{A}| = \sum_{i=1}^{n} A_{ir}a_{ir}$$

only in the fact that the column $\{c_i\}$ replaces the column $\{a_{ir}\}$ of the coefficients of x_r in $\mathbf{A}$. Thus we deduce *Cramer's rule*, which can be stated as follows:

When the determinant $|\mathbf{A}|$ *of the matrix of coefficients in a set of n linear algebraic equations in n unknowns* $x_1, \cdots, x_n$ *is not zero, that set of equations has a unique solution. The value of any* x_r *can be expressed as the ratio of two determinants, the denominator being the determinant of the matrix of coefficients, and the numerator being the determinant of the matrix obtained by replacing the column of the coefficients of* x_r *in the coefficient matrix by the column of the right-hand members.*

In the case when all right-hand members c_i are zero, the said to be *homogeneous*. In this case, one solution is clearly $x_1 = x_2 = \cdots = x_n = 0$. The preceding result then states that only possible solution if $|\mathbf{A}| \neq 0$, so that *a set of n linear homogeneous equations in n unknowns cannot possess a nontrivial solution unless the determinant of the coefficient matrix vanishes.*

We postpone the treatment of the case when $|\mathbf{A}| = 0$, as well as the case when the number of equations differs from the number of unknowns, to Sections 1.9 and 1.11.

It can be shown that the *determinant of the product* of two square matrices of the same order is equal to the *product of the determinants:**

$$|\mathbf{A}\,\mathbf{B}| = |\mathbf{A}|\,|\mathbf{B}|. \tag{32}$$

A square matrix whose determinant vanishes is called a *singular* matrix; a *nonsingular* matrix is a square matrix whose determinant is *not* zero. From (32) it then follows that *the product of two nonsingular matrices is also nonsingular.*

It is true also that if a square matrix $\mathbf{M}$ is of one of the special forms

$$\mathbf{M} = \left[\begin{array}{c|c} \mathbf{A} & \mathbf{0} \\ \hline \mathbf{C} & \mathbf{B} \end{array}\right] \quad \textit{or} \quad \mathbf{M} = \left[\begin{array}{c|c} \mathbf{A} & \mathbf{C} \\ \hline \mathbf{0} & \mathbf{B} \end{array}\right], \tag{33a}$$

where $\mathbf{A}$ and $\mathbf{B}$ are square submatrices and where $\mathbf{0}$ is a submatrix whose elements are all *zeros*, then

$$|\mathbf{M}| = |\mathbf{A}|\,|\mathbf{B}|. \tag{33b}$$

Here the dimensions of $\mathbf{A}$ and $\mathbf{B}$ need not be equal and the zero submatrix $\mathbf{0}$ of $\mathbf{M}$ need not be square.

It follows from the definitions that the determinant of the negative of a square matrix is *not* necessarily the negative of the determinant, but that one has the relationship

$$|-\mathbf{A}| = (-1)^n\,|\mathbf{A}|,$$

where n is the order of the matrix $\mathbf{A}$.

1.5. Special matrices. In this section we define certain special matrices which are of importance, and investigate some of their properties.

That matrix which is obtained from $\mathbf{A} = [a_{ij}]$ by interchanging rows and columns is called the *transpose* of $\mathbf{A}$, and is here indicated by $\mathbf{A}^T$:

$$\mathbf{A}^T = \begin{bmatrix} a_{11} & a_{21} & \cdots & a_{m1} \\ a_{12} & a_{22} & \cdots & a_{m2} \\ \cdots & \cdots & \cdots & \cdots \\ a_{1n} & a_{2n} & \cdots & a_{mn} \end{bmatrix}.$$

* While this result is in no sense profound, the existent proofs of (32) are either lengthy or indirect.

Thus the transpose of an $m \times n$ matrix is an $n \times m$ matrix. If the element in row r and column s of $\mathbf{A}$ is a_{rs}, where r may vary from 1 to m and s from 1 to n, then the element a'_{rs} in row r and column s of $\mathbf{A}^T$ is given by $a'_{rs} = a_{sr}$, where now r may vary from 1 to n and s from 1 to m.

If $\mathbf{A}$ is an $m \times l$ matrix and $\mathbf{B}$ is an $l \times n$ matrix, then both the products $\mathbf{A}\,\mathbf{B}$ and $\mathbf{B}^T\mathbf{A}^T$ exist, the former being an $m \times n$ matrix, and the latter an $n \times m$ matrix. We show next that the latter matrix is the transpose of the former. Since the element in row r and column s of the product $\mathbf{A}\,\mathbf{B} \equiv \mathbf{C}$ is given by

$$\sum_{k=1}^{l} a_{rk} b_{ks} \equiv c_{rs},$$

where r may vary from 1 to m and s from 1 to n, whereas the element c'_{rs} in row r and column s of the product $\mathbf{B}^T\,\mathbf{A}^T$ is given by

$$c'_{rs} \equiv \sum_{k=1}^{l} b'_{rk} a'_{ks} = \sum_{k=1}^{l} b_{kr} a_{sk} = c_{sr},$$

where now r may vary from 1 to n and s from 1 to m, it follows that $\mathbf{B}^T\,\mathbf{A}^T$ is indeed the transpose of $\mathbf{A}\,\mathbf{B}$.

Thus, we have shown that *the transpose of the product* $\mathbf{A}\,\mathbf{B}$ *is the product of the transposes in reverse order:*

$$(\mathbf{A}\,\mathbf{B})^T = \mathbf{B}^T\,\mathbf{A}^T. \tag{34}$$

This result will be of frequent usefulness.

When $\mathbf{A}$ is a *square* matrix, the matrix obtained from $\mathbf{A}$ by replacing each element by its cofactor and then interchanging rows and columns is called the *adjoint* of $\mathbf{A}$:

$$\text{Adj}\,\mathbf{A} = \begin{bmatrix} A_{11} & A_{21} & \cdots & A_{n1} \\ A_{12} & A_{22} & \cdots & A_{n2} \\ \cdots & \cdots & \cdots & \cdots \\ A_{1n} & A_{2n} & \cdots & A_{nn} \end{bmatrix} = [A_{ji}].$$

The adjoint of a product is found to be equal to the product of the adjoints *in the reverse order.*

When the elements of a matrix $\mathbf{A}$ are complex, we denote by $\bar{\mathbf{A}}$ the matrix obtained by replacing each element of $\mathbf{A}$ by its complex conjugate, and call the matrix $\bar{\mathbf{A}}$ the *conjugate* of $\mathbf{A}$.

The *unit matrix* $\mathbf{I}$ of order n is the *square* $n \times n$ matrix having ones in its *principal diagonal* and zeros elsewhere,

$$\mathbf{I} = \begin{bmatrix} 1 & 0 & \cdots & 0 \\ 0 & 1 & \cdots & 0 \\ \cdots & \cdots & \cdots & \cdots \\ 0 & 0 & \cdots & 1 \end{bmatrix},$$

while a *zero matrix* $\mathbf{0}$ has zeros for *all* its elements. When it is necessary to indicate the dimensions of such matrices, the notation $\mathbf{I}_n$ may be used to denote the $n \times n$ unit matrix and the notation $\mathbf{0}_{m,n}$ to denote the $m \times n$ zero matrix. It is readily verified that for any matrix $\mathbf{A}$ there follow

$$\mathbf{A}\,\mathbf{I} = \mathbf{A}, \qquad \mathbf{I}\,\mathbf{A} = \mathbf{A} \tag{35a,b}$$

and

$$\mathbf{A}\,\mathbf{0} = \mathbf{0}, \qquad \mathbf{0}\,\mathbf{A} = \mathbf{0}. \tag{36a,b}$$

Here, if $\mathbf{A}$ is an $m \times n$ matrix, the symbol $\mathbf{I}$ necessarily stands for $\mathbf{I}_n$ in (35a) and for $\mathbf{I}_m$ in (35b), while $\mathbf{0}$ stands for $\mathbf{0}_{n,s}$ on the left and $\mathbf{0}_{m,s}$ on the right in (36a), and for $\mathbf{0}_{r,m}$ on the left and $\mathbf{0}_{r,n}$ on the right in (36b), where r and s are any positive integers.

The notation of the so-called *Kronecker delta*,

$$\delta_{pq} = \begin{cases} 0 & \text{when} \quad p \neq q, \\ 1 & \text{when} \quad p = q, \end{cases} \tag{37}$$

is frequently useful. With this notation, the general term of the unit matrix is merely δ_{ij}; that is, we can write

$$\mathbf{I} = [\delta_{ij}].$$

More generally, if all elements of a square matrix except those in the principal diagonal are zeros, the matrix is said to be a *diagonal matrix.* A diagonal matrix can thus be written in the form

$$\mathbf{D} = [d_i \delta_{ij}] = [d_j \delta_{ij}]$$

where the diagonal elements, for which $i = j$, are $d_1, d_2, \cdots, d_n$. The notation

$$\mathbf{D} = \operatorname{diag}(d_1, d_2, \cdots, d_n) \tag{38}$$

is also useful.

*Pre*multiplication of a matrix $\mathbf{A}$ by $\mathbf{D}$ multiplies the ith *row* of $\mathbf{A}$ by d_i; *post*multiplication multiplies the jth *column* by d_j. These results follow from the calculations

$$\mathbf{D}\,\mathbf{A} = [d_i \delta_{ik}][a_{kj}] = \left[\sum_{k=1}^{n} d_i \delta_{ik} a_{kj}\right] = [d_i a_{ij}] \tag{39}$$

and

$$\mathbf{A}\,\mathbf{D} = [a_{ik}][d_k \delta_{kj}] = \left[\sum_{k=1}^{n} a_{ik} d_k \delta_{kj}\right] = [a_{ij} d_j] = [d_j a_{ij}]. \tag{40}$$

A diagonal matrix whose diagonal elements are all *equal* is called a *scalar matrix.* Thus, a scalar matrix must be of the form

$$\mathbf{S} = k\mathbf{I} = [k\delta_{ij}].$$

1.6. The inverse matrix. With the notation of (37), the two equations (25a) and (26a) can be combined in the form

$$\sum_{k=1}^{n} a_{ik} A_{jk} = |\mathbf{A}|\, \delta_{ij}, \tag{41a}$$

while (25b) and (26b) lead to the relation

$$\sum_{k=1}^{n} a_{kj} A_{ki} = |\mathbf{A}|\, \delta_{ij}. \tag{41b}$$

If we write temporarily

$$m_{ij} = \frac{A_{ji}}{|\mathbf{A}|}, \tag{42}$$

under the assumption that $|\mathbf{A}| \neq 0$, these equations become

$$\sum_{k=1}^{n} a_{ik} m_{kj} = \delta_{ij}, \qquad \sum_{k=1}^{n} m_{ik} a_{kj} = \delta_{ij}. \tag{43a,b}$$

Hence, reviewing the definition (20) of the matrix product, we see that these equations imply the matrix equations

$$[a_{ik}][m_{kj}] = \mathbf{I}, \qquad [m_{ik}][a_{kj}] = \mathbf{I}. \tag{44}$$

That is, the matrix $\mathbf{M} = [m_{ij}]$ has the property that

$$\mathbf{A\,M} = \mathbf{M\,A} = \mathbf{I}, \tag{45}$$

where $\mathbf{I}$ is the unit matrix. It is natural to define a matrix satisfying (45) to be an *inverse* or *reciprocal* of $\mathbf{A}$, and to write $\mathbf{M} = \mathbf{A}^{-1}$.

If $\mathbf{A}$ is not *square*, there is no matrix $\mathbf{M}$ for which both $\mathbf{A\,M}$ and $\mathbf{M\,A}$ are unit matrices (see Problem 31), and hence $\mathbf{A}$ then cannot possess an inverse. Further, *a square matrix can have only one inverse.* To prove this statement, we suppose that $\mathbf{M}'$ is any matrix such that

$$\mathbf{A\,M}' = \mathbf{I}, \tag{46}$$

where $\mathbf{A}$ is square. Then since accordingly $|\mathbf{A}|\,|\mathbf{M}'| = 1$, it follows that $\mathbf{A}$ must be *nonsingular*, and hence the associated matrix $\mathbf{M}$ exists. If we premultiply both sides of (46) by $\mathbf{M}$ and use (45) and (35), there follows

$$(\mathbf{M\,A})\mathbf{M}' = \mathbf{M\,I} \quad \textit{or} \quad \mathbf{M}' = \mathbf{M},$$

as was to be shown.

We conclude that *if and only if the matrix* $\mathbf{A} = [a_{ij}]$ *is nonsingular*, it possesses an inverse $\mathbf{A}^{-1}$ such that

$$\mathbf{A}^{-1}\,\mathbf{A} = \mathbf{A}\,\mathbf{A}^{-1} = \mathbf{I}, \tag{47}$$

and that inverse is of the form

$$\mathbf{A}^{-1} = [m_{ij}] \quad \textit{where} \quad m_{ij} = \frac{A_{ji}}{|\mathbf{A}|}. \tag{48}$$

Thus, to obtain the inverse of a nonsingular square matrix $[a_{ij}]$ we may first replace a_{ij} by its cofactor $A_{ij} = (-1)^{i+j}M_{ij}$, then interchange rows and columns and divide each element by the determinant $|a_{ij}|$. In the terminology of Section 1.5, the inverse of **A** is the adjoint of **A** divided by the determinant of **A**:

$$\mathbf{A}^{-1} = \frac{1}{|\mathbf{A}|}\,\text{Adj}\,\mathbf{A}. \tag{49}$$

This equation also implies the useful relations

$$\mathbf{A}(\text{Adj}\,\mathbf{A}) = |\mathbf{A}|\,\mathbf{I}, \qquad (\text{Adj}\,\mathbf{A})\mathbf{A} = |\mathbf{A}|\,\mathbf{I}. \tag{50a,b}$$

It may be noticed that equations (50a,b) also follow directly from (41a,b) and hence are valid even when $|\mathbf{A}| = 0$.

To determine the inverse of a *product* of nonsingular square matrices, we write

$$\mathbf{A}\,\mathbf{B} = \mathbf{C}.$$

If we premultiply both sides of this equation successively by $\mathbf{A}^{-1}$ and $\mathbf{B}^{-1}$, there follows

$$\mathbf{I} = \mathbf{B}^{-1}\,\mathbf{A}^{-1}\,\mathbf{C}$$

and hence, by postmultiplying both sides of this equation by $\mathbf{C}^{-1}$ and replacing $\mathbf{C}$ by $\mathbf{A}\,\mathbf{B}$, we obtain the rule

$$(\mathbf{A}\,\mathbf{B})^{-1} = \mathbf{B}^{-1}\,\mathbf{A}^{-1}. \tag{51}$$

To illustrate the use of the inverse matrix, we again consider the problem of solving the set of linear equations (27) under the assumption (28). In matrix notation we have

$$\mathbf{A}\,\mathbf{x} = \mathbf{c},$$

and hence, after premultiplying both sides by $\mathbf{A}^{-1}$, there follows

$$\mathbf{x} = \mathbf{A}^{-1}\,\mathbf{c} \tag{52a}$$

or

$$\begin{Bmatrix} x_1 \\ x_2 \\ \cdot \\ \cdot \\ \cdot \\ x_n \end{Bmatrix} = \frac{1}{|\mathbf{A}|}\begin{bmatrix} A_{11} & A_{21} & \cdots & A_{n1} \\ A_{12} & A_{22} & \cdots & A_{n2} \\ \cdots & \cdots & \cdots & \cdots \\ A_{1n} & A_{2n} & \cdots & A_{nn} \end{bmatrix}\begin{Bmatrix} c_1 \\ c_2 \\ \cdot \\ \cdot \\ \cdot \\ c_n \end{Bmatrix} \tag{52b}$$

or

$$x_i = \frac{1}{|\mathbf{A}|}(A_{1i}c_1 + A_{2i}c_2 + \cdots + A_{ni}c_n) \qquad (i = 1, 2, \cdots, n), \tag{52c}$$

in accordance with the expanded form of Cramer's rule.

1.7. Rank of a matrix. Before proceeding to the analytical treatment of *general* sets of linear equations, to which Cramer's rule may not apply, it is desirable to introduce an additional definition and to establish certain preliminary basic results.

We define the *rank* of any matrix **A** as *the order of the largest square submatrix of* **A** (formed by deleting certain rows and/or columns of **A**) *whose determinant does not vanish.*

Suppose now that a certain matrix **A** is of rank r. We next show that if a set of r rows of **A** containing a nonsingular $r \times r$ submatrix **R** is selected, then any *other* row of **A** is a linear combination of those r rows.

To simplify the notation, we suppose that the square array **R** of order r in the upper left corner of the matrix **A** has a nonvanishing determinant, and consider the following submatrix of **A**:

$$\mathbf{M} = \begin{bmatrix} a_{11} & \cdots & a_{1r} & a_{1s} \\ \cdots & \cdots & \cdots & \cdots \\ a_{r1} & \cdots & a_{rr} & a_{rs} \\ \\ a_{q1} & \cdots & a_{qr} & a_{qs} \end{bmatrix} = \left[\begin{array}{ccc:c} & \mathbf{R} & & a_{1s} \\ & & & \vdots \\ & & & a_{rs} \\ \hdashline a_{q1} & \cdots & a_{qr} & a_{qs} \end{array}\right] \tag{53}$$

where $s > r$ and $q > r$. Then, since the matrix **A** is of rank r, the determinant of this square submatrix must vanish for all such s and q.

Now it is possible to determine constants $\lambda_1, \lambda_2, \cdots, \lambda_r$ such that the equations

$$\left.\begin{aligned} \lambda_1 a_{11} + \lambda_2 a_{21} + \cdots + \lambda_r a_{r1} &= a_{q1}, \\ \lambda_1 a_{12} + \lambda_2 a_{22} + \cdots + \lambda_r a_{r2} &= a_{q2}, \\ \cdots\cdots\cdots\cdots\cdots\cdots\cdots\cdots \\ \lambda_1 a_{1r} + \lambda_2 a_{2r} + \cdots + \lambda_r a_{rr} &= a_{qr} \end{aligned}\right\} \tag{54}$$

are satisfied, since the coefficient determinant $|\mathbf{R}^T| = |\mathbf{R}|$ assuredly does not vanish and Cramer's rule applies. Hence, with these constants of combination, we can determine a row of elements which is a linear combination of the first r rows of **M**, and which will have its first r elements identical with the first r elements of the last row. Let the last element of that combination be denoted by a'_{qs}. In evaluating the *determinant* of **M**, we may subtract this linear combination of the first r rows from the last row without changing the value of the determinant, to obtain the result

$$|\mathbf{M}| = \begin{vmatrix} a_{11} & \cdots & a_{1r} & a_{1s} \\ \cdots & \cdots & \cdots & \cdots \\ a_{r1} & \cdots & a_{rr} & a_{rs} \\ 0 & \cdots & 0 & a_{qs} - a'_{qs} \end{vmatrix}. \tag{55}$$

But since $|\mathbf{M}|$ is equal, by the Laplace expansion, to the product of $a_{qs} - a'_{qs}$ and the determinant $|\mathbf{R}|$ which does *not* vanish, by hypothesis, and since $|\mathbf{M}| = 0$, it follows that $a'_{qs} = a_{qs}$. Hence we see that *the last row of* $\mathbf{M}$ *is a linear combination of the first r rows.* Since this is true for *any* q and s greater than r, the result can be stated as follows:

If a matrix is of rank r, and a set of r rows containing a nonsingular submatrix of order r is selected, then any other row in the matrix is a linear combination of these r rows.

The same statement is easily seen to be true, by a similar argument, if the word "row" is replaced by "column" throughout.

As a special case, we deduce that if a square matrix $\mathbf{A}$ is singular, then at least one of the rows of $\mathbf{A}$ must be a linear combination of the others, and the same statement applies to the columns. Since the converse has already been established in Section 1.4, it follows that *a square matrix is singular if and only if one of its rows is a linear combination of the others. The same statement applies to the columns.*

It is obvious that the process of interchanging rows and columns does not affect the rank of a matrix, so that *the two matrices* $\mathbf{A}$ *and* $\mathbf{A}^T$ *have the same rank.* The following section treats certain other operations on matrices which also leave the rank invariant.

1.8. Elementary operations. Associated with a set of m linear equations in n unknowns,

$$\left.\begin{aligned} a_{11}x_1 + a_{12}x_2 + \cdots + a_{1n}x_n &= c_1, \\ a_{21}x_1 + a_{22}x_2 + \cdots + a_{2n}x_n &= c_2, \\ \cdots\cdots\cdots\cdots\cdots\cdots\cdots\cdots\cdots \\ a_{m1}x_1 + a_{m2}x_2 + \cdots + a_{mn}x_n &= c_m \end{aligned}\right\}, \tag{56}$$

we consider two matrices: the $m \times n$ matrix of coefficients $[a_{ij}]$, and the $m \times (n + 1)$ matrix formed by joining to the columns of $[a_{ij}]$ the column of constants $\{c_i\}$. We refer to the former matrix as the *coefficient matrix* and to the second as the *augmented matrix.*

As was shown in Section 1.2, we may use the Gauss-Jordan reduction (renumbering certain equations or variables, if necessary) to replace (56) by an equivalent set of equations of the form

$$\left.\begin{aligned} x_1 \qquad - \alpha_{11}x_{r+1} - \cdots - \alpha_{1,n-r}x_n &= \gamma_1, \\ x_2 \quad - \alpha_{21}x_{r+1} - \cdots - \alpha_{2,n-r}x_n &= \gamma_2, \\ \cdots\cdots\cdots\cdots\cdots\cdots\cdots\cdots\cdots \\ x_r - \alpha_{r1}x_{r+1} - \cdots - \alpha_{r,n-r}x_n &= \gamma_r, \\ 0 &= \gamma_{r+1}, \\ \cdots\cdots\cdots\cdots\cdots\cdots\cdots\cdots\cdots \\ 0 &= \gamma_m \end{aligned}\right\}, \tag{57}$$

by a process which involves, in addition to possible renumbering, only the multiplication of equal quantities by equal nonzero quantities and the addition of equals to equals. Accordingly, the augmented matrix of (56) is transformed to the augmented matrix of (57), which is of the form

$$\begin{bmatrix} 1 & 0 & \cdots & 0 & -\alpha_{11} & -\alpha_{12} & \cdots & -\alpha_{1,n-r} & \gamma_1 \\ 0 & 1 & \cdots & 0 & -\alpha_{21} & -\alpha_{22} & \cdots & -\alpha_{2,n-r} & \gamma_2 \\ \cdots & \cdots & \cdots & \cdots & \cdots & \cdots & \cdots & \cdots & \cdots \\ 0 & 0 & \cdots & 1 & -\alpha_{r1} & -\alpha_{r2} & \cdots & -\alpha_{r,n-r} & \gamma_r \\ 0 & 0 & \cdots & 0 & 0 & 0 & \cdots & 0 & \gamma_{r+1} \\ \cdots & \cdots & \cdots & \cdots & \cdots & \cdots & \cdots & \cdots & \cdots \\ 0 & 0 & \cdots & 0 & 0 & 0 & \cdots & 0 & \gamma_m \end{bmatrix}, \tag{58}$$

and, at the same time, the coefficient matrix of (56) transforms into the result of deleting the last column of the matrix (58).

The steps in the reduction involve only the following so-called *elementary row and column operations:*

1. The interchange of two rows (or of two columns).
2. The multiplication of the elements of a row (or column) by a number other than zero.
3. The addition, to the elements of a row (column), of k times the corresponding elements of another row (column).

It is clear that the transformed *coefficient* matrix in the above case is of rank r, whereas if one or more of the numbers $\gamma_{r+1}, \gamma_{r+2}, \cdots, \gamma_m$ is not zero, the rank of the transformed *augmented* matrix is $r + 1$. If $\gamma_{r+1} = \gamma_{r+2} = \cdots = \gamma_m = 0$, *both* transformed matrices are of rank r.

It is next shown that *the ranks of the two matrices associated with* (56) *are the same as the ranks of the corresponding transformed matrices*, that is, that *the rank of a matrix is not changed by the elementary operations.*

We need show only that a nonvanishing determinant of *largest* order r is not reduced to zero, and that no nonvanishing determinant of higher order is *introduced*, by any such operation.

Operation 1 is equivalent to renumbering rows or columns, and obviously cannot affect over-all vanishing or nonvanishing of determinants.

Similarly, *operation 2* can only multiply certain determinants by a nonzero constant.

According to Property 7 of determinants (page 10), *operation 3* does not change the value of any determinant which involves either *both* or *neither* of the two rows (or columns) concerned. To simplify the notation in the remaining case, we again suppose that one nonsingular $r \times r$ submatrix $\mathbf{R}$

is in the upper left corner of **A**, and that k times the qth row of **A** is to be added to the ith row, with $i \leq r < q$. If we then consider the effect of this row operation on the submatrix

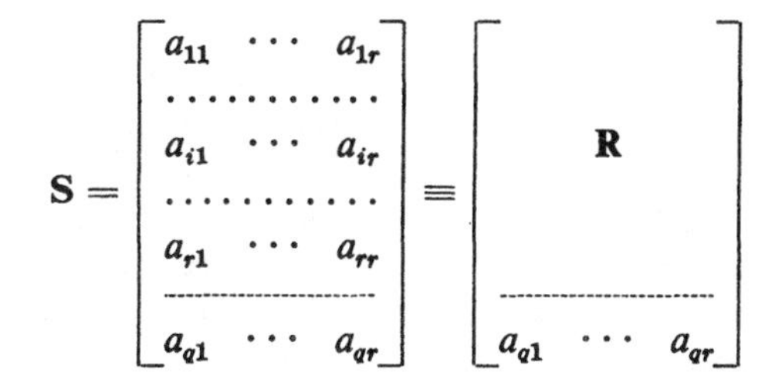

$$\mathbf{S} = \begin{bmatrix} a_{11} & \cdots & a_{1r} \\ \cdots & \cdots & \cdots \\ a_{i1} & \cdots & a_{ir} \\ \cdots & \cdots & \cdots \\ a_{r1} & \cdots & a_{rr} \\ \hline a_{q1} & \cdots & a_{qr} \end{bmatrix} \equiv \begin{bmatrix} & \mathbf{R} & \\ \hline a_{q1} & \cdots & a_{qr} \end{bmatrix}$$

we may use the last result of Section 1.7 to deduce that the rank of **A** is not *reduced* by *row* operation 3.

For either the last row of **S** is a linear combination of rows of **R** *excluding* the ith, in which case $|\mathbf{R}|$ is unchanged, or the $r \times r$ subarray of **S** which is obtained by deleting the ith row (and which is *unaffected* by the operation) is nonsingular. Thus there is at least *one* nonsingular $r \times r$ submatrix in **S** after the operation is effected.

Conversely, this operation cannot *increase* the rank of a matrix, since the reversed operation (which would be of the same type) then would *reduce* the rank of the *new* matrix.

The same argument applies to operation 3 effected on *columns*. Hence we conclude that *the elementary operations, applied to rows or to columns, do not change the rank of a matrix.*

1.9. Solvability of sets of linear equations. If we notice that no *column* operations are involved in the Gauss-Jordan reduction, except perhaps a renumbering of certain columns of the *coefficient* matrix, we conclude both that the augmented matrices of (56) and (57) are of equal rank, and also that the same is true of the coefficient matrices.

If and only if one or more of the numbers $\gamma_{r+1}, \gamma_{r+2}, \cdots, \gamma_m$ in (57) is not zero, the given set of equations possesses no solution. But if and only if this is so, the rank of the augmented matrix is greater than the rank of the coefficient matrix. Thus we deduce the following basic result:

A set of linear equations possesses a solution if and only if the rank of the augmented matrix is equal to the rank of the coefficient matrix.

If the two ranks *are* both equal to r, and if we select a set of r equations whose coefficient matrix contains a nonsingular submatrix of order r, then we may disregard all other equations, since they are implied by the r basic equations (that is, their coefficients are linear combinations of the coefficients in the r basic equations). The $n - r$ unknowns whose coefficients are *not* involved in the nonsingular $r \times r$ submatrix can be assigned arbitrary values, after which the remaining r unknowns can be determined in terms of them (by Cramer's rule or otherwise).*

In particular, *if* $r = n$ *the unknowns are determined uniquely*. Otherwise, if $n - r = d > 0$, the most general solution involves d independent arbitrary parameters.

In the *homogeneous* case, when the right-hand members of (56) are all zeros, the coefficient matrix and the augmented matrix are automatically of equal rank, and a solution *always* exists. But this fact is obvious, since such a system is always satisfied by the *trivial solution* $x_1 = x_2 = \cdots = x_n = 0$. If the rank r of the coefficient matrix is equal to the number n of unknowns, then this is the *only* solution, in accordance with the special results of Section 1.4. However, if $r < n$ (in particular, if the number of equations is less than the number of unknowns) infinitely many solutions exist, the number of independent arbitrary parameters involved being given by the difference $n - r$.

We notice that, in consequence of the *linearity* of the relevant equations, the general solution of a *nonhomogeneous* set of equations is the sum of any one *particular* solution of that set and the most general solution of the associated homogeneous set.

A case of particular interest is that of a set of n homogeneous equations in n unknowns, in which the coefficient matrix is of rank $n - 1$; that is, a set of the form

$$\sum_{k=1}^{n} a_{ik}x_k = 0 \qquad (i = 1, 2, \cdots, n) \tag{59}$$

where

$$|a_{ij}| = 0 \tag{60}$$

but where the determinant of at least one square submatrix of $\mathbf{A}$ of order $n - 1$ does not vanish. In consequence of equations (41a) and (60), these equations are satisfied by the expressions

$$x_i = CA_{si} \qquad (i = 1, 2, \cdots, n) \tag{61}$$

where C is an arbitrary constant and s may take on any value from 1 to n. Since here $d = n - r = 1$, and since (61) contains one arbitrary parameter, (61) must represent the most general solution of (59) unless, for the particular value of s chosen, all cofactors A_{si} happen to vanish. (This exception cannot exist for *all* values of s if the rank of $\mathbf{A}$ is $n - 1$.) With this reservation, the result obtained is equivalent to the statement that, in the case under consideration, *the unknowns are proportional to the cofactors of their coefficients in any row of the matrix* $[a_{ij}]$.

* In actual numerical cases, a procedure such as that of the Gauss or Gauss-Jordan reduction avoids the necessity of perhaps evaluating a large number of determinants. However, the results obtained here are of great importance in more general considerations.

1.10. Linear vector space. The preceding results have interesting and instructive interpretations in terms of so-called "vector space," which is briefly discussed in this section.

It is conventional to speak of a *one-column* matrix as a *column vector* or, more briefly, as a *vector*. (Such an array is often called a *numerical* vector, to distinguish it from a *geometrical* or *physical* vector quantity—such as a force or an acceleration—which it may *represent*.) In accordance with the notation introduced in Section 1.3, a lower-case boldface letter is used in this text to specifically denote a vector, so that we write

$$\mathbf{x} = \{x_1, x_2, \cdots, x_n\}.$$

A *one-row* matrix $[x_1, x_2, \cdots, x_n]$, which is the *transpose* of the vector $\mathbf{x}$, will be termed a *row vector* and will be denoted by $\mathbf{x}^T$ when a special symbolism is desirable.

In *two-dimensional* space, the *elements* of the vector $\{x_1, x_2\}$ can be interpreted as the *components* of $\mathbf{x}$ in the directions of rectangular coordinate (x_1 and x_2) axes. The square of the *length* of this vector is then given by $l^2 = x_1^2 + x_2^2 = \mathbf{x}^T \mathbf{x}$.* Also, if $\mathbf{u}$ and $\mathbf{v}$ are two vectors in two-dimensional space, the *scalar product* of $\mathbf{u}$ and $\mathbf{v}$ is defined to be $u_1v_1 + u_2v_2 = \mathbf{u}^T \mathbf{v} = \mathbf{v}^T \mathbf{u}$. It is seen that the scalar product $\mathbf{u}^T \mathbf{v}$ here is the equivalent, in matrix notation, of the "dot product" $\mathbf{u} \cdot \mathbf{v}$ in vector analysis. We recall that the vectors $\mathbf{u}$ and $\mathbf{v}$ are orthogonal (perpendicular) if and only if this scalar product vanishes. The vectors $\mathbf{i}_1 = \{1, 0\}$ and $\mathbf{i}_2 = \{0, 1\}$ are the orthogonal *unit vectors* ordinarily denoted by $\mathbf{i}$ and $\mathbf{j}$, respectively, in vector analysis.

The above terminology is extended by analogy to the general case of n *dimensions*. When $n > 3$, it is impossible to visualize the vectors geometrically. However, we use the language associated with space of two or three dimensions, and say that an n-dimensional rectangular coordinate system comprises n mutually orthogonal axes, that a point has n corresponding coordinates, and that a vector has n components along these axes.

The *scalar product* of two vectors $\mathbf{u}$ and $\mathbf{v}$ is defined to be

$$\mathbf{u}^T \mathbf{v} = \mathbf{v}^T \mathbf{u} = u_1v_1 + u_2v_2 + \cdots + u_nv_n \tag{62}$$

and the *square of the length* of a vector $\mathbf{u}$ is defined to be

$$l^2(\mathbf{u}) = \mathbf{u}^T \mathbf{u} = u_1^2 + u_2^2 + \cdots + u_n^2. \tag{63}$$

It is sometimes convenient to denote the scalar product by the abbreviation $(\mathbf{u}, \mathbf{v})$, so that

$$(\mathbf{u}, \mathbf{v}) = \mathbf{u}^T \mathbf{v} = \mathbf{v}^T \mathbf{u} = (\mathbf{v}, \mathbf{u}). \tag{64}$$

* A product of the form $\mathbf{x}^T\mathbf{x}$, which is truly a one-element matrix, is conventionally treated as a scalar.

The abbreviation

$$\mathbf{u}^2 \equiv (\mathbf{u}, \mathbf{u})$$

is frequently used in the special case when $\mathbf{v} = \mathbf{u}$.

Two vectors $\mathbf{u}$ and $\mathbf{v}$ are said to be *orthogonal* if their scalar product vanishes, $(\mathbf{u}, \mathbf{v}) = 0$. A *zero vector* is thus orthogonal to *all* vectors of the same dimension. A vector is said to be a *unit vector* if its length is unity, so that $(\mathbf{u}, \mathbf{u}) = 1$.

When the components of $\mathbf{u}$ and $\mathbf{v}$ are *real*, it can be shown that

$$|(\mathbf{u}, \mathbf{v})| \leqq (\mathbf{u}, \mathbf{u})^{1/2}(\mathbf{v}, \mathbf{v})^{1/2} \tag{65}$$

(see Problem 43), that is, that the magnitude of the *scalar product* of $\mathbf{u}$ and $\mathbf{v}$ is not larger than the product of the *lengths* of $\mathbf{u}$ and $\mathbf{v}$. This important result is known as the *Schwarz inequality*.

Hence the natural definition

$$\cos\theta = \frac{(\mathbf{u}, \mathbf{v})}{(\mathbf{u}, \mathbf{u})^{1/2}(\mathbf{v}, \mathbf{v})^{1/2}} \qquad (-\pi < \theta \leqq \pi) \tag{66}$$

makes $|\cos\theta| \leqq 1$ and so yields a real interpretation for the "*angle* θ between the real vectors $\mathbf{u}$ and $\mathbf{v}$" in the general n-dimensional case.

When the components of the vectors are *complex* numbers, the preceding properties of length, angle, and orthogonality usually are of limited significance and it is desirable to define more appropriate properties, which reduce to the preceding ones in the real case.

For this purpose, the *Hermitian scalar product* of a vector $\mathbf{u}$ into a vector $\mathbf{v}$ may be defined as

$$(\bar{\mathbf{u}}, \mathbf{v}) = \bar{\mathbf{u}}^T\mathbf{v} = \bar{u}_1 v_1 + \bar{u}_2 v_2 + \cdots + \bar{u}_n v_n = (\mathbf{v}, \bar{\mathbf{u}}), \tag{67}$$

and is complex and *not*, in general, equal to its conjugate $(\mathbf{u}, \bar{\mathbf{v}})$. The square of the *Hermitian length* (or *absolute length*) of a vector $\mathbf{u}$ with complex components is then defined to be the nonnegative *real* quantity

$$l_H^{\,2}(\mathbf{u}) = (\bar{\mathbf{u}}, \mathbf{u}) = \bar{\mathbf{u}}^T\mathbf{u} = \bar{u}_1 u_1 + \bar{u}_2 u_2 + \cdots + \bar{u}_n u_n, \tag{68}$$

and the *Hermitian angle* θ_H between $\mathbf{u}$ and $\mathbf{v}$ is a real quantity defined by the relation

$$\cos\theta_H = \frac{(\bar{\mathbf{u}}, \mathbf{v}) + (\bar{\mathbf{v}}, \mathbf{u})}{2(\bar{\mathbf{u}}, \mathbf{u})^{1/2}(\bar{\mathbf{v}}, \mathbf{v})^{1/2}} \qquad (-\pi < \theta_H \leqq \pi) \tag{69}$$

(see Problem 44). Two complex vectors $\mathbf{u}$ and $\mathbf{v}$ are then said to be *orthogonal in the Hermitian sense* when $(\bar{\mathbf{u}}, \mathbf{v}) = 0$, and hence also $(\bar{\mathbf{v}}, \mathbf{u}) = 0$.

When the elements involved are real, they are equal to their complex conjugates, and it is seen that (67), (68), and (69) reduce to (62), (63), and

(66). However, it should be noted, for example, that if $\mathbf{v} = \{1, i\}$, where $i^2 = -1$, then there follows $l(\mathbf{v}) = 0$ but $l_H(\mathbf{v}) = \sqrt{2}$.

A set of m vectors $\mathbf{u}_1, \mathbf{u}_2, \cdots, \mathbf{u}_m$ is said to be *linearly independent* if *no* set of constants $c_1, c_2, \ldots, c_m$, at least one of which is not zero, exists such that

$$c_1\mathbf{u}_1 + c_2\mathbf{u}_2 + \cdots + c_m\mathbf{u}_m = \mathbf{0}. \tag{70}$$

In two-dimensional space, the existence of c_1 and c_2 (not both zero) such that

$$c_1\mathbf{u}_1 + c_2\mathbf{u}_2 = \mathbf{0}$$

would imply that one of the two-dimensional vectors $\mathbf{u}_1$ and $\mathbf{u}_2$ is a scalar multiple of the other. Hence any two vectors which are not multiples of the same vector (parallel to a line) are linearly independent in two-dimensional space. Further, geometrical considerations indicate that any *three* vectors which are not *parallel to a plane* are linearly independent in three-dimensional space.

To obtain an analytical criterion for linear dependence of a set of m vectors with *real* components, we suppose that c's *do* exist, at least one of which is not zero, such that (70) is satisfied. Then, by successively forming the scalar products of $\mathbf{u}_1, \cdots, \mathbf{u}_m$ into both sides of (70), we find that the constants c_i must also satisfy the equations

$$\begin{array}{c}
c_1\mathbf{u}_1^2 + c_2(\mathbf{u}_1, \mathbf{u}_2) + \cdots + c_m(\mathbf{u}_1, \mathbf{u}_m) = 0, \\
c_1(\mathbf{u}_2, \mathbf{u}_1) + c_2\mathbf{u}_2^2 + \cdots + c_m(\mathbf{u}_2, \mathbf{u}_m) = 0, \\
\cdots\cdots\cdots\cdots\cdots\cdots\cdots\cdots\cdots\cdots\cdots \\
c_1(\mathbf{u}_m, \mathbf{u}_1) + c_2(\mathbf{u}_m, \mathbf{u}_2) + \cdots + c_m\mathbf{u}_m^2 = 0.
\end{array}$$

These conditions clearly require merely that the left-hand member of (70) be simultaneously orthogonal to $\mathbf{u}_1, \mathbf{u}_2, \cdots, \mathbf{u}_m$. But, according to Cramer's rule, this set of m equations in the m constants c_i cannot possess a nontrivial solution unless the determinant of the matrix of coefficients vanishes:

$$G \equiv \begin{vmatrix} \mathbf{u}_1^2 & (\mathbf{u}_1, \mathbf{u}_2) & \cdots & (\mathbf{u}_1, \mathbf{u}_m) \\ (\mathbf{u}_2, \mathbf{u}_1) & \mathbf{u}_2^2 & \cdots & (\mathbf{u}_2, \mathbf{u}_m) \\ \cdots & \cdots & \cdots & \cdots \\ (\mathbf{u}_m, \mathbf{u}_1) & (\mathbf{u}_m, \mathbf{u}_2) & \cdots & \mathbf{u}_m^2 \end{vmatrix} = 0. \tag{71}$$

This determinant is called the Gram determinant or *Gramian* of $\mathbf{u}_1, \cdots, \mathbf{u}_m$. Thus, if the vectors are linearly dependent the Gramian must vanish. The converse can also be shown to be true (see Problem 42). Hence it follows that *a set of real vectors is linearly dependent if and only if its Gramian vanishes.*

For vectors with *complex* components, this theorem is still true if all scalar products in the definition of the Gramian are replaced by *Hermitian* scalar products.

If $\mathbf{u}_1, \mathbf{u}_2, \cdots, \mathbf{u}_m$ are n-dimensional vectors, then the set of all vectors $\mathbf{v}$ which can be expressed in the form

$$\mathbf{v} = c_1\mathbf{u}_1 + c_2\mathbf{u}_2 + \cdots + c_m\mathbf{u}_m \tag{72}$$

is called the *vector space* generated or *spanned* by the vector set $\mathbf{u}_1, \cdots, \mathbf{u}_m$. If r and only r of the $\mathbf{u}$'s are linearly independent, the space so generated is said to be of *dimension* r. Thus the *dimension* of the generated *space* is also the *rank* of a *matrix* which has the elements of the successive generating vectors as its successive rows or columns.

When $m > r$, we see that $m - r$ of the $\mathbf{u}$'s can be expressed as linear combinations of the r *independent* $\mathbf{u}$'s, so that any vector $\mathbf{v}$ in the vector space generated by the set of m vectors also can be generated by a subset of r independent ones. Such a set of r *linearly independent* vectors is said to form a *basis* for the r-dimensional vector space which it spans.

The space of *all* n-dimensional vectors, that is, the set of all vectors having n elements, will be of principal importance in this work and will be referred to, for brevity, as *n-space*. It is clear that n-space is spanned by *any* set of n linearly independent n-dimensional vectors $\mathbf{u}_1, \cdots, \mathbf{u}_n$. For, if $\mathbf{v}$ is any n-dimensional vector, then in the n equations which equate the n components of the two members of the relation

$$\mathbf{v} = c_1\mathbf{u}_1 + c_2\mathbf{u}_2 + \cdots + c_n\mathbf{u}_n, \tag{73}$$

the matrix of coefficients of the c's has the property that no column is a linear combination of the others, and hence the determinant of the coefficient matrix cannot vanish. Thus c's can be determined so that (73) is true.

To determine the constants in (73) by an alternative method, we may form the scalar product of each $\mathbf{u}$ into the equal members of (73). The resultant set of n scalar equations always can be solved for the c's, when the $\mathbf{u}$'s are linearly independent, since the determinant of the relevant coefficient matrix is the Gramian of the $\mathbf{u}$'s, and hence does not vanish. In particular, it follows that if the n-dimensional vector $\mathbf{v}$ is *orthogonal* to each of the n linearly independent $\mathbf{u}$'s, then $\mathbf{v}$ must be the *zero* vector.*

A set of r linearly independent n-dimensional vectors is said to span an r-dimensional *subspace* of the space of *all* n-dimensional vectors, when $r < n$, and is said to be of *defect* (or *nullity*) $d = n - r$ in n-space.

Clearly, any set of n nonzero *mutually orthogonal* n-dimensional vectors is a basis in n-space, since its Gramian is the determinant of a diagonal

* In the *complex* case, the scalar products and the orthogonality are here to be defined in the *Hermitian* sense.

matrix with nonzero diagonal elements, and hence cannot vanish. An especially convenient basis comprises the particular orthogonal *unit* vectors

$$\mathbf{i}_1 = \{1, 0, 0, \cdots, 0\}, \qquad \mathbf{i}_2 = \{0, 1, 0, \cdots, 0\}, \qquad \cdots, \qquad \mathbf{i}_n = \{0, 0, 0, \cdots, 1\}, \tag{74}$$

and is sometimes called the *standard basis* of n-space.

1.11. Linear equations and vector space. We now indicate briefly the interpretation of the basic results of Section 1.9, relevant to a set of m linear algebraic equations in n unknowns, with reference to the terminology of Section 1.10.

The set of equations

$$\left.\begin{array}{l} a_{11}x_1 + a_{12}x_2 + \cdots + a_{1n}x_n = c_1, \\ a_{21}x_1 + a_{22}x_2 + \cdots + a_{2n}x_n = c_2, \\ \cdots\cdots\cdots\cdots\cdots\cdots\cdots\cdots \\ a_{m1}x_1 + a_{m2}x_2 + \cdots + a_{mn}x_n = c_m \end{array}\right\}, \tag{75}$$

which corresponds to the vector equation

$$\mathbf{A}\,\mathbf{x} = \mathbf{c}, \tag{76}$$

can also be written in the form

$$(\boldsymbol{\alpha}_i, \mathbf{x}) = c_i \qquad (i = 1, 2, \cdots, m). \tag{77}$$

Here $\mathbf{x}$ is the n-dimensional vector $\{x_1, \cdots, x_n\}$ and

$$\boldsymbol{\alpha}_i = \{a_{i1}, a_{i2}, \cdots, a_{in}\} \qquad (i = 1, 2, \cdots, m) \tag{78}$$

is the n-dimensional vector whose elements comprise the ith *row* of the coefficient matrix $\mathbf{A}$ of (75) so that, schematically,

$$\mathbf{A} = [a_{ij}] = \begin{bmatrix} -\boldsymbol{\alpha}_1{}^T \rightarrow \\ \cdots\cdots \\ -\boldsymbol{\alpha}_m{}^T \rightarrow \end{bmatrix}.$$

Thus the equations in (75) can be interpreted as prescribing the scalar product of the unknown vector $\mathbf{x}$ and the transpose of each of the m *row vectors* of $\mathbf{A}$.

In particular, when the c's are all zeros the equations (75) become *homogeneous* and require that $\mathbf{x}$ be *orthogonal* to each of the m vectors $\boldsymbol{\alpha}_i$. When the rank of $\mathbf{A}$ is equal to n, the results of Section 1.9 state that $\mathbf{x} = \mathbf{0}$ is the only solution of the associated matrix equation

$$\mathbf{A}\,\mathbf{x} = \mathbf{0}. \tag{79}$$

This result is in accordance with the fact that in this case the m vectors $\boldsymbol{\alpha}_i$ *span* all of n-space. However, when the rank of $\mathbf{A}$ is r, where $r < n$, Section 1.9 states that there is an $(n - r)$-fold infinity of vectors, each of which satisfies (79) and hence is orthogonal to each of the m vectors $\boldsymbol{\alpha}_i$. This means that when the vectors $\boldsymbol{\alpha}_1, \cdots, \boldsymbol{\alpha}_m$ span only a *subspace* of dimension $r < n$, it is possible to find a set of $d = n - r$ linearly independent nonzero vectors, say $\mathbf{u}_1, \mathbf{u}_2, \cdots, \mathbf{u}_d$, each of which is orthogonal to all the $\boldsymbol{\alpha}$'s. Any vector $\mathbf{x}$ which is a linear combination of these vectors,

$$\mathbf{x} = C_1\mathbf{u}_1 + C_2\mathbf{u}_2 + \cdots + C_d\mathbf{u}_d, \tag{80}$$

will satisfy the equation (79), and its components will satisfy (75) with $c_i = 0$.

We may notice that the basis $\mathbf{u}_1, \cdots, \mathbf{u}_{n-r}$, which generates the *solution space* (80), and a basis consisting of any r linearly independent vectors in the set $\boldsymbol{\alpha}_1, \cdots, \boldsymbol{\alpha}_m$, which generates the so-called *row space* of $\mathbf{A}$, together span all of n-space. For since the Gramian of the combination of the *two* bases just considered is the product of the separate nonvanishing Gramians [see equation (33)] it follows that their combination comprises an n-member basis for the space of *all* n-dimensional vectors. The *orthogonal complement* of a vector space is defined as the set of all vectors orthogonal to all vectors in that space. Hence we see that *a nonzero vector* $\mathbf{x}$ *satisfies* (79) *if and only if it is in the orthogonal complement of the row space of* $\mathbf{A}$.

In order to display the general solution of (75) or, equivalently, (57) in the form (80) when the right-hand members of (75) vanish, we may write $x_{r+1} = C_1, x_{r+2} = C_2, \cdots, x_n = C_{n-r}$, where the C's are arbitrary. The solution can then be written in the vector form

$$\begin{Bmatrix} x_1 \\ x_2 \\ \cdot \\ \cdot \\ \cdot \\ x_r \\ x_{r+1} \\ x_{r+2} \\ \cdot \\ \cdot \\ \cdot \\ x_n \end{Bmatrix} = C_1 \begin{Bmatrix} \alpha_{11} \\ \alpha_{21} \\ \cdot \\ \cdot \\ \cdot \\ \alpha_{r1} \\ 1 \\ 0 \\ \cdot \\ \cdot \\ \cdot \\ 0 \end{Bmatrix} + C_2 \begin{Bmatrix} \alpha_{12} \\ \alpha_{22} \\ \cdot \\ \cdot \\ \cdot \\ \alpha_{r2} \\ 0 \\ 1 \\ \cdot \\ \cdot \\ \cdot \\ 0 \end{Bmatrix} + \cdots + C_{n-r} \begin{Bmatrix} \alpha_{1,n-r} \\ \alpha_{2,n-r} \\ \cdot \\ \cdot \\ \cdot \\ \alpha_{r,n-r} \\ 0 \\ 0 \\ \cdot \\ \cdot \\ \cdot \\ 1 \end{Bmatrix}. \tag{80'}$$

It is clear from the form of the $n - r$ solution vectors that these vectors are indeed linearly independent.

In the more general case when equations (75) are *nonhomogeneous*, so that the scalar products of $\mathbf{x}$ and the vectors $\boldsymbol{\alpha}_1, \cdots, \boldsymbol{\alpha}_m$ are each to take on prescribed values, the most general vector $\mathbf{x}$ having this property is expressible as the sum of any *particular* vector having this property (if such exist) and an arbitrary linear combination of all vectors which are *orthogonal* to all the $\boldsymbol{\alpha}$'s.

It is useful to notice also that the m equations (75) can be combined into the single vector equation

$$x_1 \begin{Bmatrix} a_{11} \\ a_{21} \\ \cdot \\ \cdot \\ \cdot \\ a_{m1} \end{Bmatrix} + x_2 \begin{Bmatrix} a_{12} \\ a_{22} \\ \cdot \\ \cdot \\ \cdot \\ a_{m2} \end{Bmatrix} + \cdots + x_n \begin{Bmatrix} a_{1n} \\ a_{2n} \\ \cdot \\ \cdot \\ \cdot \\ a_{mn} \end{Bmatrix} = \begin{Bmatrix} c_1 \\ c_2 \\ \cdot \\ \cdot \\ \cdot \\ c_m \end{Bmatrix} \tag{81}$$

or

$$x_1\boldsymbol{\beta}_1 + x_2\boldsymbol{\beta}_2 + \cdots + x_n\boldsymbol{\beta}_n = \mathbf{c}, \tag{82}$$

where $\mathbf{c}$ is the m-dimensional vector $\{c_1, \cdots, c_m\}$ and

$$\boldsymbol{\beta}_j = \{a_{1j}, a_{2j}, \cdots, a_{mj}\} \qquad (j = 1, 2, \cdots, n) \tag{83}$$

is the m-dimensional vector whose elements comprise the jth *column* of the matrix $\mathbf{A}$,

$$\mathbf{A} = [a_{ij}] = \begin{bmatrix} | & & | \\ \boldsymbol{\beta}_1 & \cdots & \boldsymbol{\beta}_n \\ \downarrow & & \downarrow \end{bmatrix}.$$

Thus, with this interpretation, we see that the equations (75) are compatible if and only if $\mathbf{c}$ is representable as a linear combination of the *column vectors* comprising $\mathbf{A}$, and hence if and only if $\mathbf{c}$ is in the *column space* of $\mathbf{A}$. The components of $\mathbf{x}$, when they exist, then are to be the constants of combination in such a representation.

In some considerations, it is desirable to associate with the relation $\mathbf{A}\,\mathbf{x} = \mathbf{c}$, corresponding to the m equations (75) in n unknowns $x_1, \cdots, x_n$, the *transposed homogeneous set*

$$\left.\begin{aligned} a_{11}x_1' + a_{21}x_2' + \cdots + a_{m1}x_m' &= 0, \\ a_{12}x_1' + a_{22}x_2' + \cdots + a_{m2}x_m' &= 0, \\ \cdots\cdots\cdots\cdots\cdots\cdots\cdots\cdots& \\ a_{1n}x_1' + a_{2n}x_2' + \cdots + a_{mn}x_m' &= 0 \end{aligned}\right\} \tag{84}$$

of n *homogeneous* equations in m *new* unknowns $x_1', \cdots, x_m'$, corresponding to the relation $\mathbf{A}^T\,\mathbf{x}' = \mathbf{0}$. These equations also can be written in the forms

$$(\boldsymbol{\beta}_j, \mathbf{x}') = 0 \qquad (j = 1, 2, \cdots, n) \tag{85}$$

and

$$\sum_{i=1}^{m} x_i' \boldsymbol{\alpha}_i = \mathbf{0}, \tag{86}$$

with the notation of (78) and (83), in accordance with the fact that the row and column vectors of $\mathbf{A}$ are the column and row vectors, respectively, of $\mathbf{A}^T$.

From the relations (82) and (85), we may deduce the following useful result:

The nonhomogeneous equation $\mathbf{A}\,\mathbf{x} = \mathbf{c}$ *possesses a vector solution* $\mathbf{x}$ *if and only if the vector* $\mathbf{c}$ *is orthogonal to all vector solutions of the associated homogeneous equation* $\mathbf{A}^T\,\mathbf{x}' = \mathbf{0}$.

In order to establish this result, we notice first that if $\mathbf{A}\,\mathbf{x} = \mathbf{c}$ has a solution, then $\mathbf{c}$ is in the *column* space of $\mathbf{A}$ and hence in the *row* space of $\mathbf{A}^T$. Hence $\mathbf{c}$ then is a linear combination of the vectors $\boldsymbol{\beta}_1, \cdots, \boldsymbol{\beta}_n$, each of which is orthogonal to every $\mathbf{x}'$ which satisfies $\mathbf{A}^T\,\mathbf{x}' = \mathbf{0}$, according to (85). Thus $\mathbf{c}$ also is orthogonal to each $\mathbf{x}'$. Conversely, if $\mathbf{c}$ is orthogonal to each solution of $\mathbf{A}^T\,\mathbf{x}' = \mathbf{0}$, then $\mathbf{c}$ is in the orthogonal complement of the associated solution space. Thus $\mathbf{c}$ is in the *row* space of $\mathbf{A}^T$ and hence in the *column* space of $\mathbf{A}$, so that $x_1, \cdots, x_n$ exist such that (82) is satisfied and accordingly $\mathbf{A}\,\mathbf{x} = \mathbf{c}$ has a solution.

1.12. Characteristic-value problems. A frequently encountered problem is that of determining those values of a constant λ for which *nontrivial* solutions exist to a homogeneous set of equations of the form

$$\left.\begin{aligned} a_{11}x_1 + a_{12}x_2 + \cdots + a_{1n}x_n &= \lambda x_1, \\ a_{21}x_1 + a_{22}x_2 + \cdots + a_{2n}x_n &= \lambda x_2, \\ \cdots\cdots\cdots\cdots\cdots\cdots\cdots\cdots\cdots \\ a_{n1}x_1 + a_{n2}x_2 + \cdots + a_{nn}x_n &= \lambda x_n \end{aligned}\right\}. \tag{87}$$

Such a problem is known as a characteristic-value problem; values of λ for which nontrivial solutions exist are called *characteristic values* (also *eigenvalues* or *latent roots*) of the problem or *of the matrix* $\mathbf{A}$, and corresponding vector solutions are known as *characteristic vectors* (also *eigenvectors*) of the problem or *of the matrix* $\mathbf{A}$. A column made up of the elements of a characteristic vector is often called a *modal column*.

In many practical considerations in which such problems arise, the matrix $\mathbf{A}$ is *real and symmetric*, so that two elements which are symmetrically placed with respect to the principal diagonal are equal:

$$a_{ji} = a_{ij}. \tag{88}$$

More generally, when the coefficients are *complex* the most important cases are those in which symmetrically situated elements are *complex conjugates:*

$$a_{ji} = \bar{a}_{ij}. \tag{89}$$

Matrices having the symmetry property (89) are known as *Hermitian matrices*, and are considered in Section 1.17.

The discussion of the present section is to be restricted to *real symmetric matrices*, for which the symmetry property (88) applies. Whereas certain of the results to be obtained hold also for symmetric matrices with imaginary elements, such matrices are of limited importance in applications.

In matrix notation, equation (87) takes the form

$$\mathbf{A}\,\mathbf{x} = \lambda\,\mathbf{x} \qquad \textit{or} \qquad (\mathbf{A} - \lambda\,\mathbf{I})\mathbf{x} = \mathbf{0}, \tag{90}$$

where $\mathbf{I}$ is the unit matrix of order n. This homogeneous problem possesses nontrivial solutions if and only if the determinant of the coefficient matrix $\mathbf{A} - \lambda\,\mathbf{I}$ vanishes:

$$|\mathbf{A} - \lambda\,\mathbf{I}| \equiv \begin{vmatrix} a_{11} - \lambda & a_{12} & \cdots & a_{1n} \\ a_{21} & a_{22} - \lambda & \cdots & a_{2n} \\ \cdots & \cdots & \cdots & \cdots \\ a_{n1} & a_{n2} & \cdots & a_{nn} - \lambda \end{vmatrix} = 0. \tag{91}$$

This condition requires that λ be a root of an algebraic equation of degree n, known as the *characteristic* (or *secular*) *equation*. The n solutions $\lambda_1, \lambda_2, \cdots, \lambda_n$, which need not all be distinct, are the characteristic numbers or latent roots of the matrix $\mathbf{A}$.

Corresponding to each such value λ_k, there exists at least one vector solution (modal column) of (87) or (90), which is determined within an arbitrary multiplicative constant.* Now let λ_1 and λ_2 be two *distinct* characteristic numbers and denote corresponding characteristic vectors by $\mathbf{u}_1$ and $\mathbf{u}_2$, respectively, so that the equations

$$\mathbf{A}\,\mathbf{u}_1 = \lambda_1\mathbf{u}_1, \qquad \mathbf{A}\,\mathbf{u}_2 = \lambda_2\mathbf{u}_2 \qquad (\lambda_1 \neq \lambda_2) \tag{92a,b}$$

are satisfied. If we postmultiply the transpose of (92a) by $\mathbf{u}_2$ there follows

$$(\mathbf{A}\,\mathbf{u}_1)^T\,\mathbf{u}_2 = \lambda_1\mathbf{u}_1{}^T\,\mathbf{u}_2$$

or, using (34),

$$\mathbf{u}_1{}^T\,\mathbf{A}^T\,\mathbf{u}_2 = \lambda_1\mathbf{u}_1{}^T\,\mathbf{u}_2. \tag{93a}$$

Also, by premultiplying (92b) by $\mathbf{u}_1{}^T$, we obtain

$$\mathbf{u}_1{}^T\,\mathbf{A}\,\mathbf{u}_2 = \lambda_2\mathbf{u}_1{}^T\,\mathbf{u}_2. \tag{93b}$$

The result of subtracting (93a) from (93b), and noticing that for a *symmetric* matrix $\mathbf{A}^T = \mathbf{A}$, is then the relation

$$(\lambda_2 - \lambda_1)(\mathbf{u}_1, \mathbf{u}_2) = 0, \tag{93c}$$

* As was shown in Section 1.9, the components of this solution vector can be expressed as arbitrary multiples of the cofactors of the elements in a row of the matrix $\mathbf{A} - \lambda_k\mathbf{I}$ unless all those cofactors vanish.

and, since we have assumed that $\lambda_1 \neq \lambda_2$, we thus have the following important result:

Two characteristic vectors of a real symmetric matrix, corresponding to different characteristic numbers, are orthogonal,

$$(\mathbf{u}_1, \mathbf{u}_2) = 0. \tag{94}$$

A second basic result is that the characteristic numbers of such a matrix are always *real.* To establish this fact, we suppose that $\lambda_1 = \alpha + i\beta$ is a root of (91), where α and β are real.

Then if $\mathbf{u}_1$ is a corresponding characteristic vector, so that

$$\mathbf{A}\,\mathbf{u}_1 = \lambda_1\mathbf{u}_1,$$

there follows also

$$\mathbf{A}\,\bar{\mathbf{u}}_1 = \bar{\lambda}_1\bar{\mathbf{u}}_1,$$

by virtue of the fact that the conjugate of a product is the product of the conjugates, and of the fact that here $\mathbf{A}$ is real. Thus $\lambda_2 = \alpha - i\beta = \bar{\lambda}_1$ must also be a characteristic number with $\mathbf{u}_2 = \bar{\mathbf{u}}_1$ as an associated characteristic vector. By premultiplying the first of the relations

$$\mathbf{A}\,\mathbf{u}_1 = \lambda_1\mathbf{u}_1, \qquad \mathbf{A}\,\bar{\mathbf{u}}_1 = \bar{\lambda}_1\,\bar{\mathbf{u}}_1 \tag{95a,b}$$

by $\bar{\mathbf{u}}_1{}^T$ and postmultiplying the transpose of the second by $\mathbf{u}_1$, we obtain the relation

$$(\lambda_1 - \bar{\lambda}_1)(\bar{\mathbf{u}}_1, \mathbf{u}_1) = \bar{\mathbf{u}}_1{}^T\,\mathbf{A}\,\mathbf{u}_1 - (\mathbf{A}\,\bar{\mathbf{u}}_1)^T\,\mathbf{u}_1 = 0. \tag{96}$$

But now, since the product $(\bar{\mathbf{u}}_1, \mathbf{u}_1)$ is a *positive* quantity, it follows that $\lambda_1 - \bar{\lambda}_1 = 2i\beta$ must vanish, so that λ_1 must be real. Thus we conclude that *all characteristic numbers of a real symmetric matrix are real.*

Accordingly, all characteristic *vectors* also can be taken to be real, by rejecting permissible imaginary multiplicative factors.

If a characteristic number, say λ_1, *of a symmetric matrix is a multiple root of multiplicity* s, that is, if the left-hand member of (91) possesses the factor $(\lambda - \lambda_1)^s$, *then to* λ_1 *there correspond* s *linearly independent characteristic vectors*, any nontrivial linear combinations of which accordingly also have the same property. Proof of this important fact is postponed to Section 1.18.

The preceding statement is *not* necessarily true for *nonsymmetric* matrices, as can be seen by considering the equations

$$\left.\begin{aligned} x_1 + x_2 &= \lambda x_1, \\ -x_1 - x_2 &= \lambda x_2 \end{aligned}\right\},$$

for which

$$\mathbf{A} = \begin{bmatrix} 1 & 1 \\ -1 & -1 \end{bmatrix}.$$

Here the characteristic equation is readily found to be merely $\lambda^2 = 0$, so that $\lambda = 0$ is a characteristic number of multiplicity *two*. However, when $\lambda = 0$, the only possible solution is given by $x_1 = C_1$, $x_2 = -C_1$ and hence $\mathbf{x} = C_1\{1, -1\}$. Thus, here the double root $\lambda = 0$ corresponds to only *one* characteristic vector

As is shown in the following section, it is always possible to choose s linearly independent vectors corresponding to a characteristic number of multiplicity s in such a way that they are orthogonal *to each other*, in addition to being (automatically) orthogonal to all other characteristic vectors. Thus, if multiple roots of (91) are counted separately, we obtain always exactly n characteristic numbers, and we can determine a corresponding set of n mutually orthogonal characteristic vectors. By virtue of the results of Section 1.10, this set of vectors comprises a *basis* in n-space; that is, *any vector in n-space can be expressed as some linear combination of these n vectors.*

In particular, if each of the n orthogonal characteristic vectors has been divided by its length, and so is a *unit* vector, we say that the resultant set of vectors is an *orthonormal* set. If these vectors are denoted by $\mathbf{e}_1, \mathbf{e}_2, \cdots, \mathbf{e}_n$, so that there follows also

$$(\mathbf{e}_i, \mathbf{e}_k) = \delta_{ik}, \tag{97}$$

then the ith coefficient in the representation

$$\mathbf{v} = \sum_{k=1}^{n} \alpha_k \mathbf{e}_k \tag{98}$$

of an arbitrary n-dimensional vector $\mathbf{v}$ is easily obtained by forming the scalar product of $\mathbf{e}_i$ with the equal members of (98), in the form

$$\alpha_i = (\mathbf{e}_i, \mathbf{v}). \tag{99}$$

Consider now the *nonhomogeneous* equation

$$\mathbf{A}\,\mathbf{x} - \lambda\,\mathbf{x} = \mathbf{c}, \tag{100}$$

where $\mathbf{A}$ is a real symmetric matrix. This equation reduces to (87) or (90) when $\mathbf{c} = \mathbf{0}$. If (100) has a solution, then that solution can be expressed as a linear combination of the characteristic vectors of $\mathbf{A}$. We suppose that n orthogonal unit characteristic vectors $\mathbf{e}_1, \mathbf{e}_2, \cdots, \mathbf{e}_n$ are known, and notice that they satisfy the respective equations

$$\mathbf{A}\,\mathbf{e}_1 = \lambda_1 \mathbf{e}_1, \qquad \cdots, \qquad \mathbf{A}\,\mathbf{e}_n = \lambda_n \mathbf{e}_n. \tag{101}$$

The solution of (100) then can be assumed in the form

$$\mathbf{x} = \sum_{k=1}^{n} \alpha_k \mathbf{e}_k, \tag{102}$$

where the constants α_k are to be determined. The introduction of (102) into (100), and the use of (101), then leads to the requirement

$$\sum_{k=1}^{n}(\lambda_k - \lambda)\alpha_k \mathbf{e}_k = \mathbf{c}. \tag{103}$$

By forming the scalar product of any $\mathbf{e}_i$ with both sides of (103) we may deduce that (102) satisfies (100) if and only if the ith coefficient α_i satisfies the equation

$$(\lambda_i - \lambda)\alpha_i = (\mathbf{e}_i, \mathbf{c}) \qquad (i = 1, 2, \cdots, n). \tag{104}$$

Hence, *if λ is not a characteristic number*, the solution (102) is obtained in the form

$$\mathbf{x} = \sum_{k=1}^{n} \frac{(\mathbf{e}_k, \mathbf{c})}{\lambda_k - \lambda}\, \mathbf{e}_k. \tag{105}$$

Thus a unique solution of the nonhomogeneous problem is obtained when λ is not a characteristic number. *If $\lambda = \lambda_p$, no solution exists unless the vector* $\mathbf{c}$ *is orthogonal to the characteristic vector* (*or vectors*) *corresponding to λ_p*. In case this condition is satisfied, equation (104) shows that the corresponding coefficient (or coefficients) α_p may be chosen *arbitrarily*, so that *infinitely many* solutions then exist.

In particular, if $\lambda = 0$, equation (100) reduces to the equation

$$\mathbf{A}\,\mathbf{x} = \mathbf{c},$$

which was studied previously. This equation thus has a unique solution unless $\lambda = 0$ is a characteristic number of $\mathbf{A}$, that is, unless the equation $\mathbf{A}\,\mathbf{x} = \mathbf{0}$ has nontrivial solutions. In this exceptional situation no solution exists unless $\mathbf{c}$ is orthogonal to the vectors which satisfy $\mathbf{A}\,\mathbf{x} = \mathbf{0}$, in which case infinitely many solutions exist. This result is in accordance with the results of the preceding section, where it was shown that the requirement for the existence of a solution in the exceptional case is that $\mathbf{c}$ be orthogonal to the vectors which satisfy the equation $\mathbf{A}^T\,\mathbf{x} = \mathbf{0}$, since in the present case we have considered only a symmetric matrix, for which $\mathbf{A}^T = \mathbf{A}$.

The existence criterion obtained here, in the *more general* case when $\lambda = \lambda_p$, is also obtainable from the last result of the preceding section, by replacing $\mathbf{A}$ by $\mathbf{A} - \lambda_p\mathbf{I}$ in that result, and noticing that the latter matrix is symmetric when $\mathbf{A}$ is symmetric.

1.13. Orthogonalization of vector sets. It is often desirable, as in the preceding section, to form from a set of s linearly independent vectors $\mathbf{u}_1, \mathbf{u}_2, \cdots, \mathbf{u}_s$ an *orthogonal* set of s linear combinations of the original vectors. It is also convenient to "normalize" the vectors in such a way that each is a *unit vector*. The following procedure is a simple one, and it can be extended by analogy to other similar problems.

We first select *any one* of the original vectors, say $\mathbf{v}_1 = \mathbf{u}_1$, and divide it by its length. This is the first member of the desired set:

$$\mathbf{e}_1 = \frac{\mathbf{u}_1}{l(\mathbf{u}_1)}. \tag{106}$$

We next choose a second vector, say $\mathbf{u}_2$, from the original set and write $\mathbf{v}_2 = \mathbf{u}_2 - c\,\mathbf{e}_1$. The requirement that $\mathbf{v}_2$ be orthogonal to $\mathbf{e}_1$ leads to the determination

$$(\mathbf{e}_1, \mathbf{v}_2) = (\mathbf{e}_1, \mathbf{u}_2) - c(\mathbf{e}_1, \mathbf{e}_1) = 0$$

or

$$c = (\mathbf{e}_1, \mathbf{u}_2),$$

so that

$$\mathbf{v}_2 = \mathbf{u}_2 - (\mathbf{e}_1, \mathbf{u}_2)\mathbf{e}_1. \tag{107a}$$

Since $\mathbf{e}_1$ is a unit vector, the familiar geometrical interpretation of the scalar product in two or three dimensions leads us to say that $(\mathbf{e}_1, \mathbf{u}_2)$ is "the *scalar component* of $\mathbf{u}_2$ in the direction of $\mathbf{e}_1$," and hence that in (107a) we have "subtracted off the $\mathbf{e}_1$ component of $\mathbf{u}_2$."

The second member, $\mathbf{e}_2$, of the desired set of orthogonal unit vectors is obtained by dividing $\mathbf{v}_2$ by its length:

$$\mathbf{e}_2 = \frac{\mathbf{v}_2}{l(\mathbf{v}_2)}. \tag{107b}$$

In the third step we write $\mathbf{v}_3 = \mathbf{u}_3 - c_1\mathbf{e}_1 - c_2\mathbf{e}_2$. The requirement that $\mathbf{v}_3$ be simultaneously orthogonal to $\mathbf{e}_1$ and $\mathbf{e}_2$ then determines values of c_1 and c_2 which are in accordance with the geometrical interpretation described above, and there follows

$$\mathbf{v}_3 = \mathbf{u}_3 - (\mathbf{e}_1, \mathbf{u}_3)\mathbf{e}_1 - (\mathbf{e}_2, \mathbf{u}_3)\mathbf{e}_2, \tag{108a}$$

so that the "$\mathbf{e}_1$ and $\mathbf{e}_2$ components" of $\mathbf{u}_3$ are subtracted off. The third required vector $\mathbf{e}_3$ is then given by

$$\mathbf{e}_3 = \frac{\mathbf{v}_3}{l(\mathbf{v}_3)}. \tag{108b}$$

A continuation of this process finally determines the sth member of the required set in the form

$$\mathbf{e}_s = \frac{\mathbf{v}_s}{l(\mathbf{v}_s)} \quad \textit{where} \quad \mathbf{v}_s = \mathbf{u}_s - \sum_{k=1}^{s-1} (\mathbf{e}_k, \mathbf{u}_s)\mathbf{e}_k. \tag{109}$$

This method, which is often called the *Gram-Schmidt orthogonalization procedure*, would fail if and only if at some stage $\mathbf{v}_r = \mathbf{0}$. But this would mean that $\mathbf{u}_r$ is a linear combination of $\mathbf{e}_1, \mathbf{e}_2, \cdots, \mathbf{e}_{r-1}$, and hence also a linear combination of $\mathbf{u}_1, \mathbf{u}_2, \cdots, \mathbf{u}_{r-1}$, in contradiction with the hypothesis that the set of $\mathbf{u}$'s is linearly independent.

It is seen that this procedure permits the determination of an *orthonormal basis* (that is, a basis comprising *mutually orthogonal unit vectors*) for a vector space when *any* set of spanning vectors is known.

When the vectors $\mathbf{u}_1, \cdots, \mathbf{u}_s$ are *complex*, the same procedure clearly applies if *Hermitian* products and lengths are used throughout.

1.14. Quadratic forms. A homogeneous expression of second degree, of the form

$$A \equiv a_{11}x_1^2 + a_{22}x_2^2 + \cdots + a_{nn}x_n^2 \\ + 2a_{12}x_1x_2 + 2a_{13}x_1x_3 + \cdots + 2a_{n-1,n}x_{n-1}x_n, \quad (110)$$

is called a *quadratic form* in $x_1, x_2, \cdots, x_n$. It will be supposed here that the elements a_{ij} and the variables x_i are *real.* In two-dimensional space the equation $A =$ constant represents a second-degree curve (conic) with center at the origin, while in three-dimensional space the equation $A =$ constant represents a central *quadric surface* with center at the origin. Many problems associated with such forms are intimately related to problems associated with sets of linear equations.

We may notice first that if we write

$$y_i = \frac{1}{2}\frac{\partial A}{\partial x_i} \qquad (i = 1, 2, \cdots, n),$$

we obtain the equations

$$\left.\begin{aligned} a_{11}x_1 + a_{12}x_2 + \cdots + a_{1n}x_n &= y_1, \\ a_{12}x_1 + a_{22}x_2 + \cdots + a_{2n}x_n &= y_2, \\ \cdots\cdots\cdots\cdots\cdots&\cdots\cdots \\ a_{1n}x_1 + a_{2n}x_2 + \cdots + a_{nn}x_n &= y_n \end{aligned}\right\}. \quad (111)$$

This set of equations can be written in the form

$$\mathbf{A}\,\mathbf{x} = \mathbf{y}, \quad (112)$$

where $\mathbf{A} = [a_{ij}]$ is a *symmetric* matrix. That is, the elements satisfy the symmetry condition

$$a_{ji} = a_{ij}. \quad (113)$$

On the other hand, it is easily seen that (110) is equivalent to the relation $A \equiv (\mathbf{x}, \mathbf{y})$; that is, (110) can be written in the form

$$A \equiv \mathbf{x}^T\,\mathbf{A}\,\mathbf{x}. \quad (114)$$

In many cases it is desirable to express $x_1, x_2, \cdots, x_n$ as linear combinations of new real variables $x_1', x_2', \cdots, x_n'$ in such a way that A is reduced to a linear combination of only *squares* of the new variables, the cross-product terms being eliminated. A form of this type is said to be a *canonical form.*

Let the vector $\mathbf{x}$ be expressed in terms of $\mathbf{x}'$ by the equation

$$\mathbf{x} = \mathbf{Q}\,\mathbf{x}', \tag{115}$$

where $\mathbf{Q}$ is a square matrix of order n. The introduction of (115) into (114) then gives

$$A = (\mathbf{Q}\,\mathbf{x}')^T\,\mathbf{A}\,\mathbf{Q}\,\mathbf{x}' = \mathbf{x}'^T\,\mathbf{Q}^T\,\mathbf{A}\,\mathbf{Q}\,\mathbf{x}' \tag{116}$$

or

$$A = \mathbf{x}'^T\,\mathbf{A}'\,\mathbf{x}', \tag{117}$$

where the new matrix $\mathbf{A}'$ is defined by the equation

$$\mathbf{A}' = \mathbf{Q}^T\,\mathbf{A}\,\mathbf{Q}. \tag{118}$$

Thus we see that, if A is to involve only squares of the variables x_i', the matrix $\mathbf{Q}$ in (115) must be chosen so that $\mathbf{Q}^T\,\mathbf{A}\,\mathbf{Q}$ is a *diagonal matrix;* that is, so that all elements for which $i \neq j$ vanish.

We show next that if the characteristic numbers and corresponding characteristic vectors of the real symmetric matrix $\mathbf{A}$ are known, a matrix $\mathbf{Q}$ having this property can be very easily constructed. Suppose that the characteristic numbers of $\mathbf{A}$ are $\lambda_1, \lambda_2, \cdots, \lambda_n$, repeated roots of the characteristic equation being numbered separately, and denote the corresponding members of the orthogonalized set of n characteristic unit vectors by $\mathbf{e}_1, \mathbf{e}_2, \cdots, \mathbf{e}_n$. We then have the relations

$$\mathbf{A}\,\mathbf{e}_1 = \lambda_1\mathbf{e}_1, \qquad \cdots, \qquad \mathbf{A}\,\mathbf{e}_n = \lambda_n\mathbf{e}_n. \tag{119}$$

Let a matrix $\mathbf{Q}$ be constructed in such a way that the elements of the unit vectors $\mathbf{e}_1, \mathbf{e}_2, \cdots, \mathbf{e}_n$ are the elements of the successive *columns* of $\mathbf{Q}$:

$$\mathbf{Q} = \begin{bmatrix} e_{11} & e_{21} & \cdots & e_{n1} \\ e_{12} & e_{22} & \cdots & e_{n2} \\ \cdots & \cdots & \cdots & \cdots \\ e_{1n} & e_{2n} & \cdots & e_{nn} \end{bmatrix}. \tag{120}$$

Then, if use is made of (119), it is easily seen that

$$\mathbf{A}\,\mathbf{Q} = \begin{bmatrix} \lambda_1 e_{11} & \lambda_2 e_{21} & \cdots & \lambda_n e_{n1} \\ \lambda_1 e_{12} & \lambda_2 e_{22} & \cdots & \lambda_n e_{n2} \\ \cdots & \cdots & \cdots & \cdots \\ \lambda_1 e_{1n} & \lambda_2 e_{2n} & \cdots & \lambda_n e_{nn} \end{bmatrix} \tag{121a}$$

or

$$\mathbf{A}\,\mathbf{Q} = \mathbf{Q}\cdot\begin{bmatrix} \lambda_1 & 0 & \cdots & 0 \\ 0 & \lambda_2 & \cdots & 0 \\ \cdots & \cdots & \cdots & \cdots \\ 0 & 0 & \cdots & \lambda_n \end{bmatrix}. \tag{121b}$$

This relation follows directly from the fact that the product of $\mathbf{A}$ into the kth column of $\mathbf{Q}$ is the kth column of the right-hand member of (121a). Since the vectors $\mathbf{e}_1, \cdots, \mathbf{e}_n$ are linearly independent, it follows that $|\mathbf{Q}| \neq 0$. Thus the inverse $\mathbf{Q}^{-1}$ exists, and by premultiplying the equal members of (121b) by $\mathbf{Q}^{-1}$ we obtain the result

$$\mathbf{Q}^{-1}\,\mathbf{A}\,\mathbf{Q} = [\lambda_i \delta_{ij}]. \tag{122}$$

Hence, the matrix $\mathbf{A}$ is diagonalized by the indicated operations, the diagonal elements being merely the characteristic numbers of $\mathbf{A}$.

However, the *desired* diagonalization (118) was to be of the form $\mathbf{Q}^T\,\mathbf{A}\,\mathbf{Q}$. Thus, the matrix $\mathbf{Q}$ defined by (120) is not acceptable for present purposes unless it can be shown that

$$\mathbf{Q}^T = \mathbf{Q}^{-1} \qquad or \qquad \mathbf{Q}^T\,\mathbf{Q} = \mathbf{I}. \tag{123}$$

But the typical term p_{ij} of the product $\mathbf{Q}^T\,\mathbf{Q}$ is of the form

$$p_{ij} = \sum_{k=1}^{n} e_{ik} e_{jk},$$

where

$$\mathbf{e}_i = \{e_{i1}, e_{i2}, \cdots, e_{in}\},$$

and since the $\mathbf{e}$'s are *orthogonal* the indicated sum *vanishes* unless $i = j$, in which case the sum is *unity* since the $\mathbf{e}$'s are *unit vectors.*

Hence there follows $p_{ij} = \delta_{ij}$; that is, $\mathbf{Q}^T\,\mathbf{Q} = [\delta_{ij}] = \mathbf{I}$, as is required by (123). Further, since $|\mathbf{Q}| = |\mathbf{Q}^T|$, we may obtain from (123) the useful result

$$|\mathbf{Q}|^2 = 1: \qquad |\mathbf{Q}| = \pm 1. \tag{124}$$

It follows that the matrix $\mathbf{Q}$ defined by (120) does indeed have the property that *the quadratic form*

$$A = \mathbf{x}^T\,\mathbf{A}\,\mathbf{x} \tag{125}$$

is reduced by the change in variables

$$\mathbf{x} = \mathbf{Q}\,\mathbf{x}' \tag{126}$$

to the form

$$A = \mathbf{x}'^T\,\mathbf{A}'\,\mathbf{x}' \qquad where \qquad \mathbf{A}' = [\lambda_i \delta_{ij}],$$

that is, to the form

$$A = \lambda_1 x_1'^2 + \lambda_2 x_2'^2 + \cdots + \lambda_n x_n'^2, \tag{127}$$

where the numbers λ_i are the characteristic numbers of $\mathbf{A}$.

A matrix whose columns comprise the elements of n linearly independent characteristic vectors of a given matrix $\mathbf{A}$, of order n, is called a *modal matrix* of $\mathbf{A}$. In particular, when those n vectors are mutually orthogonal and of unit length, it is convenient to say that the modal matrix is *orthonormal.* Thus the matrix $\mathbf{Q}$ is an *orthonormal modal matrix* of $\mathbf{A}$.

We notice that if $\lambda = 0$ is an s-fold root of the characteristic equation the form (127) has only $n - s$ nonvanishing terms. It is shown in Section 1.18 that this situation arises if and only if the symmetric matrix $\mathbf{A}$ is of *rank* $r = n - s$.

The new variables x_i' are related to the original ones, in accordance with (115) and (123), by the equation

$$\mathbf{x}' = \mathbf{Q}^{-1}\,\mathbf{x} = \mathbf{Q}^T\,\mathbf{x}, \tag{128}$$

and hence are of the form

$$\left.\begin{array}{l} x_1' = e_{11}x_1 + e_{12}x_2 + \cdots + e_{1n}x_n, \\ \cdots\cdots\cdots\cdots\cdots\cdots\cdots\cdots\cdots \\ x_n' = e_{n1}x_1 + e_{n2}x_2 + \cdots + e_{nn}x_n \end{array}\right\} \tag{129}$$

or

$$x_i' = (\mathbf{e}_i, \mathbf{x}) \qquad (i = 1, 2, \cdots, n). \tag{129'}$$

When all the characteristic numbers of $\mathbf{A}$ are *distinct*, the orthonormal modal matrix $\mathbf{Q}$ is uniquely determined except for the ordering of the columns and the arbitrary algebraic sign associated with each column. However, if a root is of multiplicity s, the corresponding s orthogonalized unit vectors can be chosen in infinitely many ways, as was shown in Section 1.13.

We remark that the modal matrix $\mathbf{Q}$ specified by (120) is not the *only* matrix which can be used in (126) to reduce a quadratic form to a sum of squares. However, it is the only such matrix which possesses the useful property that

$$\mathbf{Q}^T = \mathbf{Q}^{-1}.$$

A matrix having this property is called an *orthogonal matrix.*

It is easily seen that *a square matrix is an orthogonal matrix if and only if its columns comprise the elements of real mutually orthogonal unit vectors.*

1.15. A numerical example. To illustrate the preceding reduction in a specific numerical case, we consider the quadratic form

$$A = 25x_1^{\,2} + 34x_2^{\,2} + 41x_3^{\,2} - 24x_2x_3.$$

The corresponding matrix $\mathbf{A}$ is then of the form

$$\mathbf{A} = \begin{bmatrix} 25 & 0 & 0 \\ 0 & 34 & -12 \\ 0 & -12 & 41 \end{bmatrix}$$

and the equations $\mathbf{A}\,\mathbf{x} - \lambda\,\mathbf{x} = \mathbf{0}$ become

$$\begin{aligned} (25 - \lambda)x_1 \qquad\qquad\qquad\qquad\qquad &= 0, \\ (34 - \lambda)x_2 \qquad -12x_3 &= 0, \\ -12x_2 + (41 - \lambda)x_3 &= 0. \end{aligned}$$

The characteristic equation $|\mathbf{A} - \lambda\,\mathbf{I}| = 0$ then takes the form

$$(25 - \lambda)(\lambda^2 - 75\lambda + 1250) = 0,$$

from which the characteristic numbers are determined:

$$\lambda_1 = \lambda_2 = 25, \qquad \lambda_3 = 50.$$

When $\lambda = \lambda_1 = \lambda_2 = 25$, the equations $\mathbf{A}\,\mathbf{x} - \lambda\,\mathbf{x} = \mathbf{0}$ become

$$\begin{aligned} 0 &= 0, \\ 9x_2 - 12x_3 &= 0, \\ -12x_2 + 16x_3 &= 0, \end{aligned}$$

with the general solution $x_1 = C_1$, $x_2 = C_2$, $x_3 = \frac{3}{4}C_2$. In vector form we may write $\mathbf{x} = C_1\mathbf{u}_1 + C_2\mathbf{u}_2$, where $\mathbf{u}_1 = \{1, 0, 0\}$ and $\mathbf{u}_2 = \{0, 1, \frac{3}{4}\}$. Since it happens that $\mathbf{u}_1$ and $\mathbf{u}_2$ are orthogonal, we need only divide them by their lengths $l_1 = 1$ and $l_2 = \frac{5}{4}$ to obtain the two orthogonal unit characteristic vectors

$$\mathbf{e}_1 = \{1, 0, 0\}, \qquad \mathbf{e}_2 = \{0, \tfrac{4}{5}, \tfrac{3}{5}\}.$$

In a similar way, a unit characteristic vector corresponding to $\lambda = \lambda_3 = 50$ is found to be

$$\mathbf{e}_3 = \{0, \tfrac{3}{5}, -\tfrac{4}{5}\}.$$

Hence the orthonormal modal matrix $\mathbf{Q}$ of equation (120) can be taken in the form

$$\mathbf{Q} = \begin{bmatrix} 1 & 0 & 0 \\ 0 & \frac{4}{5} & \frac{3}{5} \\ 0 & \frac{3}{5} & -\frac{4}{5} \end{bmatrix}$$

and the new coordinates defined by (129) are then given by

$$\begin{aligned} x_1' &= x_1, \\ x_2' &= \tfrac{4}{5}x_2 + \tfrac{3}{5}x_3, \\ x_3' &= \tfrac{3}{5}x_2 - \tfrac{4}{5}x_3. \end{aligned}$$

With this choice of the new coordinates, (127) states that the quadratic form under consideration takes the form

$$A = 25x_1'^2 + 25x_2'^2 + 50x_3'^2.$$

In particular, it follows that the quadric surface with the equation

$$25x_1^2 + 34x_2^2 + 41x_3^2 - 24x_2x_3 = 25$$

takes the standard form

$$x_1'^2 + x_2'^2 + 2x_3'^2 = 1$$

with the introduction of the new coordinates. It is shown in Section 1.21 that *the new* x_1'-x_n' *coordinate system defined by* (126) *is also a rectangular system when* $\mathbf{Q}$ *is an orthogonal matrix* and that *length and angle then are preserved by the transformation.* Hence the quadric surface just considered is an *oblate spheroid* with semiaxes of length 1, 1, $\sqrt{2}/2$.

In the present example it happens that the chosen matrix $\mathbf{Q}$ is *symmetric*, and hence also is such that $\mathbf{Q} = \mathbf{Q}^T = \mathbf{Q}^{-1}$.

It may be noticed that, by the usual method of "completing squares," we may, for example, also reduce the form A as follows:

$$\begin{aligned} A &= 25x_1^{\,2} + 34[x_2^{\,2} - \tfrac{24}{34}x_2x_3 + (\tfrac{12}{34})^2x_3^{\,2}] + (41 - \tfrac{144}{34})x_3^{\,2} \\ &= 25x_1^{\,2} + 34(x_2 - \tfrac{6}{17}x_3)^2 + \tfrac{625}{17}x_3^{\,2}. \end{aligned}$$

Hence, if we introduce new variables by the relations

$$\begin{aligned} x_1' &= x_1, \\ x_2' &= \qquad x_2 - \tfrac{6}{17}x_3, \\ x_3' &= \qquad\qquad\quad x_3, \end{aligned}$$

we can reduce A to the form

$$A = 25x_1'^2 + 34x_2'^2 + \tfrac{625}{17}x_3'^2.$$

However, here the matrix $\mathbf{Q}$ for which $\mathbf{x} = \mathbf{Q}\,\mathbf{x}'$, and which takes the *triangular* form

$$\mathbf{Q} = \begin{bmatrix} 1 & 0 & 0 \\ 0 & 1 & \tfrac{6}{17} \\ 0 & 0 & 1 \end{bmatrix},$$

is *not* an orthogonal matrix. Consequently, as is shown in Section 1.21, the new $x_1'x_2'x_3'$ coordinate system is *not* a rectangular system in this case; that is, the new coordinate axes are *not mutually perpendicular*. Nevertheless, the matrix $\mathbf{Q}$ does have the property that $\mathbf{Q}^T\,\mathbf{A}\,\mathbf{Q}$ is a diagonal matrix.

1.16. Equivalent matrices and transformations. Two matrices $\mathbf{A}$ and $\mathbf{B}$ which can be obtained from each other by a finite number of successive applications of the *elementary operations* (Section 1.8) to rows and/or columns are said to be *equivalent* (but not necessarily *equal*) *matrices.*

It can be shown that any such sequence of operations on the *rows* of $\mathbf{A}$ can be effected by *pre*multiplying $\mathbf{A}$ by some nonsingular matrix $\mathbf{P}$, while corresponding operations on *columns* can always be effected by *post*-multiplying $\mathbf{A}$ by a nonsingular matrix $\mathbf{Q}$. This result is a consequence of the easily established fact that an elementary operation on rows (columns) of $\mathbf{A}$ may be accomplished by first performing that operation on the *unit* matrix **I** of appropriate order, and then premultiplying (postmultiplying) **A** by the resultant matrix (see Problems 32 and 33).

The converse of the preceding statement is also true (see Problem 34); that is, *the matrices* **A** *and* **B** *are equivalent if and only if nonsingular matrices* **P** *and* **Q** *exist such that* $\mathbf{B} = \mathbf{P\,A\,Q}$.

Since the elementary operations do not change the rank of a matrix, it follows that *two equivalent matrices have the same rank.*

Transformations of the form $\mathbf{P\,A\,Q}$ are classified according to restrictions imposed on **P** and **Q**. Thus if $\mathbf{P} = \mathbf{Q}^T = \mathbf{Q}^{-1}$, as in the reduction of Section 1.14, the transformation is called an *orthogonal* transformation. If only $\mathbf{P} = \mathbf{Q}^T$, as is required by equation (118), the resulting transformation $\mathbf{Q}^T\,\mathbf{A}\,\mathbf{Q}$ is called a *congruence* transformation, whereas a transformation of the form $\mathbf{Q}^{-1}\,\mathbf{A}\,\mathbf{Q}$, for which $\mathbf{P} = \mathbf{Q}^{-1}$, is called a *similarity* transformation. This terminology is motivated by certain geometrical considerations. We notice that *an orthogonal transformation is both a congruence and a similarity transformation.*

Conjunctive and *unitary* transformations, which are of importance in dealing with matrices of *complex* elements, are defined in the following section.

1.17. Hermitian matrices. We now consider a matrix with *complex* elements which satisfy the relation

$$a_{ji} = \bar{a}_{ij}. \tag{130}$$

Such a matrix is hence of the special form

$$\mathbf{H} = \begin{bmatrix} a_{11} & a_{12} & a_{13} & \cdots & a_{1n} \\ \bar{a}_{12} & a_{22} & a_{23} & \cdots & a_{2n} \\ \bar{a}_{13} & \bar{a}_{23} & a_{33} & \cdots & a_{3n} \\ \cdots & \cdots & \cdots & \cdots & \cdots \\ \bar{a}_{1n} & \bar{a}_{2n} & \bar{a}_{3n} & \cdots & a_{nn} \end{bmatrix}, \tag{131}$$

and is known as a *Hermitian* matrix. Thus a Hermitian matrix has the property that two elements situated symmetrically with respect to the principal diagonal are *complex conjugates*. In particular, (130) requires that the elements in the principal diagonal ($i = j$) be *real.*

We see that the *conjugate* of the matrix **H**, obtained by replacing each element by its complex conjugate, and denoted by $\overline{\mathbf{H}}$, is equal to the transpose of **H**:

$$\mathbf{H}^T = \overline{\mathbf{H}}. \tag{132}$$

The product

$$H \equiv \bar{\mathbf{x}}^T\,\mathbf{H}\,\mathbf{x} \tag{133}$$

is known as a *Hermitian form.* In two dimensions, the general Hermitian form is thus given by

$$\begin{aligned} H &\equiv [\bar{x}_1 \quad \bar{x}_2]\begin{bmatrix} a_{11} & a_{12} \\ \bar{a}_{12} & a_{22} \end{bmatrix}\begin{Bmatrix} x_1 \\ x_2 \end{Bmatrix} \\ &\equiv a_{11}\bar{x}_1 x_1 + (a_{12}\bar{x}_1 x_2 + \bar{a}_{12}\bar{x}_2 x_1) + a_{22}\bar{x}_2 x_2. \end{aligned} \tag{134}$$

Although the elements a_{ij} and variables x_i may be complex, *the values assumed by a Hermitian form are always real.* To establish this fact, we recall first that *the conjugate of a product of complex quantities is equal to the product of the conjugates.* Thus, if H were imaginary (nonreal), and given by (133), then its conjugate $\bar{H}$ would be given by

$$\bar{H} = \mathbf{x}^T \, \bar{\mathbf{H}} \, \bar{\mathbf{x}} = \mathbf{x}^T \, \mathbf{H}^T \, \bar{\mathbf{x}} = (\mathbf{H} \, \mathbf{x})^T \, \bar{\mathbf{x}} = \bar{\mathbf{x}}^T \, (\mathbf{H} \, \mathbf{x}) = H. \tag{135}$$

But $\bar{H} = H$ only if H is real, as was to be shown.

Also, we can show that the characteristic numbers of a Hermitian matrix are real. For if $\mathbf{u}_1$ is a characteristic vector corresponding to λ_1, we must have

$$\mathbf{H} \, \mathbf{u}_1 = \lambda_1 \mathbf{u}_1, \tag{136}$$

and hence also, after premultiplying both sides by $\bar{\mathbf{u}}_1^T$,

$$\bar{\mathbf{u}}_1^T \, \mathbf{H} \, \mathbf{u}_1 = \lambda_1 \bar{\mathbf{u}}_1^T \, \mathbf{u}_1. \tag{137}$$

But since $\bar{\mathbf{u}}_1^T \, \mathbf{H} \, \mathbf{u}_1$ and $\bar{\mathbf{u}}_1^T \, \mathbf{u}_1$ are both real, and $\bar{\mathbf{u}}_1^T \, \mathbf{u}_1 \neq 0$, λ_1 must also be real.

Further, let $\mathbf{u}_2$ be a characteristic vector corresponding to a second characteristic number $\lambda_2 \neq \lambda_1$, so that

$$\mathbf{H} \, \mathbf{u}_2 = \lambda_2 \mathbf{u}_2. \tag{138}$$

If the transposed conjugate of (136) is postmultiplied by $\mathbf{u}_2$, there follows

$$(\bar{\mathbf{H}} \, \bar{\mathbf{u}}_1)^T \, \mathbf{u}_2 = \lambda_1 \bar{\mathbf{u}}_1^T \, \mathbf{u}_2,$$

while premultiplication of (138) by $\bar{\mathbf{u}}_1^T$ leads to the relation

$$\bar{\mathbf{u}}_1^T \, \mathbf{H} \, \mathbf{u}_2 = \lambda_2 \bar{\mathbf{u}}_1^T \, \mathbf{u}_2.$$

By subtracting these equations from each other, and using equations (34) and (132), there follows

$$\begin{aligned} (\lambda_2 - \lambda_1)\bar{\mathbf{u}}_1^T \, \mathbf{u}_2 &= \bar{\mathbf{u}}_1^T \, \mathbf{H} \, \mathbf{u}_2 - (\bar{\mathbf{H}} \, \bar{\mathbf{u}}_1)^T \, \mathbf{u}_2 \\ &= \bar{\mathbf{u}}_1^T \, \mathbf{H} \, \mathbf{u}_2 - \bar{\mathbf{u}}_1^T \, \bar{\mathbf{H}}^T \, \mathbf{u}_2 \\ &= 0. \end{aligned}$$

Hence we conclude that *two characteristic vectors of a Hermitian matrix, corresponding to different characteristic numbers, are orthogonal in the Hermitian sense:*

$$(\bar{\mathbf{u}}_1, \mathbf{u}_2) \equiv \bar{\mathbf{u}}_1{}^T \mathbf{u}_2 = 0. \tag{139}$$

The vectors $\mathbf{u}_i$ then can be divided by their Hermitian lengths

$$l_H(\mathbf{u}_i) = (\bar{\mathbf{u}}_i, \mathbf{u}_i)^{1/2},$$

to give a set of orthogonal *unit* vectors $\mathbf{e}_i$ corresponding to successive non-repeated roots of the characteristic equation. Corresponding to a root of multiplicity s there exists a set of s linearly independent characteristic vectors (see Section 1.18), which can be orthogonalized and reduced to absolute length unity, by a procedure completely analogous to that given in Section 1.13. Thus we may again obtain a set of n mutually orthogonal unit characteristic vectors $\mathbf{e}_1, \mathbf{e}_2, \cdots, \mathbf{e}_n$, orthogonality and length being defined in the Hermitian sense.

It then follows that any complex n-dimensional vector $\mathbf{v}$ can be expressed in the form

$$\mathbf{v} = \sum_{k=1}^{n} \alpha_k \mathbf{e}_k, \tag{140}$$

where the kth coefficient is given by the formula

$$\alpha_k = (\bar{\mathbf{e}}_k, \mathbf{v}). \tag{141}$$

In order to solve the equation

$$\mathbf{H}\,\mathbf{x} - \lambda\,\mathbf{x} = \mathbf{c}, \tag{142}$$

we may assume the expansion

$$\mathbf{x} = \sum_{k=1}^{n} \alpha_k \mathbf{e}_k, \tag{143}$$

as in the real case, so that (142) takes the form

$$\sum_{k=1}^{n} (\lambda_k - \lambda) \alpha_k \mathbf{e}_k = \mathbf{c}$$

and there follows

$$(\lambda_k - \lambda)\alpha_k = (\bar{\mathbf{e}}_k, \mathbf{c}).$$

Thus, if λ is not a characteristic number, the solution becomes

$$\mathbf{x} = \sum_{k=1}^{n} \frac{(\bar{\mathbf{e}}_k, \mathbf{c})}{\lambda_k - \lambda}\, \mathbf{e}_k \tag{144}$$

in analogy with (105). If $\lambda = \lambda_p$, no solution exists unless $\mathbf{c}$ is such that $(\bar{\mathbf{e}}_p, \mathbf{c}) = 0$, in which case α_p is arbitrary, and infinitely many solutions exist.

The reduction of a Hermitian form to a sum of the *canonical form*

$$H = \lambda_1 \bar{x}_1' x_1' + \lambda_2 \bar{x}_2' x_2' + \cdots + \lambda_n \bar{x}_n' x_n' \tag{145}$$

may be accomplished by a method analogous to that of Section 1.14. Thus, if we write

$$\mathbf{x} = \mathbf{U}\,\mathbf{x}', \tag{146}$$

the form H of (133) becomes

$$H = (\bar{\mathbf{U}}\,\bar{\mathbf{x}}')^T\,\mathbf{H}\,\mathbf{U}\,\mathbf{x}' = \bar{\mathbf{x}}'^T(\bar{\mathbf{U}}^T\,\mathbf{H}\,\mathbf{U})\mathbf{x}'. \tag{147}$$

This form will be identified with (145) if and only if $\mathbf{U}$ is such that

$$\bar{\mathbf{U}}^T\,\mathbf{H}\,\mathbf{U} = [\lambda_i\delta_{ij}]. \tag{148}$$

As in Section 1.13, a permissible choice of $\mathbf{U}$ consists of an orthonormal modal matrix formed by arranging n orthogonal unit characteristic vectors of $\mathbf{H}$ as its columns, in which case the scalars λ_i are the characteristic numbers of $\mathbf{H}$. For the matrix $\mathbf{U}$ it is found that

$$\bar{\mathbf{U}}^T = \mathbf{U}^{-1} \qquad \mathit{or} \qquad \bar{\mathbf{U}}^T\,\mathbf{U} = \mathbf{I}. \tag{149}$$

A matrix $\mathbf{U}$ having the property (149) is called a *unitary* (or *Hermitian orthogonal*) *matrix*, and the product $\bar{\mathbf{U}}^T\,\mathbf{H}\,\mathbf{U}$ is then called a *unitary transformation* of $\mathbf{H}$. More generally, a transformation of the form $\bar{\mathbf{U}}^T\,\mathbf{H}\,\mathbf{U}$, where $\mathbf{U}$ does not necessarily satisfy (149), is called a *conjunctive transformation* of $\mathbf{H}$.

It is easily seen that *a square matrix is unitary if and only if its columns comprise the elements of mutually orthogonal unit vectors, length and orthogonality being defined in the Hermitian sense.*

1.18. Multiple characteristic numbers of symmetric matrices. We next establish the assertion, made in Section 1.12, that a real symmetric matrix $\mathbf{A}$ with a characteristic number λ_1 of multiplicity s has an s-parameter family of characteristic vectors corresponding to λ_1.

For this purpose, suppose first that $\mathbf{A}$ is a symmetric $n \times n$ matrix such that the characteristic polynomial

$$F(\lambda) = |\mathbf{A} - \lambda\,\mathbf{I}|$$

has $(\lambda - \lambda_1)^2$ as a factor, and let $\mathbf{e}_1$ be *one* unit characteristic vector of $\mathbf{A}$ corresponding to λ_1. Then, if $\mathbf{Q}$ is *any* $n \times n$ *orthogonal* matrix having $\mathbf{e}_1$ as its *first* column vector, there follows

$$\begin{aligned}\mathbf{Q}^T\,\mathbf{A}\,\mathbf{Q} &= \begin{bmatrix} -\mathbf{e}_1{}^T\!\rightarrow \\ \cdots\cdots \\ \cdots\cdots \end{bmatrix} \mathbf{A} \begin{bmatrix} \mid & \vdots & \vdots \\ \mathbf{e}_1 & \vdots & \vdots \\ \downarrow & \vdots & \vdots \end{bmatrix} \\ &= \begin{bmatrix} -\mathbf{e}_1{}^T\!\rightarrow \\ \cdots\cdots \\ \cdots\cdots \end{bmatrix} \begin{bmatrix} \mid & \vdots & \vdots \\ \lambda_1\mathbf{e}_1 & \vdots & \vdots \\ \downarrow & \vdots & \vdots \end{bmatrix}.\end{aligned} \tag{150}$$

Since each vector whose elements comprise a row of $\mathbf{Q}^T$, except the first, is orthogonal to the column $\lambda_1\mathbf{e}_1$, each element of the first column of the product except the leading element will vanish. Thus the result will be of the form

$$\mathbf{Q}^T\,\mathbf{A}\,\mathbf{Q} = \begin{bmatrix} \lambda_1 & \alpha_{12} & \cdots & \alpha_{1n} \\ 0 & \alpha_{22} & \cdots & \alpha_{2n} \\ \cdots & \cdots & \cdots & \cdots \\ 0 & \alpha_{n2} & \cdots & \alpha_{nn} \end{bmatrix} \tag{151}$$

where the α's are certain constants, depending upon the remaining columns of $\mathbf{Q}$. But, since the symmetry of $\mathbf{A}$ implies the symmetry of $\mathbf{Q}^T\,\mathbf{A}\,\mathbf{Q}$, the elements $\alpha_{12}, \cdots, \alpha_{1n}$ *also* must vanish.

Hence, if $\mathbf{Q}$ is *any* orthogonal matrix having $\mathbf{e}_1$ as its first column vector, the product $\mathbf{Q}^T\,\mathbf{A}\,\mathbf{Q} = \mathbf{Q}^{-1}\,\mathbf{A}\,\mathbf{Q}$ is of the form

$$\mathbf{Q}^{-1}\,\mathbf{A}\,\mathbf{Q} = \begin{bmatrix} \lambda_1 & 0 & \cdots & 0 \\ 0 & \alpha_{22} & \cdots & \alpha_{2n} \\ \cdots & \cdots & \cdots & \cdots \\ 0 & \alpha_{2n} & \cdots & \alpha_{nn} \end{bmatrix} \tag{152}$$

and also there follows

$$\mathbf{Q}^{-1}\,\mathbf{A}\,\mathbf{Q} - \lambda\,\mathbf{I} = \begin{bmatrix} \lambda_1 - \lambda & 0 & \cdots & 0 \\ 0 & \alpha_{22} - \lambda & \cdots & \alpha_{2n} \\ \cdots & \cdots & \cdots & \cdots \\ 0 & \alpha_{2n} & \cdots & \alpha_{nn} - \lambda \end{bmatrix}. \tag{153}$$

Next we notice that, in consequence of the relation

$$\mathbf{Q}^{-1}\,\mathbf{A}\,\mathbf{Q} - \lambda\,\mathbf{I} = \mathbf{Q}^{-1}(\mathbf{A} - \lambda\,\mathbf{I})\mathbf{Q}, \tag{154}$$

there follows also

$$|\mathbf{Q}^{-1}\,\mathbf{A}\,\mathbf{Q} - \lambda\,\mathbf{I}| = |\mathbf{A} - \lambda\,\mathbf{I}| \tag{155}$$

and hence (153) implies the relation

$$|\mathbf{A} - \lambda\,\mathbf{I}| = \begin{vmatrix} \lambda_1 - \lambda & 0 & \cdots & 0 \\ 0 & \alpha_{22} - \lambda & \cdots & \alpha_{2n} \\ \cdots & \cdots & \cdots & \cdots \\ 0 & \alpha_{2n} & \cdots & \alpha_{nn} - \lambda \end{vmatrix}$$

$$= (\lambda_1 - \lambda) \begin{vmatrix} \alpha_{22} - \lambda & \cdots & \alpha_{2n} \\ \cdots & \cdots & \cdots \\ \alpha_{2n} & \cdots & \alpha_{nn} - \lambda \end{vmatrix}. \tag{156}$$

But since, by hypothesis, the left-hand member of (156) has $(\lambda_1 - \lambda)^2$ as a factor, it follows that the coefficient of $(\lambda_1 - \lambda)$ in the right-hand member has $(\lambda_1 - \lambda)$ as a factor, and hence *vanishes when* $\lambda = \lambda_1$. Thus *the matrix* $\mathbf{Q}^{-1}\,\mathbf{A}\,\mathbf{Q} - \lambda\,\mathbf{I}$ *in* (153) *is of rank* $n - 2$ *or less when* $\lambda = \lambda_1$.

Finally, since (154) states that the matrix $\mathbf{Q}^{-1}\,\mathbf{A}\,\mathbf{Q} - \lambda_1\mathbf{I}$ is *similar* to the matrix $\mathbf{A} - \lambda_1\mathbf{I}$, we deduce that *the rank of the matrix* $\mathbf{A} - \lambda_1\mathbf{I}$ *is not greater than* $n - 2$ *when* λ_1 *is a characteristic number of* $\mathbf{A}$ *of multiplicity two or more*, that is, that *the matrix equation* $\mathbf{A}\,\mathbf{x} = \lambda_1\mathbf{x}$ *has at least a two-parameter family of solutions in that case.*

If the multiplicity of λ_1 is *greater* than two, then by taking $\mathbf{Q}$ to be any orthogonal matrix having *two* orthogonal unit vectors $\mathbf{e}_1$ and $\mathbf{e}_2$, both corresponding to λ_1, as its first *two* column vectors, we deduce in a completely analogous way that

$$\mathbf{Q}^{-1}\,\mathbf{A}\,\mathbf{Q} - \lambda\,\mathbf{I} = \begin{bmatrix} \lambda_1 - \lambda & 0 & 0 & \cdots & 0 \\ 0 & \lambda_1 - \lambda & 0 & \cdots & 0 \\ 0 & 0 & \alpha_{33} - \lambda & \cdots & \alpha_{3n} \\ \cdots & \cdots & \cdots & \cdots & \cdots \\ 0 & 0 & \alpha_{3n} & \cdots & \alpha_{nn} - \lambda \end{bmatrix} \tag{157}$$

and the same argument leads to the conclusion that the matrix $\mathbf{A} - \lambda\,\mathbf{I}$ is of rank not greater than $n - 3$ when $\lambda = \lambda_1$, so that at least a *three*-parameter family of corresponding characteristic vectors exists when the multiplicity of λ_1 is at least three.

By inductive reasoning, we thus deduce that if λ_1 is a characteristic number of a symmetric matrix $\mathbf{A}$, of multiplicity s, then the rank of the matrix $\mathbf{A} - \lambda\,\mathbf{I}$ is not greater than $n - s$ when $\lambda = \lambda_1$, so that *at least* s linearly independent characteristic vectors corresponding to λ_1 exist. However, the rank also cannot be *less* than $n - s$, for if this were so, more than s linearly independent characteristic vectors would correspond to λ_1, in which case the total number of linearly independent characteristic vectors corresponding to *all* characteristic numbers would exceed the dimension of n-space.

Thus we deduce the desired result:

If λ_1 *is a characteristic number of multiplicity* s, *of a real symmetric matrix* $\mathbf{A}$ *of order* n, *then the rank of the matrix* $\mathbf{A} - \lambda\,\mathbf{I}$ *is exactly* $n - s$ *when* $\lambda = \lambda_1$; *that is, there exist exactly* s *linearly independent corresponding characteristic vectors.*

This statement does not apply *in general* to a *nonsymmetric* matrix, as was shown by an example in Section 1.12. However, an argument analogous to that given above shows that *the same statement applies to Hermitian matrices.*

1.19. Definite forms. If the real quadratic form $\mathbf{x}^T\mathbf{A}\,\mathbf{x}$, associated with a real symmetric matrix $\mathbf{A}$, is *nonnegative* for all real values of the variables x_i, and is *zero* only if *each* of those n variables is zero, then that quadratic form is said to be *positive definite.* It is then conventional to say also that the *matrix* $\mathbf{A}$ is positive definite.

Similarly, a *Hermitian* matrix $\mathbf{H}$ is said to be positive definite if the associated *Hermitian* form $\bar{\mathbf{x}}^T \mathbf{H} \mathbf{x}$ is nonnegative for any real or complex vector $\mathbf{x}$, and vanishes only when $\mathbf{x} = \mathbf{0}$.

If a real quadratic form $A = \mathbf{x}^T \mathbf{A} \mathbf{x}$ is reducible by a transformation of the form $\mathbf{x} = \mathbf{Q} \mathbf{x}'$, where $\mathbf{Q}$ is a nonsingular real matrix, to the sum of *squares* of the n new variables, each with a *positive* coefficient, then it is clear that A is a positive definite form relative to the real variables $x_1', \cdots, x_n'$. But from the relation $\mathbf{x}' = \mathbf{Q}^{-1} \mathbf{x}$, which is a consequence of the assumed nonvanishing of $|\mathbf{Q}|$, we see that a real vector $\mathbf{x}$ then corresponds always to a real vector $\mathbf{x}'$, and that the vectors $\mathbf{x} = \mathbf{0}$ and $\mathbf{x}' = \mathbf{0}$ then correspond uniquely. Hence it follows in this case that A is also positive definite relative to the *original* real variables $x_1, \cdots, x_n$.

Similarly, if a Hermitian form is reducible by a nonsingular complex transformation to the canonical form (145), wherein all coefficients are positive, the form is then nonnegative for any *complex* values of the variables, and is zero if and only if all the n variables vanish.

We notice that if the coefficients of the squares of any of the n variables are zero, then the vanishing of the form does not imply the vanishing of *those* variables, and hence the form is then *not* positive definite relative to the entire set of n variables.

It then follows from the results of preceding sections [see equations (127) and (145)] that *a real quadratic* (*or Hermitian*) *form is positive definite if and only if the characteristic numbers of the associated real symmetric* (*or Hermitian*) *matrix are all positive.*

A form is often said to be *positive semidefinite* when it takes on only *nonnegative* values for all permissible values of the variables, but *vanishes* for *some* nonzero values of the variables. The preceding argument leads easily to the fact that *a quadratic* (*or Hermitian*) *form is positive semidefinite if and only if the associated symmetric* (*or Hermitian*) *matrix is singular and possesses no negative characteristic numbers.*

Positive definite forms are of particular importance in applications, and are found to possess certain useful properties. In particular, we show next that if at least one of the *two* real quadratic forms

$$A = \mathbf{x}^T \mathbf{A} \mathbf{x}, \qquad B = \mathbf{x}^T \mathbf{B} \mathbf{x} \tag{158a,b}$$

is *positive definite*, then it is always possible to reduce the two forms *simultaneously* to linear combinations of only squares of new variables, that is, to canonical forms, by a nonsingular real transformation. For this purpose, suppose that the form B is positive definite. Then, by proceeding exactly as in Section 1.14, we first set

$$\mathbf{x} = \mathbf{Q} \mathbf{y}, \tag{159}$$

where $\mathbf{Q}$ is an *orthonormal modal matrix* of $\mathbf{B}$, defined in that section, and so reduce B to the form

$$B = \mu_1 y_1^2 + \mu_2 y_2^2 + \cdots + \mu_n y_n^2, \tag{160}$$

where here μ_i is written for the ith characteristic number of the symmetric matrix $\mathbf{B}$. Since B is positive definite, the μ's are all positive. Hence we may make the real substitution

$$\eta_i = \sqrt{\mu_i}\, y_i \qquad (i = 1, 2, \cdots, n), \tag{161}$$

and thus reduce (160) to the form

$$B = \eta_1^2 + \eta_2^2 + \cdots + \eta_n^2 = \boldsymbol{\eta}^T \boldsymbol{\eta}. \tag{162}$$

At the same time, the substitution (159) reduces A to the form

$$A = (\mathbf{Q}\,\mathbf{y})^T \mathbf{A}\,\mathbf{Q}\,\mathbf{y} = \mathbf{y}^T (\mathbf{Q}^T \mathbf{A}\,\mathbf{Q})\,\mathbf{y} \tag{163}$$

and the subsequent substitution (161) reduces this form to the expression

$$A = \boldsymbol{\eta}^T(\mathbf{Q}'^T \mathbf{A}\,\mathbf{Q}')\boldsymbol{\eta}, \tag{164}$$

where $\mathbf{Q}'$ is a matrix obtained from $\mathbf{Q}$ by dividing each element of the ith column of $\mathbf{Q}$ by $\sqrt{\mu_i}$. Hence, if we write

$$\mathbf{G} = \mathbf{Q}'^T \mathbf{A}\,\mathbf{Q}', \tag{165}$$

equation (164) takes the form

$$A = \boldsymbol{\eta}^T \mathbf{G}\,\boldsymbol{\eta}. \tag{166}$$

Now $\mathbf{G}$ is a symmetric matrix, since

$$\mathbf{G}^T = (\mathbf{Q}'^T \mathbf{A}\,\mathbf{Q}')^T = \mathbf{Q}'^T \mathbf{A}^T \mathbf{Q}' = \mathbf{Q}'^T \mathbf{A}\,\mathbf{Q}' = \mathbf{G}. \tag{167}$$

Hence we may reduce (166) to canonical form by setting

$$\boldsymbol{\eta} = \mathbf{R}\,\boldsymbol{\alpha}, \tag{168}$$

where $\mathbf{R}$ is made up of the characteristic vectors of $\mathbf{G}$ just as $\mathbf{Q}$ is formed from those of $\mathbf{B}$, and (166) is reduced to the form

$$A = \lambda_1 \alpha_1^2 + \lambda_2 \alpha_2^2 + \cdots + \lambda_n \alpha_n^2 \tag{169}$$

where λ_i is the ith characteristic number of the matrix $\mathbf{G}$.

At the same time, the final substitution (168) reduces (162) to

$$B = \boldsymbol{\eta}^T \boldsymbol{\eta} = (\mathbf{R}\,\boldsymbol{\alpha})^T(\mathbf{R}\,\boldsymbol{\alpha}) = \boldsymbol{\alpha}^T \mathbf{R}^T \mathbf{R}\,\boldsymbol{\alpha}. \tag{170}$$

But since the matrix $\mathbf{R}$ is an *orthogonal* matrix, there follows $\mathbf{R}^T \mathbf{R} = \mathbf{I}$, and hence we have the result

$$B = \boldsymbol{\alpha}^T \boldsymbol{\alpha} = \alpha_1^2 + \alpha_2^2 + \cdots + \alpha_n^2. \tag{171}$$

Thus, finally, with the substitution

$$\mathbf{x} = \mathbf{Q}\,\mathbf{y} = \mathbf{Q}'\,\boldsymbol{\eta} = \mathbf{Q}'\,\mathbf{R}\,\boldsymbol{\alpha}, \tag{172}$$

the two forms (158a,b) are simultaneously reduced to the canonical forms (169) and (171).

If we define the diagonal matrix

$$\mathbf{M} = \begin{bmatrix} m_1 & 0 & \cdots & 0 \\ 0 & m_2 & \cdots & 0 \\ \cdots & \cdots & \cdots & \cdots \\ 0 & 0 & \cdots & m_n \end{bmatrix}, \qquad m_i = \frac{1}{\sqrt{\mu_i}}, \tag{173}$$

it follows that

$$\mathbf{Q}' = \mathbf{Q}\,\mathbf{M} \tag{174}$$

and (172) becomes

$$\mathbf{x} = \mathbf{Q}\,\mathbf{M}\,\mathbf{R}\,\boldsymbol{\alpha}. \tag{175}$$

Since $\mathbf{Q}$ and $\mathbf{R}$ are orthogonal matrices, with determinants equal to unity in absolute value [see equation (124)], and since clearly $|\mathbf{M}| \neq 0$, it follows that the transformation (172) is indeed nonsingular.

In certain applications to dynamical problems (see Section 2.14) the positive definite form B (*kinetic energy*) involves the time derivative $d\mathbf{x}/dt$ in place of $\mathbf{x}$, whereas the form A (*potential energy*) involves only $\mathbf{x}$ itself. The above reduction is still applicable, however, since $\mathbf{x}$ and $d\mathbf{x}/dt$ are transformed in the same way at each step of the process.

Another method of accomplishing the same reduction, which is usually more conveniently applied in practice, is presented in Section 1.25 [see equations (265)–(267)].

1.20. Discriminants and invariants. It is frequently of importance to determine whether a quadratic or Hermitian form which involves cross-product terms is or is not a *positive definite* form, without reducing it to a canonical form or determining the characteristic numbers of the associated matrix. This problem is to be considered in the present section.

If we write the characteristic equation $|\mathbf{A} - \lambda\,\mathbf{I}| = 0$ of a square matrix $\mathbf{A}$ in the form

$$\begin{vmatrix} a_{11} - \lambda & a_{12} & \cdots & a_{1n} \\ a_{21} & a_{22} - \lambda & \cdots & a_{2n} \\ \cdots & \cdots & \cdots & \cdots \\ a_{n1} & a_{n2} & \cdots & a_{nn} - \lambda \end{vmatrix}$$

$$\equiv (-1)^n[\lambda^n - \beta_1\lambda^{n-1} + \beta_2\lambda^{n-2} - \cdots + (-1)^n\beta_n] = 0, \tag{176}$$

and denote the n roots of this equation as $\lambda_1, \lambda_2, \cdots, \lambda_n$, numbering multiple roots separately, it follows that

$$\lambda^n - \beta_1\lambda^{n-1} + \beta_2\lambda^{n-2} - \cdots + (-1)^n\beta_n \equiv (\lambda - \lambda_1)(\lambda - \lambda_2)\cdots(\lambda - \lambda_n). \tag{177}$$

By comparing coefficients of λ in the two sides of (177) it can be shown that

$$\left.\begin{aligned} \beta_1 &= \lambda_1 + \lambda_2 + \cdots + \lambda_n, \\ \beta_2 &= \lambda_1\lambda_2 + \lambda_1\lambda_3 + \cdots + \lambda_{n-1}\lambda_n, \\ \beta_3 &= \lambda_1\lambda_2\lambda_3 + \cdots + \lambda_{n-2}\lambda_{n-1}\lambda_n, \\ &\cdots\cdots\cdots\cdots\cdots\cdots\cdots\cdots \\ \beta_n &= \lambda_1\lambda_2\lambda_3\cdots\lambda_n \end{aligned}\right\} . \tag{178}$$

Now, for either a *real symmetric* or a *Hermitian* matrix, we have shown that the roots of (176) are all *real.* Hence, by Descartes' rule of signs, we see that in such cases *the roots of the characteristic equation* (176) *are all positive if and only if the quantities* $\beta_1, \beta_2, \cdots, \beta_n$ *are all positive.*

From (176) it follows that β_n is the value of $|\mathbf{A} - \lambda\,\mathbf{I}|$ when $\lambda = 0$; that is, β_n is the value of the determinant of $\mathbf{A}$:

$$\beta_n = |a_{ij}|. \tag{179}$$

Further, it is easily seen that the coefficient of λ^{n-1} in the expansion of the determinant in (176) is merely

$$(-1)^{n+1}(a_{11} + a_{22} + \cdots + a_{nn});$$

that is, β_1 is the sum of the *diagonal elements* of $\mathbf{A}$:

$$\beta_1 = a_{11} + a_{22} + \cdots + a_{nn} = \sum_{k=1}^{n} a_{kk}. \tag{180}$$

This sum is called the *trace* of $\mathbf{A}$.

More generally, it can be shown that β_i *is the sum of all determinants formed from square arrays of order i whose principal diagonals lie along the principal diagonal of* $\mathbf{A}$. Such determinants are called the *principal minors* of $\mathbf{A}$.

Thus it follows that *a real quadratic* (*or Hermitian*) *form is positive definite if and only if the sums* β_i, *relevant to the associated symmetric* (*or Hermitian*) *matrix, are all positive.*

In illustration, the real quadratic form

$$A = a_{11}x_1^{\,2} + a_{22}x_2^{\,2} + a_{33}x_3^{\,2} + 2a_{12}x_1x_2 + 2a_{23}x_2x_3 + 2a_{13}x_1x_3 \tag{181}$$

in three dimensions, which is associated with the real matrix

$$\mathbf{A} = \begin{bmatrix} a_{11} & a_{12} & a_{13} \\ a_{12} & a_{22} & a_{23} \\ a_{13} & a_{23} & a_{33} \end{bmatrix}, \tag{182}$$

is positive definite if and only if the three conditions

$$a_{11} + a_{22} + a_{33} > 0, \tag{183a}$$

$$(a_{11}a_{22} - a_{12}^2) + (a_{22}a_{33} - a_{23}^2) + (a_{11}a_{33} - a_{13}^2) > 0, \tag{183b}$$

$$|a_{ij}| > 0, \tag{183c}$$

are satisfied.

It is readily verified by direct expansion that the determinant of the *symmetric* matrix (182) can be written in the form

$$|a_{ij}| = \frac{(a_{11}a_{22} - a_{12}^2)(a_{11}a_{33} - a_{13}^2) - (a_{11}a_{23} - a_{12}a_{13})^2}{a_{11}}, \tag{184}$$

and also in two further equivalent forms obtained by cyclic permutation of the subscripts.

Suppose that we require only that

$$a_{11} > 0, \qquad a_{11}a_{22} - a_{12}^2 > 0, \qquad |a_{ij}| > 0. \tag{185a,b,c}$$

It then follows from (185a,b) that we must have $a_{22} > 0$, and also, by referring to (184), we see that (185a,b,c) imply that $a_{11}a_{33} - a_{13}^2 > 0$. This relation, together with (185a), in turn implies that $a_{33} > 0$. By considering the permutation of the right-hand member of (184) in which $1 \rightarrow 2$, $2 \rightarrow 3$, $3 \rightarrow 1$, we then deduce similarly that (185a,b,c) also imply the inequality $a_{22}a_{33} - a_{23}^2 > 0$. Thus it follows that *the three conditions* (185) *imply the three conditions* (183).

By considering the conditions that (181) still be positive definite when, first, only *one* variable differs from zero and when, second, only *two* variables differ from zero, it is easily seen that *each* diagonal term a_{ii} must be positive and also that *each* principal minor of second order must be positive. Hence these conditions imply and must be implied by either the conditions (183) or the more convenient conditions (185).

More generally, if for *any* real symmetric (or Hermitian) matrix $\mathbf{A}$ we define the mth *discriminant* Δ_m to be the determinant of the submatrix $\mathbf{D}_m$ obtained by deleting all elements which do not simultaneously lie in the first m rows and columns of $\mathbf{A}$, it can be shown that *the real symmetric* (*or Hermitian*) *matrix* $\mathbf{A}$, *and the corresponding quadratic* (*or Hermitian*) *form, is positive definite if and only if each of the n discriminants* Δ_m *is positive.* If and only if this is so, *all* the principal minors of $\mathbf{A}$ are positive.

To establish the sufficiency of this criterion, we need only prove that, if $\mathbf{D}_m$ is positive definite and $\Delta_{m+1} \equiv |\mathbf{D}_{m+1}|$ is positive, then $\mathbf{D}_{m+1}$ is also positive definite. Suppose, on the contrary, that $\mathbf{D}_{m+1}$ is *not* positive definite. Then, since $|\mathbf{D}_{m+1}|$ is the *product* of the characteristic numbers of $\mathbf{D}_{m+1}$, it follows that an *even* number of these characteristic numbers must be negative. Let γ_1 and γ_2 be two such numbers, and denote by $\mathbf{u}_1$ and $\mathbf{u}_2$ corresponding orthogonal unit characteristic vectors of $\mathbf{D}_{m+1}$, length and orthogonality being defined in the Hermitian sense. If we define the $(m + 1)$-dimensional vector

$$\mathbf{x}^* = c_1\mathbf{u}_1 + c_2\mathbf{u}_2,$$

where at least one of the c's does not vanish, and notice that then we have $\mathbf{D}_{m+1}\,\mathbf{u}_1 = \gamma_1\mathbf{u}_1$ and $\mathbf{D}_{m+1}\,\mathbf{u}_2 = \gamma_2\mathbf{u}_2$, there follows easily

$$\bar{\mathbf{x}}^{*T}\,\mathbf{D}_{m+1}\,\mathbf{x}^* = \bar{c}_1c_1\gamma_1 + \bar{c}_2c_2\gamma_2 < 0,$$

for any c_1 and c_2. Thus the vector $\mathbf{x}^*$ renders the Hermitian form associated with $\mathbf{D}_{m+1}$ negative. Now let c_1 and c_2 be related in such a way that the component x^*_{m+1} vanishes. If we notice that the Hermitian form $\bar{\mathbf{x}}^T\,\mathbf{D}_{m+1}\,\mathbf{x}$ reduces to the form $\bar{\mathbf{x}}^T\,\mathbf{D}_m\,\mathbf{x}$ when $x_{m+1} = 0$, we conclude that the m-dimensional vector made up of the first m components of the $\mathbf{x}^*$ so determined renders the Hermitian form associated with $\mathbf{D}_m$ negative. Since $\mathbf{D}_m$ is positive definite, this situation is impossible, and the desired contradiction is obtained.

The specialization of the preceding argument to the case of a real symmetric matrix, and its associated real quadratic form, is obtained by deleting the bars indicating complex conjugates. In this case, $\mathbf{u}_1$ and $\mathbf{u}_2$ are real, and the constants c_1 and c_2 are also to be real.

Whereas the requirements that a form or matrix be positive definite thus need not be stated in terms of the *sums* β_i, these sums nevertheless are of considerable importance in themselves. We see from (178) that each β_i is a symmetric function, of degree i, of the characteristic numbers of $\mathbf{A}$. Also, it follows from (176) that *for any two square matrices* $\mathbf{A}$ *and* $\mathbf{B}$ *such that* $|\mathbf{A} - \lambda\,\mathbf{I}| = |\mathbf{B} - \lambda\,\mathbf{I}|$ *for all values of* λ, *the n quantities* β_i *are the same.*

In order to determine conditions under which this situation exists, let $\mathbf{A}$ and $\mathbf{B}$ be two equivalent matrices. This means that nonsingular matrices $\mathbf{P}$ and $\mathbf{Q}$ exist such that $\mathbf{B} = \mathbf{P\,A\,Q}$. Hence we have, for any value of λ,

$$\begin{aligned}\mathbf{B} - \lambda\,\mathbf{I} &= \mathbf{P\,A\,Q} - \lambda\,\mathbf{I}\\ &= \mathbf{P}(\mathbf{A} - \lambda\,\mathbf{P}^{-1}\,\mathbf{Q}^{-1})\mathbf{Q}\end{aligned}$$

and also

$$|\mathbf{B} - \lambda\,\mathbf{I}| = |\mathbf{P}|\;|\mathbf{Q}|\;|\mathbf{A} - \lambda\,\mathbf{P}^{-1}\,\mathbf{Q}^{-1}|. \tag{186}$$

Thus, if $\mathbf{P}$ and $\mathbf{Q}$ are such that $\mathbf{P}^{-1}\,\mathbf{Q}^{-1} = \mathbf{I}$ or $\mathbf{P} = \mathbf{Q}^{-1}$, so that

$$\mathbf{B} = \mathbf{Q}^{-1}\,\mathbf{A}\,\mathbf{Q}, \tag{187}$$

there follows $\mathbf{P}\,\mathbf{Q} = \mathbf{I}$, and hence $|\mathbf{P}|\,|\mathbf{Q}| = 1$, and (186) takes the form

$$|\mathbf{B} - \lambda\,\mathbf{I}| = |\mathbf{A} - \lambda\,\mathbf{I}|, \tag{188}$$

for all values of λ.

A transformation of the form (187) has been defined as a *similarity transformation*, and the matrices **A** and **B** are said to be *similar*. Since (188) states that **A** and **B** have the same characteristic equation, it follows that *the quantities β_i are invariant under* (unchanged by) *any similarity transformation*. This result has important consequences in many physical considerations.

Since *orthogonal* and *unitary* transformations are special types of similarity transformations, in which also $\mathbf{Q}^{-1} = \mathbf{Q}^T$ and $\mathbf{Q}^{-1} = \overline{\mathbf{Q}}^T$, respectively, the preceding statement applies to them.

1.21. Coordinate transformations. The elements of an n-dimensional *numerical vector* $\mathbf{x} = \{x_1, x_2, \cdots, x_n\}$ may be interpreted as the *components* of a certain *geometrical vector* $\mathscr{X}$ (such as a *force* or an *acceleration* or a *displacement* from an origin to a point in a geometrical n-space) in the directions of the n basic mutually orthogonal unit vectors

$$\mathbf{i}_1 = \{1, 0, \cdots, 0\}, \qquad \cdots, \qquad \mathbf{i}_n = \{0, 0, \cdots, 1\}$$

which lie along the axes of reference rectangular coordinates $x_1, x_2, \cdots, x_n$ in n-space. This interpretation corresponds to the relationship*

$$\mathbf{x} = \{x_1, x_2, \cdots, x_n\} = \sum_{k=1}^{n} x_k \mathbf{i}_k. \tag{189}$$

Now let a *new* coordinate system in n-space be so chosen that the unit vectors $\mathbf{i}'_1, \mathbf{i}'_2, \cdots, \mathbf{i}'_n$ in the directions of the axes of the new coordinates $x'_1, x'_2, \cdots, x'_n$ are linearly independent and are related to the original unit vectors by the equations

$$\left.\begin{aligned} \mathbf{i}'_1 &= q_{11}\mathbf{i}_1 + q_{21}\mathbf{i}_2 + \cdots + q_{n1}\mathbf{i}_n, \\ &\cdots\cdots\cdots\cdots\cdots\cdots\cdots\cdots \\ \mathbf{i}'_n &= q_{1n}\mathbf{i}_1 + q_{2n}\mathbf{i}_2 + \cdots + q_{nn}\mathbf{i}_n \end{aligned}\right\} \tag{190}$$

or

$$\mathbf{i}'_k = \sum_{r=1}^{n} q_{rk}\mathbf{i}_r. \tag{191}$$

The geometrical vector $\mathscr{X}$ then can be specified by its components $x'_1, x'_2, \cdots, x'_n$ along the new axes. If we denote the numerical vector comprising this array of components by $\mathbf{x}'$, there follows

$$\mathbf{x}' = x'_1\mathbf{i}'_1 + x'_2\mathbf{i}'_2 + \cdots + x'_n\mathbf{i}'_n = \sum_{k=1}^{n} x'_k\mathbf{i}'_k. \tag{192}$$

* In nongeometric terms, we can say that **x** is the *numerical vector* whose elements are the constants of combination in the representation (189) of an *abstract vector* $\mathscr{X}$ in terms of the *standard basis* $\mathbf{i}_1, \mathbf{i}_2, \cdots, \mathbf{i}_n$.

To determine the new components in terms of the original ones, we transform the representation $\mathbf{x}'$ to the representation $\mathbf{x}$ by introducing (191) into (192):

$$\mathbf{x} = \sum_{k=1}^{n} \sum_{r=1}^{n} x_k' q_{rk} \mathbf{i}_r = \sum_{r=1}^{n} \left(\sum_{k=1}^{n} q_{rk} x_k' \right) \mathbf{i}_r. \tag{193}$$

Then, since the vectors $\mathbf{i}_r$ are mutually orthogonal, their respective coefficients in (189) and (193) must be equal, so that

$$x_r = \sum_{k=1}^{n} q_{rk} x_k'. \tag{194}$$

Thus, if we write $\mathbf{x}' = \{x_1', x_2', \cdots, x_n'\}$ for the numerical vector comprising the components of $\mathscr{X}$ in the directions of the new coordinate axes specified by (190), there follows

$$\mathbf{x} = \mathbf{Q}\,\mathbf{x}', \tag{195}$$

where $\mathbf{Q}$ is the *transformation matrix*

$$\mathbf{Q} = \begin{bmatrix} q_{11} & q_{12} & \cdots & q_{1n} \\ q_{21} & q_{22} & \cdots & q_{2n} \\ \cdots & \cdots & \cdots & \cdots \\ q_{n1} & q_{n2} & \cdots & q_{nn} \end{bmatrix}, \tag{196}$$

of which the coefficient matrix in (190) is the *transpose*. We notice that each *column* of (196) contains *the components of a new unit vector along the original coordinate axes.*

Here we interpret the matrix $\mathbf{Q}$ of (195) as relating the components of a geometrical vector along the original coordinate axes to the components of the *same* geometrical vector along the new coordinate axes. In other considerations we may suppose that no *change of axes* is involved, and that an equation of the form (195) merely transforms one numerical vector into another one, both vectors then being referred to the same axes. Which interpretation is to be attached to such an equation in practice clearly depends upon the nature of the problem involved.*

In order that the new unit vectors be linearly independent, and hence span n-space, the matrix $\mathbf{Q}$ must be nonsingular. Hence $\mathbf{Q}^{-1}$ then exists,

* In many other applications the two number sets $\{x_1, \cdots, x_n\}$ and $\{x_1', \cdots, x_n'\}$ are not inherently subject to *any* geometrical interpretations (for example, they could represent the costs of n manufactured items under two different production schedules) and (195) is merely a compact way of stating that the two sets of numbers are related by a certain set of linear algebraic equations. In *abstract* terms, (195) relates the components of an abstract vector $\mathscr{X}$, relative to the *standard* basis $\mathbf{i}_1, \cdots, \mathbf{i}_n$, to the components of $\mathscr{X}$, relative to a *new* basis $\mathbf{i}_1', \cdots, \mathbf{i}_n'$, when the two bases are related by (191).

and we have also, from (195),

$$\mathbf{x}' = \mathbf{Q}^{-1}\,\mathbf{x}. \tag{197}$$

Suppose now that two numerical vectors **x** and **y**, representing $\mathscr{X}$ and $\mathscr{Y}$, respectively, are related by an equation of the form

$$\mathbf{y} = \mathbf{A}\,\mathbf{x}, \tag{198}$$

when the components refer to the original coordinate frame, and that we require the relationship between the two corresponding numerical vectors $\mathbf{x}'$ and $\mathbf{y}'$ whose components are referred to a new coordinate frame (190). (We may, for example, imagine that **y** and $\mathbf{y}'$ represent a *force* and that **x** and $\mathbf{x}'$ represent an *acceleration*. In *Newtonian* mechanics, the matrix **A** then would be a *scalar* matrix.) By replacing **x** by $\mathbf{Q}\,\mathbf{x}'$ and **y** by $\mathbf{Q}\,\mathbf{y}'$, we obtain the relation

$$\mathbf{Q}\,\mathbf{y}' = \mathbf{A}\,\mathbf{Q}\,\mathbf{x}'$$

from (198) and hence, after premultiplying both sides by $\mathbf{Q}^{-1}$, we deduce the desired result

$$\mathbf{y}' = (\mathbf{Q}^{-1}\,\mathbf{A}\,\mathbf{Q})\,\mathbf{x}'. \tag{199}$$

Thus we see that the matrix relating $\mathbf{x}'$ and $\mathbf{y}'$ is obtained from that relating **x** and **y** by a *similarity transformation*. In particular, it follows that the invariance properties discussed at the close of the preceding section apply in the present case. That is, the quantities β_i of that section, pertaining to the matrix **A** of (198), are invariant under a nonsingular coordinate transformation. This result is of great importance in many applications.

If the *new* unit vectors are *mutually orthogonal*, we readily obtain from (196) the result

$$\mathbf{Q}^T\,\mathbf{Q} = \mathbf{I} \qquad \textit{or} \qquad \mathbf{Q}^T = \mathbf{Q}^{-1}, \tag{200}$$

so that **Q** then is an *orthogonal* matrix. Thus, *a transformation from one set of orthogonal axes to another is accomplished by an orthogonal transformation.* We may verify that in such a transformation the length of the vector representation in the new system is the same as the length of the representation in the original system (that is, there is no change in *scale*). For if $\mathbf{x} = \mathbf{Q}\,\mathbf{x}'$ there follows

$$l^2 = \mathbf{x}^T\,\mathbf{x} = (\mathbf{Q}\,\mathbf{x}')^T\,\mathbf{Q}\,\mathbf{x}' = \mathbf{x}'^T\,\mathbf{Q}^T\,\mathbf{Q}\,\mathbf{x}' = \mathbf{x}'^T\,\mathbf{x}' = l'^2. \tag{201}$$

Also, the magnitude of the scalar product of the numerical vectors representing two quantities is the same in both systems (that is, the magnitude of an "angle" is also preserved), since

$$(\mathbf{x}, \mathbf{y}) = \mathbf{x}^T\,\mathbf{y} = \mathbf{x}'^T\,\mathbf{Q}^T\,\mathbf{Q}\,\mathbf{y}' = \mathbf{x}'^T\,\mathbf{y}' = (\mathbf{x}', \mathbf{y}'). \tag{202}$$

As these results suggest, it can be shown that any orthogonal transformation in n-space can be interpreted as a combination of *rotations* and *reflections.**

In particular, reference to Section 1.14 shows that, when $\mathbf{A}$ is an $n \times n$ real symmetric matrix, the real quadratic form

$$A = \mathbf{x}^T \mathbf{A}\, \mathbf{x} \tag{203}$$

in the real variables $x_1, \cdots, x_n$ can be reduced to the canonical form

$$A = \lambda_1 x_1'^2 + \cdots + \lambda_n x_n'^2 \tag{204}$$

by a real *coordinate transformation*, comprising only rotations and/or reflections, in which *the axes of the new coordinates $x_1', \cdots, x_n'$ are made to coincide with the respective mutually orthogonal characteristic vectors of the matrix* $\mathbf{A}$.

1.22. Functions of symmetric matrices. In this section, we again restrict attention to real *symmetric* matrices, which are of principal interest in applications. We notice first that, as is easily shown, *the sum of two symmetric matrices of the same order is also symmetric*, while *the product of two symmetric matrices of the same order is symmetric if those matrices are commutative.*

Positive integral powers of any square matrix $\mathbf{A}$ are defined by iteration:

$$\mathbf{A}^2 = \mathbf{A}\,\mathbf{A}, \quad \mathbf{A}^3 = \mathbf{A}\,\mathbf{A}^2, \quad \cdots, \quad \mathbf{A}^{n+1} = \mathbf{A}\,\mathbf{A}^n, \quad \cdots. \tag{205}$$

In consequence of this definition, there follows also

$$\mathbf{A}^r\,\mathbf{A}^s = \mathbf{A}^s\,\mathbf{A}^r = \mathbf{A}^{r+s}, \tag{206}$$

when r and s are positive integers. *Negative integral* powers are defined only for *nonsingular* matrices, for which a unique inverse $\mathbf{A}^{-1}$ exists, and are then defined by the relation

$$\mathbf{A}^{-n} = (\mathbf{A}^{-1})^n. \tag{207}$$

If we define also

$$\mathbf{A}^0 = \mathbf{I}, \tag{208}$$

then (206) applies to any nonsingular matrix, for *any* integers r and s. It is clear that *any integral power of a symmetric matrix is also symmetric.*

Polynomial functions of $\mathbf{A}$ are then defined as linear combinations of a finite number of nonnegative integral powers of $\mathbf{A}$. Any polynomial in $\mathbf{A}$ hence can be expressed as a symmetric matrix of the same order as $\mathbf{A}$.

* When two real vectors $\mathbf{x}$ and $\mathbf{y}$ undergo the *same real transformation*, so that $\mathbf{x} = \mathbf{Q}\,\mathbf{x}'$ and $\mathbf{y} = \mathbf{Q}\,\mathbf{y}'$, the two sets of variables which comprise their components are said to be *cogredient*. When the vectors are transformed separately, so that $\mathbf{x} = \mathbf{P}\,\mathbf{x}'$ and $\mathbf{y} = \mathbf{Q}\,\mathbf{y}'$, in such a way that the condition $(\mathbf{x}, \mathbf{y}) = (\mathbf{x}', \mathbf{y}')$ is satisfied, the two sets of variables are said to be *contragredient*. Equation (202) states that when two vectors undergo the same *orthogonal* transformation their components are *both cogredient and contragredient.*

Suppose now that $\mathbf{A}$ is of order n, and let its characteristic numbers be denoted by $\lambda_1, \lambda_2, \cdots, \lambda_n$ (not necessarily distinct), with corresponding orthogonalized characteristic vectors $\mathbf{u}_1, \mathbf{u}_2, \cdots, \mathbf{u}_n$, so that

$$\mathbf{A}\,\mathbf{u}_i = \lambda_i \mathbf{u}_i \tag{209}$$

for $i = 1, 2, \cdots, n$. If we multiply both sides of (209) by $\mathbf{A}$, and use (209) to simplify the resulting right-hand member, there follows

$$\mathbf{A}^2\,\mathbf{u}_i = \lambda_i \mathbf{A}\,\mathbf{u}_i = \lambda_i^2 \mathbf{u}_i, \tag{210}$$

and, by repeating this process, we deduce from (209) the relation

$$\mathbf{A}^r\,\mathbf{u}_i = \lambda_i^r \mathbf{u}_i \tag{211}$$

for any positive integer r. Similarly, if $\mathbf{A}$ is *nonsingular*, the result of multiplying both sides of (209) by $\mathbf{A}^{-1}$ becomes

$$\mathbf{A}^{-1}\,\mathbf{u}_i = \lambda_i^{-1} \mathbf{u}_i \tag{212}$$

and, by iteration, we find that (211) is then true for any integer r.

Thus we deduce that *if* λ_i *is a characteristic number of* $\mathbf{A}$, *with a corresponding characteristic vector* $\mathbf{u}_i$, *then* λ_i^r *is a characteristic number of* $\mathbf{A}^r$, *with the same characteristic vector* $\mathbf{u}_i$. For a *symmetric* matrix of order n, there are exactly n linearly independent characteristic vectors. Hence it follows in this case that $\mathbf{A}^r$ cannot possess *additional* characteristic numbers or linearly independent characteristic vectors.

It should be noticed, however, that $\mathbf{A}^r$ *may* have characteristic vectors which are *not* possessed by $\mathbf{A}$. As a simple example, it is seen that the matrix

$$\mathbf{A} = \begin{bmatrix} 1 & 0 \\ 0 & -1 \end{bmatrix}$$

has the characteristic numbers $\lambda_1 = 1$ and $\lambda_2 = -1$, and corresponding characteristic vectors $\mathbf{u}_1 = \{1, 0\}$ and $\mathbf{u}_2 = \{0, 1\}$. Here the matrix $\mathbf{A}^2 = \mathbf{I}$ has the *double* characteristic number $\lambda_1^2 = \lambda_2^2 = 1$ and it is obvious that *all* nonzero vectors in two-space are characteristic vectors of $\mathbf{A}^2$. Clearly, all such vectors are indeed linear combinations of $\mathbf{u}_1$ and $\mathbf{u}_2$.

Next, consider any polynomial in $\mathbf{A}$, of degree m, of the form

$$P(\mathbf{A}) = \alpha_0 \mathbf{A}^m + \alpha_1 \mathbf{A}^{m-1} + \cdots + \alpha_{m-1}\mathbf{A} + \alpha_m \mathbf{I}. \tag{213}$$

If we consider the product of the matrix $P(\mathbf{A})$ with any characteristic vector of $\mathbf{A}$, and use (211), we obtain the relation

$$P(\mathbf{A})\,\mathbf{u}_i = \alpha_0 \lambda_i^m \mathbf{u}_i + \alpha_1 \lambda_i^{m-1} \mathbf{u}_i + \cdots + \alpha_{m-1}\lambda_i \mathbf{u}_i + \alpha_m \mathbf{u}_i$$

or

$$P(\mathbf{A})\,\mathbf{u}_i = P(\lambda_i)\,\mathbf{u}_i. \tag{214}$$

Hence it follows that the equation

$$[P(\mathbf{A}) - \mu\,\mathbf{I}]\,\mathbf{x} = \mathbf{0} \tag{215}$$

possesses a nontrivial solution when $\mu = P(\lambda_i)$, and that a solution of (215) in this case is an arbitrary multiple of $\mathbf{u}_i$. But, as in the preceding argument, no additional linearly independent solutions of (215) can exist. Thus we find that *if* $\mathbf{A}$ *is symmetric, then all characteristic vectors of* $\mathbf{A}$ *also belong to* $P(\mathbf{A})$, and also, *if the characteristic numbers of* $\mathbf{A}$ *are* $\lambda_1, \cdots, \lambda_n$, *then those of* $P(\mathbf{A})$ *are* $P(\lambda_1), \cdots, P(\lambda_n)$.

Let the λ-polynomial $|\mathbf{A} - \lambda\,\mathbf{I}|$, the vanishing of which determines the characteristic numbers of $\mathbf{A}$, be denoted by $F(\lambda)$:

$$F(\lambda) = |\mathbf{A} - \lambda\,\mathbf{I}| \equiv (-1)^n[\lambda^n - \beta_1\lambda^{n-1} + \cdots + (-1)^n\beta_n]. \tag{216}$$

Then $F(\lambda)$ is a polynomial in λ, of degree n, which vanishes when $\lambda = \lambda_i$,

$$F(\lambda_i) = 0 \qquad (i = 1, 2, \cdots, n). \tag{217}$$

If now we identify the polynomial function P with the function F, equation (214) becomes

$$F(\mathbf{A})\,\mathbf{u}_i = \mathbf{0} \qquad (i = 1, 2, \cdots, n). \tag{218}$$

Thus if we write temporarily $\mathbf{B} = F(\mathbf{A})$, it follows that the equation $\mathbf{B}\,\mathbf{x} = \mathbf{0}$ possesses the n linearly independent solutions $\mathbf{x} = \mathbf{u}_1, \cdots, \mathbf{u}_n$. But since $\mathbf{B}$ is a square matrix of order n, the results of Section 1.9 show that $\mathbf{B}$ must be of rank $n - n = 0$. Hence $\mathbf{B} = F(\mathbf{A})$ must be a *zero matrix*, and it follows that*

$$F(\mathbf{A}) = \mathbf{0}. \tag{219}$$

That is, *if the characteristic equation of a symmetric matrix* $\mathbf{A}$ *is* $F(\lambda) = 0$, *then the matrix* $\mathbf{A}$ *satisfies the equation* $F(\mathbf{A}) = \mathbf{0}$.

This curious and useful result is known as the *Cayley-Hamilton theorem*, and is often stated briefly as follows: "A matrix satisfies its own characteristic equation."

It is important to notice that in deducing (219) from (218) we have made use of the fact that a *symmetric* matrix always has n linearly independent characteristic vectors. Since this statement does not apply to nonsymmetric matrices with repeated characteristic numbers, the preceding proof does not apply in such cases. However, it can be proved by somewhat less direct methods that *the Cayley-Hamilton theorem is true for any square matrix.*

* It should be noticed that $F(\mathbf{M})$ is defined to be the *matrix* obtained by replacing λ by $\mathbf{M}$ in the *polynomial* $F(\lambda) = (-1)^n\lambda^n + \cdots$, with the understanding that λ^0 is to be replaced by $\mathbf{I}$ in the constant term. The interpretation $F(\mathbf{M}) = |\mathbf{A} - \mathbf{M}\,\mathbf{I}| = |\mathbf{A} - \mathbf{M}|$ is *not* intended and obviously would *not* lead to (219).

If $F(\lambda)$ possesses a factor $(\lambda - \lambda_r)^s$, where $s > 1$, so that λ_r is of multiplicity s, the same argument shows that the *symmetric* matrix $\mathbf{A}$ also satisfies the *reduced characteristic equation* $G(\mathbf{A}) = \mathbf{0}$, where $G(\lambda) = F(\lambda)/(\lambda - \lambda_r)^{s-1}$. (The matrix $\mathbf{A}$ considered in Section 1.15 may be used as an illustration.) This statement is *not* necessarily true if $\mathbf{A}$ is nonsymmetric, as may be illustrated by the matrix considered on page 32.

As a verification of the theorem, we notice that, for the matrix

$$\mathbf{A} = \begin{bmatrix} 2 & 1 \\ 1 & 2 \end{bmatrix}, \tag{220}$$

we have

$$F(\lambda) = \begin{vmatrix} 2-\lambda & 1 \\ 1 & 2-\lambda \end{vmatrix} = \lambda^2 - 4\lambda + 3, \tag{221}$$

and the equation $\mathbf{A}^2 - 4\mathbf{A} + 3\mathbf{I} = \mathbf{0}$ becomes

$$\begin{bmatrix} 5 & 4 \\ 4 & 5 \end{bmatrix} - \begin{bmatrix} 8 & 4 \\ 4 & 8 \end{bmatrix} + \begin{bmatrix} 3 & 0 \\ 0 & 3 \end{bmatrix} = \begin{bmatrix} 0 & 0 \\ 0 & 0 \end{bmatrix}.$$

We notice that this theorem permits any positive integral power of a matrix $\mathbf{A}$, and hence *any polynomial* in $\mathbf{A}$, to be expressed as a linear combination of the matrices $\mathbf{I}, \mathbf{A}, \mathbf{A}^2, \cdots, \mathbf{A}^{n-1}$, where n is the order of $\mathbf{A}$.

Thus, for the matrix (220) considered above, we have the successive results

$$\begin{aligned} \mathbf{A}^2 &= 4\mathbf{A} - 3\mathbf{I}, \\ \mathbf{A}^3 &= 4\mathbf{A}^2 - 3\mathbf{A} = 4(4\mathbf{A} - 3\mathbf{I}) - 3\mathbf{A} = 13\mathbf{A} - 12\mathbf{I}, \end{aligned} \tag{222}$$

and so forth. In addition, we obtain the relation

$$\mathbf{A} - 4\mathbf{I} + 3\mathbf{A}^{-1} = \mathbf{0}.$$

Hence we deduce that

$$\mathbf{A}^{-1} = -\tfrac{1}{3}\mathbf{A} + \tfrac{4}{3}\mathbf{I},$$

and obtain successive *negative* integral powers of $\mathbf{A}$ by successive multiplications and simplifications.

A convenient determination of the constants of combination, in the case of a general polynomial, is afforded by a result next to be obtained under the assumption that all the characteristic numbers of $\mathbf{A}$ are *distinct*. In place of directly determining the constants involved in the representation $P(\mathbf{A}) = c_1\mathbf{A}^{n-1} + c_2\mathbf{A}^{n-2} + \cdots + c_n\mathbf{I}$, it is desirable for present purposes to assume the equivalent form

$$\begin{aligned} P(\mathbf{A}) = C_1[(\mathbf{A} - \lambda_2\mathbf{I})(\mathbf{A} - \lambda_3\mathbf{I}) \cdots (\mathbf{A} - \lambda_n\mathbf{I})] \\ + C_2[(\mathbf{A} - \lambda_1\mathbf{I})(\mathbf{A} - \lambda_3\mathbf{I}) \cdots (\mathbf{A} - \lambda_n\mathbf{I})] + \cdots \\ + C_n[(\mathbf{A} - \lambda_1\mathbf{I})(\mathbf{A} - \lambda_2\mathbf{I}) \cdots (\mathbf{A} - \lambda_{n-1}\mathbf{I})], \end{aligned} \tag{223}$$

where each bracketed quantity, and hence also the complete right-hand side, is clearly a polynomial of degree $n - 1$ in $\mathbf{A}$. To determine the n C's, we postmultiply the equal members of (223) successively by each of the n characteristic vectors $\mathbf{u}_1, \cdots, \mathbf{u}_n$ of the matrix $\mathbf{A}$.

If both members are postmultiplied by $\mathbf{u}_k$, and use is made of the relation $\mathbf{A}\,\mathbf{u}_k = \lambda_k \mathbf{u}_k$, it is found that the coefficients of all C's except C_k then contain the factor $(\lambda_k - \lambda_k)$, and hence vanish. Thus there follows, after a simple calculation,

$$P(\mathbf{A})\,\mathbf{u}_k = C_k[(\lambda_k - \lambda_1) \cdots (\lambda_k - \lambda_{k-1})(\lambda_k - \lambda_{k+1}) \cdots (\lambda_k - \lambda_n)]\mathbf{u}_k, \quad (224)$$

for $k = 1, 2, \cdots, n$. But reference to equation (214) then shows that the coefficient of $\mathbf{u}_k$ on the right must be equal to $P(\lambda_k)$. Thus *if the characteristic numbers of the matrix* $\mathbf{A}$ *are all distinct*, we obtain the result

$$C_k = \frac{P(\lambda_k)}{\prod_{r \neq k} (\lambda_k - \lambda_r)} \qquad (k = 1, 2, \cdots, n), \quad (225)$$

where the notation $\prod_{r \neq k}$ denotes the product of those factors for which r takes on the values 1 through n, excluding k. If this result is introduced into (223), the desired representation is obtained in the form

$$P(\mathbf{A}) = \sum_{k=1}^{n} P(\lambda_k) \mathrm{Z}_k(\mathbf{A}), \quad (226)$$

with the convenient abbreviation

$$\mathrm{Z}_k(\mathbf{A}) = \frac{\prod_{r \neq k} (\mathbf{A} - \lambda_r \mathbf{I})}{\prod_{r \neq k} (\lambda_k - \lambda_r)} \qquad (k = 1, 2, \cdots, n). \quad (227)$$

Cases in which certain characteristic numbers are repeated require special treatment.*

To verify this result in the case of the matrix (220), we notice that $\lambda_1 = 3$ and $\lambda_2 = 1$. To evaluate $P(\mathbf{A}) = \mathbf{A}^3$, we first calculate

$$\mathrm{Z}_1(\mathbf{A}) = \frac{\mathbf{A} - \lambda_2 \mathbf{I}}{3 - 1} = \frac{1}{2}(\mathbf{A} - \mathbf{I}), \qquad \mathrm{Z}_2(\mathbf{A}) = \frac{\mathbf{A} - \lambda_1 \mathbf{I}}{1 - 3} = -\frac{1}{2}(\mathbf{A} - 3\mathbf{I}).$$

Hence, with $P(3) = 27$ and $P(1) = 1$, there follows

$$\mathbf{A}^3 = \tfrac{27}{2}(\mathbf{A} - \mathbf{I}) - \tfrac{1}{2}(\mathbf{A} - 3\mathbf{I}) = 13\mathbf{A} - 12\mathbf{I},$$

in accordance with (222). The usefulness of (226) clearly would be better illustrated in the calculation of $\mathbf{A}^{100}$.

* See Reference 7. It can be shown that this representation (with appropriate modifications for repeated characteristic numbers) is valid for *any* square matrix $\mathbf{A}$.

It should be noticed that the matrices $Z_k(\mathbf{A})$ depend only on $\mathbf{A}$, and are *not* dependent upon the form of the polynomial P chosen. The result (226) is known as *Sylvester's formula*.

Having defined polynomial functions, we may next define other functions of $\mathbf{A}$ by *infinite series* such as

$$\sum_{m=0}^{\infty} \alpha_m \mathbf{A}^m = \lim_{M\to\infty} \sum_{m=0}^{M} \alpha_m \mathbf{A}^m, \tag{228}$$

for those matrices for which the indicated limit exists. We omit discussion of the convergence of such series. However, if $\mathbf{A}$ is of order n, it is clear that the sum of M terms of the series can be expressed as a polynomial of maximum degree $n - 1$ in $\mathbf{A}$, regardless of the value of M, in consequence of the preceding results. Thus we see that *if the series converges, the function represented by the series must also be so expressible, and hence must be determinable from* (226) *if the characteristic numbers of* $\mathbf{A}$ *are distinct.*

In particular, it can be shown that the series in the right-hand member of the relation

$$e^{\mathbf{A}} = \sum_{m=0}^{\infty} \frac{\mathbf{A}^m}{m!} \tag{229}$$

converges for *any* square matrix $\mathbf{A}$, and hence may serve to *define* $e^{\mathbf{A}}$. Suppose that $\mathbf{A}$ is a matrix of order *two*, with distinct characteristic numbers λ_1 and λ_2. Then (227) gives

$$Z_1(\mathbf{A}) = \frac{\mathbf{A} - \lambda_2 \mathbf{I}}{\lambda_1 - \lambda_2}, \qquad Z_2(\mathbf{A}) = \frac{\mathbf{A} - \lambda_1 \mathbf{I}}{\lambda_2 - \lambda_1},$$

and from (226) we obtain the evaluation

$$e^{\mathbf{A}} = \frac{1}{\lambda_1 - \lambda_2} [(e^{\lambda_1} - e^{\lambda_2})\mathbf{A} - (\lambda_2 e^{\lambda_1} - \lambda_1 e^{\lambda_2})\mathbf{I}]. \tag{230}$$

The corresponding evaluation when $\lambda_1 = \lambda_2$ can be obtained from this result as the limiting form when $\lambda_2 \to \lambda_1$ (see Problem 86).

1.23. Numerical solution of characteristic-value problems. In the process of dealing with a characteristic-value problem of the form

$$\mathbf{A}\,\mathbf{x} = \lambda\,\mathbf{x}, \tag{231}$$

it is necessary first to determine roots of the characteristic equation

$$|\mathbf{A} - \lambda\,\mathbf{I}| = 0, \tag{232}$$

and then, for each such value of λ, to obtain a nontrivial solution vector of (231). If $\mathbf{A}$ is of order n, equation (232) is an algebraic equation of the same degree in λ, and the numerical determination of the characteristic numbers generally involves considerable labor when $n > 2$. Further, the actual *expansion* of (232) may be tedious in such cases.

In this section we outline a numerical iterative method which avoids these steps, and which is frequently useful in practice. This method is analogous to a method, associated with the names of *Vianello* and *Stodola*, which is applied to corresponding problems involving differential equations.*

Suppose first that the *dominant* characteristic number, that is, the characteristic number with largest magnitude, is required. To start the procedure, we choose an initial nonzero approximation to the corresponding characteristic vector, say $\mathbf{x}^{(1)}$. In the absence of advance knowledge as to the nature of this vector, we may, for example, start with the vector $\{0, 0, \cdots, 1\}$ or $\{1, 1, \cdots, 1\}$. This initial approximation is then introduced into the *left*-hand member of (231). If we then set

$$\mathbf{y}^{(1)} = \mathbf{A}\,\mathbf{x}^{(1)}, \tag{233}$$

the requirement that (231) be approximately satisfied becomes

$$\mathbf{y}^{(1)} \approx \lambda\,\mathbf{x}^{(1)}. \tag{234}$$

If the respective components of $\mathbf{x}^{(1)}$ and $\mathbf{y}^{(1)}$ are nearly in a constant ratio, we may expect that the approximation $\mathbf{x}^{(1)}$ is good, and that this ratio is an approximation to the true value of λ.

It is rather conventional to choose $\mathbf{x}^{(1)}$ in such a way that one component is *unity*, and to choose, as a first approximation to the dominant characteristic value of λ, the corresponding component of $\mathbf{y}^{(1)}$. A more efficient determination is outlined in the following section [see equations (250a,b)].

A convenient multiple of $\mathbf{y}^{(1)}$ is then taken as the next approximation $\mathbf{x}^{(2)}$, and the process is repeated until satisfactory agreement between successive approximations is obtained. As will be shown, in the case when $\mathbf{A}$ is real and symmetric, this method will lead inevitably to the *dominant* characteristic value of λ and to the corresponding characteristic vector, unless the vector $\mathbf{x}^{(1)}$ happens to be exactly orthogonal to that vector, except in the unusual case when the *negative* of the dominant characteristic number is *also* a characteristic number.

Analytically, if the rth approximation is denoted by $\mathbf{x}^{(r)}$, the iteration can be specified by the relations

$$\mathbf{y}^{(r)} = \mathbf{A}\,\mathbf{x}^{(r)}, \qquad \mathbf{x}^{(r+1)} = \alpha_r \mathbf{y}^{(r)} \qquad (r = 1, 2, \cdots), \tag{235a,b}$$

where α_r is a conveniently chosen nonzero multiplicative constant, and the assertion then is that, in general,

$$\mathbf{y}^{(r)} \sim \lambda_n \mathbf{x}^{(r)}, \qquad \mathbf{x}^{(r)} \to c\,\mathbf{e}_n \qquad (r \to \infty) \tag{236a,b}$$

* See Reference 9.

where λ_n is the characteristic number of $\mathbf{A}$ which is *of largest magnitude*, $\mathbf{e}_n$ is the corresponding unit characteristic vector, and c is a constant depending upon the choice of the α's. The introduction of the arbitrarily chosen α_r in the rth cycle permits one to take the length of $\mathbf{x}^{(r+1)}$ to be of the order of unity, and so to prevent the lengths of successive approximation vectors from growing unboundedly or tending to zero.

If the *smallest* characteristic value of λ is required, we may first transform (231) to the equation

$$\mathbf{x} = \lambda\, \mathbf{A}^{-1}\, \mathbf{x}.$$

With the notations

$$\mathbf{M} = \mathbf{A}^{-1}, \qquad \kappa = \frac{1}{\lambda} \tag{237a,b}$$

this equation takes the form

$$\mathbf{M}\,\mathbf{x} = \kappa\,\mathbf{x}. \tag{238}$$

The *largest* characteristic value of κ for this equation then can be determined by the iterative method, and its reciprocal is the *smallest* characteristic value of λ for (231).*

Analytically, if the rth approximation is again denoted by $\mathbf{x}^{(r)}$, and if we now denote $\mathbf{M}\,\mathbf{x}^{(r)} \equiv \mathbf{A}^{-1}\,\mathbf{x}^{(r)}$ by $\mathbf{y}^{(r)}$, it follows that also $\mathbf{x}^{(r)} = \mathbf{A}\,\mathbf{y}^{(r)}$, so that we may specify this iteration by the relations

$$\mathbf{x}^{(r)} = \mathbf{A}\,\mathbf{y}^{(r)}, \qquad \mathbf{x}^{(r+1)} = \alpha_r \mathbf{y}^{(r)} \qquad (r = 1, 2, \cdots) \tag{239a,b}$$

and the assertion in this case is that, in general,

$$\mathbf{y}^{(r)} \sim \frac{1}{\lambda_1}\mathbf{x}^{(r)}, \qquad \mathbf{x}^{(r)} \to c\,\mathbf{e}_1 \qquad (r \to \infty) \tag{240a,b}$$

where λ_1 is the characteristic number of $\mathbf{A}$ *of smallest magnitude* and $\mathbf{e}_1$ is the corresponding unit characteristic vector. As before, we begin by choosing the elements of the vector $\mathbf{x}^{(1)}$, but here we next determine the elements of $\mathbf{y}^{(1)}$ in such a way that $\mathbf{A}\,\mathbf{y}^{(1)} = \mathbf{x}^{(1)}$, whereas in approximating λ_n we determine $\mathbf{y}^{(1)}$ such that $\mathbf{y}^{(1)} = \mathbf{A}\,\mathbf{x}^{(1)}$. Clearly, the computation of the elements of the inverse matrix $\mathbf{M} \equiv \mathbf{A}^{-1}$ for the purpose of transforming (231) to (238) may be avoided at the expense of solving the simultaneous equations corresponding to (239a),

$$\left.\begin{array}{c} a_{11}y_1^{(r)} + \cdots + a_{1n}y_n^{(r)} = x_1^{(r)}, \\ \cdots\cdots\cdots\cdots\cdots\cdots\cdots\cdots \\ a_{1n}y_1^{(r)} + \cdots + a_{nn}y_n^{(r)} = x_n^{(r)} \end{array}\right\}, \tag{239a$'$}$$

by an appropriate numerical method in each cycle of the iteration.

* See also Problem 124.

This procedure clearly fails if $\mathbf{A}$ is *singular*, that is, if $\lambda = 0$ is a characteristic number of (231). A method which is useful in this case is presented in the following section (page 68).

To illustrate the basic procedure, we seek the largest characteristic value of λ for the system

$$\left.\begin{aligned} x_1 + x_2 + x_3 &= \lambda x_1, \\ x_1 + 2x_2 + 2x_3 &= \lambda x_2, \\ x_1 + 2x_2 + 3x_3 &= \lambda x_3 \end{aligned}\right\}. \tag{241}$$

With the initial approximation $\mathbf{x}^{(1)} = \{1, 1, 1\}$, there follows

$$\mathbf{y}^{(1)} = \begin{bmatrix} 1 & 1 & 1 \\ 1 & 2 & 2 \\ 1 & 2 & 3 \end{bmatrix} \begin{Bmatrix} 1 \\ 1 \\ 1 \end{Bmatrix} = \begin{Bmatrix} 3 \\ 5 \\ 6 \end{Bmatrix} = 6 \begin{Bmatrix} \frac{1}{2} \\ \frac{5}{6} \\ 1 \end{Bmatrix}. \tag{242}$$

If we determine λ such that the x_3 components of $\lambda\,\mathbf{x}^{(1)}$ and $\mathbf{y}^{(1)}$ are equal, we have $\lambda^{(1)} = 6$. Next, with $\mathbf{x}^{(2)} = \{\frac{1}{2}, \frac{5}{6}, 1\}$, there follows

$$\mathbf{y}^{(2)} = \begin{bmatrix} 1 & 1 & 1 \\ 1 & 2 & 2 \\ 1 & 2 & 3 \end{bmatrix} \begin{Bmatrix} \frac{1}{2} \\ \frac{5}{6} \\ 1 \end{Bmatrix} = \begin{Bmatrix} \frac{7}{3} \\ \frac{25}{6} \\ \frac{31}{6} \end{Bmatrix} = \frac{31}{6} \begin{Bmatrix} \frac{14}{31} \\ \frac{25}{31} \\ 1 \end{Bmatrix}. \tag{243}$$

The second approximation to the dominant characteristic number is then $\lambda^{(2)} = \frac{31}{6} \doteq 5.17$. The third step then gives

$$\mathbf{y}^{(3)} = \begin{bmatrix} 1 & 1 & 1 \\ 1 & 2 & 2 \\ 1 & 2 & 3 \end{bmatrix} \begin{Bmatrix} \frac{14}{31} \\ \frac{25}{31} \\ 1 \end{Bmatrix} = \begin{Bmatrix} \frac{70}{31} \\ \frac{126}{31} \\ \frac{157}{31} \end{Bmatrix} = \frac{157}{31} \begin{Bmatrix} \frac{70}{157} \\ \frac{126}{157} \\ 1 \end{Bmatrix} \tag{244}$$

and also $\lambda^{(3)} = \frac{157}{31} \doteq 5.06$. The ratios $x_1 : x_2 : x_3$ according to the four approximations are $(1:1:1)$, $(0.500:0.833:1)$, and $(0.446:0.803:1)$. The next cycle leads to the value $\lambda^{(4)} \doteq 5.05$, and to the ratios $0.445:0.802:1$, which hence may be expected to be accurate to three significant figures.

1.24. Additional techniques. In order to improve and extend the procedure just outlined, in the case when $\mathbf{A}$ is *real and symmetric*, it is desirable to consider the analytical basis of the precedure in that case.* For this purpose, we may suppose that $\mathbf{e}_1, \mathbf{e}_2, \cdots, \mathbf{e}_n$ are the true orthogonalized characteristic unit vectors of the problem (231), corresponding to the characteristic numbers $\lambda_1, \lambda_2, \cdots, \lambda_n$, arranged in increasing order of

* Certain other cases are considered in Sections 1.25 and 1.26.

magnitude. If the initial assumption $\mathbf{x}^{(1)}$ is imagined to be expressed in the form

$$\mathbf{x}^{(1)} = \sum_{k=1}^{n} c_k \mathbf{e}_k, \tag{245a}$$

then the vector $\mathbf{y}^{(1)} = \mathbf{A}\,\mathbf{x}^{(1)}$ accordingly must be given by

$$\mathbf{y}^{(1)} = \sum_{k=1}^{n} c_k \mathbf{A}\,\mathbf{e}_k = \sum_{k=1}^{n} \lambda_k c_k \mathbf{e}_k. \tag{245b}$$

Next, if a multiple of $\mathbf{y}^{(1)}$, say $\alpha_1 \mathbf{y}^{(1)}$, is taken to be $\mathbf{x}^{(2)}$, there then follows similarly

$$\mathbf{x}^{(2)} = \alpha_1 \sum_{k=1}^{n} \lambda_k c_k \mathbf{e}_k, \qquad \mathbf{y}^{(2)} = \alpha_1 \sum_{k=1}^{n} \lambda_k^{\,2} c_k \mathbf{e}_k. \tag{246a,b}$$

More generally, after r steps we have

$$\mathbf{x}^{(r)} = \alpha'_{r-1} \sum_{k=1}^{n} \lambda_k^{\,r-1} c_k \mathbf{e}_k = \alpha'_{r-1} \lambda_n^{\,r-1} \sum_{k=1}^{n} \left(\frac{\lambda_k}{\lambda_n}\right)^{r-1} c_k \mathbf{e}_k$$

or

$$\mathbf{x}^{(r)} = \alpha'_{r-1} \lambda_n^{\,r-1} \left[c_n \mathbf{e}_n + \left(\frac{\lambda_{n-1}}{\lambda_n}\right)^{r-1} c_{n-1} \mathbf{e}_{n-1} + \cdots + \left(\frac{\lambda_1}{\lambda_n}\right)^{r-1} c_1 \mathbf{e}_1 \right] \tag{247a}$$

where $\alpha'_{r-1} = \alpha_1 \alpha_2 \cdots \alpha_{r-1}$, and, correspondingly,

$$\mathbf{y}^{(r)} = \alpha'_{r-1} \lambda_n^{\,r} \left[c_n \mathbf{e}_n + \left(\frac{\lambda_{n-1}}{\lambda_n}\right)^{r} c_{n-1} \mathbf{e}_{n-1} + \cdots + \left(\frac{\lambda_1}{\lambda_n}\right)^{r} c_1 \mathbf{e}_1 \right]. \tag{247b}$$

Since λ_n is the dominant characteristic number, the powers $(\lambda_k/\lambda_n)^r$ tend to zero when $k \neq n$ if $|\lambda_n| > |\lambda_{n-1}| \geqq \cdots$, and the expressions tend to multiples of $\mathbf{e}_n$ as r increases except in the very special case when the initial assumption $\mathbf{x}^{(1)}$ happens to be exactly orthogonal to $\mathbf{e}_n$, so that $c_n = 0$.

If λ_n is a multiple root of the characteristic equation, it is easily seen that the process will still lead to *one* corresponding characteristic vector. The case when λ_n and $-\lambda_n$ are both characteristic numbers requires special treatment.* However, in most practical cases the characteristic numbers are all nonnegative.

The *rate* of convergence of the method clearly depends upon the magnitude of the ratio of the two largest characteristic numbers. In case this ratio is near unity, and the convergence rate is slow, the matrix $\mathbf{A}$ may be first raised to an integral power p. The characteristic numbers of the new matrix $\mathbf{A}^p$ are then $\lambda_1^{\,p}, \cdots, \lambda_n^{\,p}$, and the ratio of the dominant and subdominant numbers is increased.

* It is apparent from (247a) that if $\lambda_{n-1} = -\lambda_n$ the sequence $\mathbf{x}^{(1)}, \mathbf{x}^{(3)}, \cdots$ converges to a multiple of $c_n \mathbf{e}_n + c_{n-1} \mathbf{e}_{n-1}$, whereas the sequence $\mathbf{x}^{(2)}, \mathbf{x}^{(4)}, \cdots$ converges to a multiple of $c_n \mathbf{e}_n - c_{n-1} \mathbf{e}_{n-1}$ (see Problem 96).

We may notice from (247a,b) that, if at any stage of the iteration the true vector $\mathbf{e}_n$ were known, the condition

$$(\mathbf{e}_n, \mathbf{y}^{(r)}) = \lambda(\mathbf{e}_n, \mathbf{x}^{(r)}) \tag{248}$$

would lead to the relation

$$\alpha'_{r-1}\lambda_n{}^r c_n = \lambda\alpha'_{r-1}\lambda_n{}^{r-1}c_n \qquad \text{or} \qquad \lambda = \lambda_n, \tag{249}$$

and hence would determine λ_n *exactly*. Clearly, any multiple of $\mathbf{e}_n$ would serve the same purpose. Thus it may be expected that a reasonably good approximation to λ_n would be obtained by replacing $\mathbf{e}_n$ by a convenient multiple of either the approximation $\mathbf{x}^{(r)}$ or the better approximation $\mathbf{y}^{(r)}$ in (248). This procedure gives the alternative formulas

$$(\mathbf{x}^{(r)}, \mathbf{y}^{(r)}) \approx \lambda_n(\mathbf{x}^{(r)}, \mathbf{x}^{(r)}) \tag{250a}$$

or

$$(\mathbf{y}^{(r)}, \mathbf{y}^{(r)}) \approx \lambda_n(\mathbf{x}^{(r)}, \mathbf{y}^{(r)}), \tag{250b}$$

of which the second is in general the more nearly accurate. It can be shown that the approximation given by (250a) is always conservative in absolute value (when $\mathbf{A}$ is real and symmetric). The same is true of that given by (250b) if the matrix $\mathbf{A}$ is also positive definite (see Problem 119).

We list in the following table the results of applying (A) the method of the preceding section, (B) the formula of (250a), and (C) the formula of (250b), to the illustrative example:

r	(A)	(B)	(C)
1	6.000	4.667	5.000
2	5.167	5.043	5.048
3	5.065	5.049	5.049
4	5.051	5.049	5.049

It may be seen that if (250a) or (250b) is used, the successive approximations to λ_n converge more rapidly than do the approximations to $\mathbf{e}_n$. This statement is generally true. Thus these formulas are useful in those cases when an accurate value of the dominant characteristic number is required, but comparable accuracy in the determination of the corresponding characteristic vector is not needed.

To obtain the *smallest* characteristic number of (241), we may write $\kappa = 1/\lambda$, resolve the equations in the form

$$\left.\begin{aligned} 2x_1 - x_2 \qquad &= \kappa x_1, \\ -x_1 + 2x_2 - x_3 &= \kappa x_2, \\ -x_2 + x_3 &= \kappa x_3 \end{aligned}\right\}, \tag{251}$$

and determine the largest characteristic value of κ by the preceding methods, or we may proceed equivalently by making use of (239) and (240).

Suppose now that *one* characteristic vector, say one which corresponds to a dominant characteristic number, is known *exactly*. Then for a *symmetric* matrix, all other characteristic vectors may be considered to be orthogonal to $\mathbf{e}_n$.* Hence, if we impose the constraint

$$(\mathbf{e}_n, \mathbf{x}) = 0 \tag{252}$$

on the problem (231), the resultant problem will possess those characteristic numbers and corresponding characteristic vectors which are in addition to λ_n and $\mathbf{e}_n$. But (252) permits one of the components, say x_r, to be expressed as a linear combination of the others. Hence we may eliminate x_r from the scalar equations corresponding to (231), disregard the rth resulting equation, and obtain a set of $n - 1$ equations involving only $n - 1$ components. The dominant characteristic number, and a corresponding characteristic vector, are then obtained as before, the component x_r being determined finally from (252).

Whereas the coefficient matrix associated with the new set of $n - 1$ equations is generally nonsymmetric, the convergence of the iterative method is assured in this case by results to be obtained in Section 1.26.

In particular, in the case when $|\mathbf{A}| = 0$ so that $\lambda = 0$ is a characteristic number of $\mathbf{A}$, we may replace $\mathbf{e}_n$ in (252) by the corresponding characteristic vector. Unless $\lambda = 0$ is of multiplicity greater than one, the corresponding reduced set of equations can then be inverted for the purpose of determining the smallest nonzero characteristic number. In the more general case, a number of unknowns equal to the multiplicity of the number $\lambda = 0$ must be eliminated in this way.

The procedure may be repeated until the solution is concluded or until only two components remain, at which stage the characteristic equation is quadratic in λ and the analysis can be conveniently completed without matrix iteration. Thus, if $\mathbf{A}$ is of order three, only one iterative process is needed. If $\mathbf{A}$ is of order four, we may conveniently determine the largest and smallest characteristic numbers and their corresponding vectors. The conditions $(\mathbf{e}_1, \mathbf{x}) = 0$ and $(\mathbf{e}_4, \mathbf{x}) = 0$ then permit the elimination of two components, and the reduction of the problem to one involving only the two remaining components.

In practice, the determination of the primary characteristic vector is only approximately effected. It is found that the numerical determination of a subdominant characteristic vector will often involve repeated subtraction of nearly equal quantities, particularly if the two relevant characteristic

* If the characteristic numbers are distinct, this *must* be so; otherwise, we may impose this condition without loss of generality.

numbers are nearly equal. In such cases, it may be necessary to calculate the components of the dominant characteristic vector to a degree of accuracy much higher than that required for the subdominant characteristic vector.

To illustrate the reduction in the preceding example, we notice that the dominant characteristic vector is given by {0.445, 0.802, 1} to three significant figures. Hence (252) here becomes

$$0.445x_1 + 0.802x_2 + x_3 = 0. \tag{253}$$

If we eliminate x_3 between (253) and (241), and notice that the third equation is then a consequence of the first two (to the three significant figures retained), we obtain the reduced problem

$$\left.\begin{aligned} 0.555x_1 + 0.198x_2 &= \lambda x_1, \\ 0.110x_1 + 0.396x_2 &= \lambda x_2 \end{aligned}\right\}. \tag{254}$$

The dominant characteristic number of (254), and the two components of the corresponding characteristic vector, can then be obtained by matrix iteration, if this is desired, the component x_3 being determined in terms of them by (253). Otherwise, since the characteristic equation of (254) is quadratic, that equation can be solved by the quadratic formula, and the ratio of the x_1 and x_2 components of the corresponding characteristic vectors can be obtained directly.

1.25. Generalized characteristic-value problems. In some applications we encounter characteristic-value problems of the more general form

$$\mathbf{A}\,\mathbf{x} = \lambda\,\mathbf{B}\,\mathbf{x}, \tag{255}$$

where **A** and **B** are real square matrices of order n. Such a problem reduces to the type considered previously when $\mathbf{B} = \mathbf{I}$. The characteristic equation corresponding to (255) is of the form

$$|\mathbf{A} - \lambda\,\mathbf{B}| = 0. \tag{256}$$

In the important practical cases in which both **A** and **B** are *symmetric*, so that $\mathbf{A}^T = \mathbf{A}$ and $\mathbf{B}^T = \mathbf{B}$, we next establish a useful generalization of the results of Section 1.12. If λ_1 and λ_2 are distinct characteristic numbers corresponding, respectively, to the characteristic vectors $\mathbf{u}_1$ and $\mathbf{u}_2$, there follows

$$\mathbf{A}\,\mathbf{u}_1 = \lambda_1\mathbf{B}\,\mathbf{u}_1, \qquad \mathbf{A}\,\mathbf{u}_2 = \lambda_2\mathbf{B}\,\mathbf{u}_2$$

and hence also

$$(\mathbf{A}\,\mathbf{u}_1)^T\,\mathbf{u}_2 = \lambda_1(\mathbf{B}\,\mathbf{u}_1)^T\,\mathbf{u}_2, \qquad \mathbf{u}_1{}^T\,\mathbf{A}\,\mathbf{u}_2 = \lambda_2\mathbf{u}_1{}^T\,\mathbf{B}\,\mathbf{u}_2,$$

or, making use of the symmetry in **A** and **B**,

$$\mathbf{u}_1{}^T\,\mathbf{A}\,\mathbf{u}_2 = \lambda_1\mathbf{u}_1{}^T\,\mathbf{B}\,\mathbf{u}_2, \qquad \mathbf{u}_1{}^T\,\mathbf{A}\,\mathbf{u}_2 = \lambda_2\mathbf{u}_1{}^T\,\mathbf{B}\,\mathbf{u}_2. \tag{257}$$

By subtracting the first equation from the second in (257), we then obtain the relation

$$(\lambda_2 - \lambda_1)\mathbf{u}_1^T \mathbf{B} \mathbf{u}_2 = 0. \tag{258}$$

Thus, since $\lambda_1 \neq \lambda_2$ by assumption, we conclude that $\mathbf{u}_1^T \mathbf{B} \mathbf{u}_2 = 0$. That is, *if* $\mathbf{u}_1$ *and* $\mathbf{u}_2$ *are characteristic vectors, corresponding to two distinct characteristic numbers of the problem* $\mathbf{A}\mathbf{x} = \lambda \mathbf{B}\mathbf{x}$, *where* $\mathbf{A}$ *and* $\mathbf{B}$ *are symmetric, there follows*

$$\mathbf{u}_1^T \mathbf{B} \mathbf{u}_2 = 0. \tag{259}$$

It is convenient to speak of the left-hand member of (259) as the *scalar product of* $\mathbf{u}_1$ *and* $\mathbf{u}_2$ *relative to* $\mathbf{B}$, and to say that when (259) is satisfied *the vectors* $\mathbf{u}_1$ *and* $\mathbf{u}_2$ *are orthogonal relative to* $\mathbf{B}$. The ordinary type of orthogonality is thus relative to the *unit matrix* $\mathbf{I}$.

In consequence of (259) and (257), we deduce that *the vectors* $\mathbf{u}_1$ *and* $\mathbf{u}_2$ *are also orthogonal relative to the matrix* $\mathbf{A}$.

The left-hand member of (259) is conveniently denoted by $(\mathbf{u}_1, \mathbf{u}_2)_\mathbf{B}$. More generally, we write

$$(\mathbf{u}, \mathbf{v})_\mathbf{B} \equiv \mathbf{u}^T \mathbf{B} \mathbf{v} = \mathbf{v}^T \mathbf{B} \mathbf{u} \tag{260}$$

for the scalar product of $\mathbf{u}$ and $\mathbf{v}$ relative to a *symmetric* matrix $\mathbf{B}$. In particular, when $\mathbf{v} = \mathbf{u}$ we define the product

$$l_\mathbf{B}^2(\mathbf{u}) \equiv (\mathbf{u}, \mathbf{u})_\mathbf{B} \equiv \mathbf{u}^T \mathbf{B} \mathbf{u} \tag{261}$$

to be the square of the *generalized length* of $\mathbf{u}$, relative to $\mathbf{B}$. In order that this quantity be necessarily *positive* when $\mathbf{u}$ is real, except only when $\mathbf{u}$ is a *zero vector*, the matrix $\mathbf{B}$ must be *positive definite*. This is the case which most frequently arises in practice.*

In the remainder of this section, we assume that $\mathbf{A}$ *is real and symmetric and* $\mathbf{B}$ *real, symmetric, and positive definite*. In particular, this implies that $\mathbf{B}$ is *nonsingular*. The generalized length of a real vector, relative to $\mathbf{B}$, is then real and positive unless the vector is a zero vector, in which case its generalized length is zero.

Further it is easily shown that *when* $\mathbf{B}$ *is real and positive definite, any set of nonzero real vectors which are mutually orthogonal relative to* $\mathbf{B}$ *is a linearly independent set* (see Problem 103).

By a method analogous to that used in Section 1.12, it is then easily shown that *the characteristic numbers of* (255) *are real*. Further, by an argument similar to that used in Section 1.18, it can be shown that *to a characteristic number of multiplicity s there correspond s linearly independent characteristic vectors*. Then, by methods completely analogous to those of

* Usually at least *one* of the matrices $\mathbf{A}$ and $\mathbf{B}$ is positive definite. If $\mathbf{A}$ is positive definite, we may replace λ by $1/\lambda'$ and interchange the roles of $\mathbf{A}$ and $\mathbf{B}$ throughout this section.

Section 1.13, this set can be orthogonalized relative to $\mathbf{B}$, and normalized in such a way that each vector possesses generalized length unity. It is seen that the condition $|\mathbf{B}| \neq 0$ guarantees that the characteristic equation (256) be of degree n. Hence, in the case under consideration, we may always obtain a set of n mutually orthogonal unit characteristic vectors $\mathbf{e}_1, \mathbf{e}_2, \cdots, \mathbf{e}_n$, such that

$$(\mathbf{e}_i, \mathbf{e}_j)_{\mathbf{B}} = \delta_{ij}. \tag{262}$$

An *orthonormal modal matrix* $\mathbf{M}$, associated with (255), may now be defined as the matrix having the components of the kth vector of the set as the elements of its kth *column*. Then, in consequence of the relation

$$\mathbf{A}\,\mathbf{e}_i = \lambda_i \mathbf{B}\,\mathbf{e}_i \qquad (i = 1, 2, \cdots, n),$$

there follows

$$\mathbf{A\,M} = \mathbf{B\,M}\begin{bmatrix} \lambda_1 & 0 & \cdots & 0 \\ 0 & \lambda_2 & \cdots & 0 \\ \cdots & \cdots & \cdots & \cdots \\ 0 & 0 & \cdots & \lambda_n \end{bmatrix} \equiv \mathbf{B\,M\,D}, \tag{263}$$

and also, by virtue of (262),

$$\mathbf{M}^T\,\mathbf{B\,M} = \mathbf{I}. \tag{264}$$

(See Problem 41.)

We may now verify the fact that, with the change of variables

$$\mathbf{x} = \mathbf{M}\,\boldsymbol{\alpha}, \tag{265}$$

the *two* quadratic forms

$$A = \mathbf{x}^T\,\mathbf{A}\,\mathbf{x}, \qquad B = \mathbf{x}^T\,\mathbf{B}\,\mathbf{x} \tag{266a,b}$$

are reduced *simultaneously* to the canonical forms

$$A = \boldsymbol{\alpha}^T\,\mathbf{D}\,\boldsymbol{\alpha} = \lambda_1\alpha_1^{\,2} + \lambda_2\alpha_2^{\,2} + \cdots + \lambda_n\alpha_n^{\,2}, \tag{267a}$$

$$B = \boldsymbol{\alpha}^T\,\boldsymbol{\alpha} = \alpha_1^{\,2} + \alpha_2^{\,2} + \cdots + \alpha_n^{\,2}. \tag{267b}$$

For the substitution of (265) into (266b), and the use of (264), gives immediately

$$B = \boldsymbol{\alpha}^T\,\mathbf{M}^T\,\mathbf{B\,M}\,\boldsymbol{\alpha} = \boldsymbol{\alpha}^T\,\boldsymbol{\alpha},$$

in accordance with (267b), whereas the substitution of (265) into (266a) gives

$$A = \boldsymbol{\alpha}^T\,\mathbf{M}^T\,\mathbf{A\,M}\,\boldsymbol{\alpha}$$

and the use of (263) and (264) leads to the result

$$A = \boldsymbol{\alpha}^T\,\mathbf{M}^T\,\mathbf{B\,M\,D}\,\boldsymbol{\alpha} = \boldsymbol{\alpha}^T\,\mathbf{D}\,\boldsymbol{\alpha},$$

in accordance with (267a).

From (264) it follows that

$$|\mathbf{M}| = \pm \frac{1}{\sqrt{|\mathbf{B}|}},$$

so that the transformation (265) is *nonsingular*. Further, since (264) leads to the relation $\mathbf{M}^{-1} = \mathbf{M}^T \mathbf{B}$, the inversion of (265) may be conveniently effected by use of the equation

$$\boldsymbol{\alpha} = \mathbf{M}^T \mathbf{B}\, \mathbf{x}. \tag{268}$$

Equations (267a,b) are identical with equations (169) and (171) of Section 1.19. It is important to notice that the coefficients λ_i in (267) are the roots of the equation $|\mathbf{A} - \lambda \mathbf{B}| = 0$, and hence are *real*, under the present restrictions on $\mathbf{A}$ and $\mathbf{B}$.

If the matrices $\mathbf{A}$ *and* $\mathbf{B}$ *are both positive definite, the characteristic numbers* λ_i *are also necessarily positive.* This result is established by noticing that the relation

$$\mathbf{A}\, \mathbf{e}_i = \lambda_i \mathbf{B}\, \mathbf{e}_i$$

implies the relation

$$\mathbf{e}_i^T \mathbf{A}\, \mathbf{e}_i = \lambda_i \mathbf{e}_i^T \mathbf{B}\, \mathbf{e}_i.$$

Since both $\mathbf{e}_i^T \mathbf{A}\, \mathbf{e}_i$ and $\mathbf{e}_i^T \mathbf{B}\, \mathbf{e}_i$ are positive when $\mathbf{A}$ and $\mathbf{B}$ are positive definite (and $\mathbf{e}_i \neq \mathbf{0}$), the same is true of λ_i.

The preceding results will be of importance in Section 2.14 of the following chapter.

Since the vectors $\mathbf{e}_1, \mathbf{e}_2, \cdots, \mathbf{e}_n$ are *linearly independent* (see Problem 103), it follows that any vector $\mathbf{v}$ in n-space can be expressed as a linear combination of these vectors, of the form

$$\mathbf{v} = \alpha_1 \mathbf{e}_1 + \alpha_2 \mathbf{e}_2 + \cdots + \alpha_n \mathbf{e}_n = \sum_{k=1}^{n} \alpha_k \mathbf{e}_k. \tag{269}$$

In order to evaluate any coefficient α_r, we merely form the generalized scalar product of $\mathbf{e}_r$ into both sides of (269), and use (262) to obtain the result

$$\alpha_r = (\mathbf{e}_r, \mathbf{v})_{\mathbf{B}} \equiv \mathbf{e}_r^T \mathbf{B}\, \mathbf{v} \qquad (r = 1, 2, \cdots, n). \tag{270}$$

The case of most common occurrence in practice is that in which $\mathbf{B}$ is a *diagonal matrix* $\mathbf{G}$, say

$$\mathbf{G} = \begin{bmatrix} g_1 & 0 & \cdots & 0 \\ 0 & g_2 & \cdots & 0 \\ \cdots & \cdots & \cdots & \cdots \\ 0 & 0 & \cdots & g_n \end{bmatrix}, \tag{271}$$

so that the equations $\mathbf{A}\,\mathbf{x} = \lambda\,\mathbf{G}\,\mathbf{x}$ take the special form

$$\left.\begin{array}{l} a_{11}x_1 + a_{12}x_2 + \cdots + a_{1n}x_n = \lambda g_1 x_1, \\ \cdots\cdots\cdots\cdots\cdots\cdots\cdots\cdots\cdots \\ a_{n1}x_1 + a_{n2}x_2 + \cdots + a_{nn}x_n = \lambda g_n x_n \end{array}\right\}, \tag{272}$$

where $a_{ji} = a_{ij}$.

The generalized scalar product $(\mathbf{x}, \mathbf{y})_G$ then takes the form

$$(\mathbf{x}, \mathbf{y})_G = g_1 x_1 y_1 + g_2 x_2 y_2 + \cdots + g_n x_n y_n, \tag{273}$$

while the generalized length of $\mathbf{x}$ is given by

$$l_G^2(\mathbf{x}) = (\mathbf{x}, \mathbf{x})_G = g_1 x_1^2 + g_2 x_2^2 + \cdots + g_n x_n^2. \tag{274}$$

The condition that $\mathbf{G}$ be positive definite requires that the diagonal elements be positive:

$$g_i > 0 \qquad (i = 1, 2, \cdots, n). \tag{275}$$

It may be noticed that in certain cases a set of equations of the matrix form $\mathbf{A}'\,\mathbf{x} = \lambda\,\mathbf{x}$, where $\mathbf{A}'$ is a *nonsymmetric* square matrix, can be reduced to a set of the matrix form $\mathbf{A}\,\mathbf{x} = \lambda\,\mathbf{G}\,\mathbf{x}$, where $\mathbf{A}$ is symmetric and $\mathbf{G}$ is a diagonal matrix with positive diagonal elements, by multiplying the ith equation of the original set by a suitably chosen *positive* constant g_i. When $\mathbf{A}'$ is of order *two*, this reduction is clearly always possible if $a'_{12}a'_{21} > 0$; it is possible in other cases only when the coefficients satisfy certain compatibility conditions. If and only if such a reduction is possible, $\mathbf{A}'$ can be expressed as a product $\mathbf{D}\,\mathbf{A}$, where $\mathbf{D} \equiv \mathbf{G}^{-1}$ is a *diagonal* matrix with positive diagonal elements, and $\mathbf{A}$ is symmetric.

To conclude this section, we indicate the extension of the numerical methods of the preceding sections to the treatment of a characteristic-value problem of the form

$$\mathbf{A}\,\mathbf{x} = \lambda\,\mathbf{B}\,\mathbf{x}, \tag{276}$$

where again $\mathbf{A}$ is a symmetric matrix of order n, and $\mathbf{B}$ is a positive definite, symmetric, matrix of the same order.

Since, by assumption, $\mathbf{B}$ is nonsingular, equation (276) can be reduced to the form

$$\mathbf{B}^{-1}\,\mathbf{A}\,\mathbf{x} = \lambda\,\mathbf{x}, \tag{277}$$

which is of the type considered previously. However, the matrix $\mathbf{B}^{-1}\,\mathbf{A}$ will now *not* be symmetric, in general. In the case of (272) the reduction to the form (277) involves only division of both sides of the ith equation by g_i.

Whether or not the transformation of (276) to (277) is effected, we may define sequences of vectors $\mathbf{x}^{(1)}, \mathbf{x}^{(2)}, \cdots$ and $\mathbf{y}^{(1)}, \mathbf{y}^{(2)}, \cdots$ by the relations

$$\mathbf{A}\,\mathbf{x}^{(r)} = \mathbf{B}\,\mathbf{y}^{(r)}, \qquad \mathbf{x}^{(r+1)} = \alpha_r \mathbf{y}^{(r)} \qquad (r = 1, 2, \cdots), \tag{278a,b}$$

where α_r is a conveniently chosen nonzero constant and where $\mathbf{x}^{(1)}$ is to be *chosen* to start the iteration, in the hope that there will follow

$$\mathbf{y}^{(r)} \sim \lambda_n \mathbf{x}^{(r)}, \qquad \mathbf{x}^{(r)} \to c\, \mathbf{e}_n \qquad (r \to \infty) \tag{279a,b}$$

in general, where λ_n is the relevant characteristic number of *largest* magnitude. Here the vector $\mathbf{y}^{(r)}$ is to be obtained from $\mathbf{x}^{(r)}$, in accordance with (278a), either by making use of the matrix $\mathbf{B}^{-1}$ or by a numerical solution of the associated set of scalar equations in each iteration.

In order to investigate the convergence of the iterative procedure in this case, let the normalized characteristic unit vectors be denoted by $\mathbf{e}_1, \mathbf{e}_2, \cdots, \mathbf{e}_n$, corresponding, respectively, to $\lambda_1, \lambda_2, \cdots, \lambda_n$, so that

$$\mathbf{A}\, \mathbf{e}_r = \lambda_r \mathbf{B}\, \mathbf{e}_r. \tag{280}$$

The initial approximation $\mathbf{x}^{(1)}$ can be imagined to be expressed as a linear combination of these vectors, in the form

$$\mathbf{x}^{(1)} = \sum_{k=1}^{n} c_k \mathbf{e}_k. \tag{281a}$$

There then follows

$$\mathbf{y}^{(1)} \equiv \mathbf{B}^{-1}\mathbf{A}\,\mathbf{x}^{(1)} = \sum_{k=1}^{n} c_k \mathbf{B}^{-1}\mathbf{A}\,\mathbf{e}_k = \sum_{k=1}^{n} c_k \mathbf{B}^{-1}\lambda_k \mathbf{B}\,\mathbf{e}_k$$

or

$$\mathbf{y}^{(1)} \equiv \mathbf{B}^{-1}\mathbf{A}\,\mathbf{x}^{(1)} = \sum_{k=1}^{n} \lambda_k c_k \mathbf{e}_k. \tag{281b}$$

By comparing (281a,b) with (245a,b) of the preceding section, we see that the arguments presented in that section again apply here, to show that successive approximations will indeed converge to a multiple of the dominant vector $\mathbf{e}_n$ when $|\lambda_n| > |\lambda_{n-1}| \geqq \cdots$, if $c_n \neq 0$.

In this case, however, it is seen that the requirement

$$(\mathbf{e}_n, \mathbf{y}^{(r)})_{\mathbf{B}} = \lambda(\mathbf{e}_n, \mathbf{x}^{(r)})_{\mathbf{B}},$$

in place of (248), would give $\lambda = \lambda_n$ exactly. Hence (250a,b) here should be modified to the alternative conditions

$$(\mathbf{x}^{(r)}, \mathbf{y}^{(r)})_{\mathbf{B}} \approx \lambda_n (\mathbf{x}^{(r)}, \mathbf{x}^{(r)})_{\mathbf{B}} \tag{282a}$$

or, better,

$$(\mathbf{y}^{(r)}, \mathbf{y}^{(r)})_{\mathbf{B}} \approx \lambda_n (\mathbf{x}^{(r)}, \mathbf{y}^{(r)})_{\mathbf{B}}. \tag{282b}$$

Similarly, equation (252) must be replaced by the relation

$$(\mathbf{e}_n, \mathbf{x})_{\mathbf{B}} = 0, \tag{283}$$

which permits reduction of the order of the system when one characteristic vector has been obtained.

The same statements apply to the inversion of (277),

$$\frac{1}{\lambda}\mathbf{x} = \mathbf{A}^{-1}\mathbf{B}\,\mathbf{x},$$

which may be used in determining the smallest characteristic value of λ when $\mathbf{A}$ is nonsingular.* Here the relations (278) and (279) are to be replaced by the relations

$$\mathbf{A}\,\mathbf{y}^{(r)} = \mathbf{B}\,\mathbf{x}^{(r)}, \qquad \mathbf{x}^{(r+1)} = \alpha_r \mathbf{y}^{(r)} \qquad (r = 1, 2, \cdots) \qquad \text{(284a,b)}$$

and

$$\mathbf{y}^{(r)} \sim \frac{1}{\lambda_1}\mathbf{x}^{(r)}, \qquad \mathbf{x}^{(r)} \to c\,\mathbf{e}_1 \qquad (r \to \infty), \qquad \text{(285a,b)}$$

where the iteration is to be initiated by choosing $\mathbf{x}^{(1)}$ and determining $\mathbf{y}^{(1)}$, either by making use of $\mathbf{A}^{-1}$ or by a direct numerical method.

1.26. Characteristic numbers of nonsymmetric matrices. Whereas the characteristic equation $|\mathbf{A} - \lambda\,\mathbf{I}| = 0$ of a nonsymmetric square matrix $\mathbf{A}$ of order n is of degree n, we have seen that when the roots of this equation are not *distinct* the total number of linearly independent characteristic vectors may be less than n. In the present section we exclude the exceptional cases, which rarely occur in practice, and suppose that the n characteristic numbers of $\mathbf{A}$ are *real and distinct*. The corresponding characteristic vectors are then linearly independent.

In order to establish this fact, we assume the contrary and deduce a contradiction. Suppose that the characteristic numbers $\lambda_1, \cdots, \lambda_n$ are all distinct, and denote the corresponding characteristic vectors by $\mathbf{u}_1, \cdots, \mathbf{u}_n$. We then have the relations $\mathbf{A}\,\mathbf{u}_i = \lambda_i \mathbf{u}_i$ for $i = 1, 2, \cdots, n$. *Assume* that the first r characteristic vectors are linearly independent, but that

$$\mathbf{u}_{r+1} = \sum_{k=1}^{r} c_k \mathbf{u}_k,$$

where at least one c_k is not zero since $\mathbf{u}_{r+1} \neq \mathbf{0}$. By premultiplying the equal members of this relation by $\mathbf{A}$, there then follows

$$\lambda_{r+1}\mathbf{u}_{r+1} = \sum_{k=1}^{r} c_k \lambda_k \mathbf{u}_k,$$

and hence also, by comparing these relations,

$$\sum_{k=1}^{r} c_k(\lambda_{r+1} - \lambda_k)\mathbf{u}_k = \mathbf{0}.$$

But, since $\mathbf{u}_1, \cdots, \mathbf{u}_r$ are linearly independent, the coefficient of *each* $\mathbf{u}_k$ must vanish. Since at least one c_k is not zero, at least one λ_k must equal λ_{r+1}, in contradiction with the hypothesis that the λ's are distinct.

In correspondence with the characteristic-value problem

$$\mathbf{A}\,\mathbf{x} = \lambda\,\mathbf{x}, \qquad \text{(286)}$$

* See also Problem 125.

we may consider the problem

$$\mathbf{A}^T \mathbf{x}' = \lambda \mathbf{x}', \tag{287}$$

associated with the *transpose* of **A**. By virtue of the fact that the two matrices $\mathbf{A} - \lambda \mathbf{I}$ and $\mathbf{A}^T - \lambda \mathbf{I}$ differ only in that rows and columns are interchanged, their determinants possess the same expansion, so that (286) *and* (287) *possess the same characteristic numbers.* Let λ_1 and λ_2 denote any two distinct characteristic numbers, and let corresponding solutions of (286) and (287) be denoted by $\mathbf{u}_1$, $\mathbf{u}_2$ and $\mathbf{u}_1'$, $\mathbf{u}_2'$, respectively. We then have the relations

$$\mathbf{A}\,\mathbf{u}_1 = \lambda_1 \mathbf{u}_1, \qquad \mathbf{A}^T\,\mathbf{u}_2' = \lambda_2 \mathbf{u}_2',$$

from which there follows

$$(\lambda_2 - \lambda_1)\mathbf{u}_1{}^T\,\mathbf{u}_2' = 0. \tag{288}$$

Hence we conclude that *any characteristic vector of* (286) *is orthogonal to any characteristic vector of* (287) *which corresponds to a different characteristic number:*

$$(\mathbf{u}_i, \mathbf{u}_j') = 0 \qquad (\lambda_i \neq \lambda_j). \tag{289}$$

This property permits the generalization of the methods of Sections 1.23 and 1.24 to the more general case considered here. While the problems considered in Section 1.25 are included in this generalization, the methods given in that section are usually preferable when they are applicable.

In particular, it is seen that the coefficients in the representation

$$\mathbf{v} = \sum_{k=1}^{n} \alpha_k \mathbf{u}_k \tag{290a}$$

are determined by forming the scalar product of $\mathbf{u}_k'$ with the two members of this equation, in the form

$$\alpha_k(\mathbf{u}_k, \mathbf{u}_k') = (\mathbf{v}, \mathbf{u}_k'). \tag{290b}$$

A development analogous to that of Section 1.24 then shows that the matrix iteration procedure again converges in this case to the characteristic vector corresponding to the dominant characteristic number of **A**. In fact, such a development shows that the convergence of this procedure is insured if the matrix **A** possesses n linearly independent characteristic vectors, and only *real* characteristic numbers (which need not be *distinct*).* While formulas analogous to (250a,b) can be devised for more accurate estimates of λ_n, their use involves a considerable increase in calculation.

* By a somewhat more involved analysis, which may be based on the generalized Sylvester formula mentioned in Section 1.22 (cf. Problem 89), it can be shown that the iterative procedure converges to the dominant characteristic number of any real matrix *if that number is real, and if no unequal characteristic number has equal absolute value.* If the dominant number is *repeated,* however, the rate of convergence is not always *exponential* and the procedure is often not practical unless the multiplicity of that number is somehow known in advance. Modifications (similar to those of Problem 96) which are useful in this special situation, as well as in the case when the dominant numbers are conjugate complex, are given in Reference 7.

The essential modification in procedure is involved in the calculation of subdominant characteristic quantities. In the more general case considered here, the constraint condition (252) must be replaced by the equation

$$(\mathbf{u}_n', \mathbf{x}) = 0. \tag{291}$$

Thus after the (approximate) determination of the dominant characteristic number λ_n, and the corresponding vector solution $\mathbf{u}_n$, a vector $\mathbf{u}_n'$ satisfying the related equation $\mathbf{A}^T \mathbf{x}' = \lambda_n \mathbf{x}'$ must be determined (see Problem 108). The constraint (291), which corresponds to the fact that all other characteristic vectors of $\mathbf{A}$ are orthogonal to $\mathbf{u}_n'$, then permits the elimination of one of the unknowns in the system of equations (and the neglect of one of the resulting equations) so that the order of the system is reduced by unity.

It may be noticed that the relation

$$\mathbf{A}^T \mathbf{u}_r' = \lambda_r \mathbf{u}_r'$$

can also be written in the form

$$\mathbf{u}_r'^T \mathbf{A} = \lambda_r \mathbf{u}_r'^T \tag{292}$$

Thus, if $\mathbf{u}_r'$ is a characteristic vector of $\mathbf{A}^T$, corresponding to λ_r, then the *row vector* $\mathbf{u}_r'^T$ is such that *post*multiplication by the matrix $\mathbf{A}$ multiplies $\mathbf{u}_r'^T$ by the scalar λ_r, whereas the *column vector* $\mathbf{u}_r$ is such that *pre*multiplication by $\mathbf{A}$ multiplies $\mathbf{u}_r$ by λ_r.

If, for *any* $n \times n$ matrix $\mathbf{A}$ with n independent characteristic vectors, a *modal matrix* $\mathbf{M}$ is formed, in such a way that the components of n successive linearly independent characteristic vectors of $\mathbf{A}$ comprise successive columns of $\mathbf{M}$, the matrix $\mathbf{A}$ can be *diagonalized* by the *similarity* transformation $\mathbf{M}^{-1} \mathbf{A} \mathbf{M}$, the resultant diagonal elements being the characteristic numbers of $\mathbf{A}$ (see Problem 109). However, in consequence of the fact that the n characteristic vectors are generally not orthogonal, it follows that $\mathbf{M}^{-1} \neq \mathbf{M}^T$, in general, so that the matrix $\mathbf{M}$ generally is *not* an *orthogonal matrix*.

If certain characteristic numbers of a nonsymmetric and non-Hermitian matrix are repeated, there may be less than n linearly independent characteristic vectors, so that complete diagonalization in this way is impossible. In any case, it can be shown that *any* square matrix can be transformed by a *similarity* transformation (which is not necessarily orthogonal) to a *canonical matrix* with the following properties:

1. All elements *below* the principal diagonal are zero.
2. The diagonal elements are the characteristic numbers of the matrix.
3. All elements *above* the principal diagonal are zero *except* possibly those elements which are adjacent to *two* equal diagonal elements.
4. The latter elements are each either zero or unity.

A matrix having these four properties is known as a *Jordan canonical matrix*.

In illustration, for a matrix of order five for which $\lambda_1 = \lambda_2 = \lambda_3$ and $\lambda_4 = \lambda_5$, but $\lambda_1 \neq \lambda_4$, this canonical form would be

$$\begin{bmatrix} \lambda_1 & \alpha_1 & 0 & 0 & 0 \\ 0 & \lambda_1 & \alpha_2 & 0 & 0 \\ 0 & 0 & \lambda_1 & 0 & 0 \\ 0 & 0 & 0 & \lambda_4 & \alpha_3 \\ 0 & 0 & 0 & 0 & \lambda_4 \end{bmatrix},$$

where each of the elements α_1, α_2, and α_3 is either unity or zero, according as λ_1 corresponds to one, two, or three independent characteristic vectors, and λ_4 to one or two independent characteristic vectors. Reductions to certain other standard forms have also been studied.*

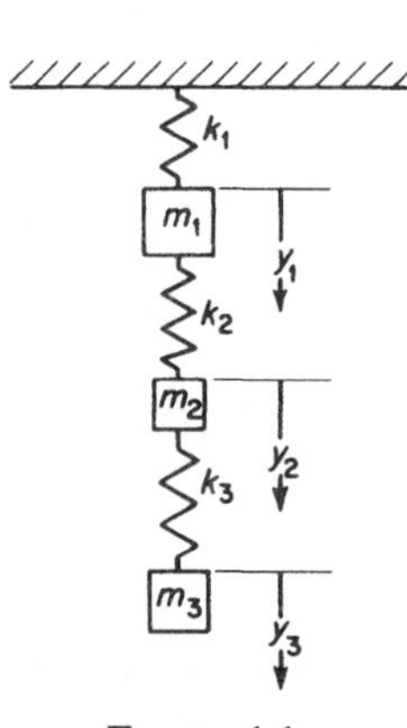

FIGURE 1.1

To conclude this section, we repeat the following comments made in Section 1.22:

The Cayley-Hamilton theorem [see equation (219)] *is true for any square matrix* **A**. *Sylvester's formula* [see equations (226) and (227)] *is valid for any square matrix* **A** *with no repeated characteristic numbers, and accordingly can be applied to the evaluation of any function* $f(\mathbf{A})$ *which is representable as a convergent series of powers of* **A**.

1.27. A physical application. Several applications of the preceding methods will be found in the chapters which follow. In this section, we present one such application to a mechanical problem.

We consider the problem of determining the natural modes of free vibration of the mechanical system indicated in Figure 1.1, in which the masses m_1, m_2, and m_3 are connected in series to a fixed support, by linear springs with spring constants k_1, k_2, and k_3. The effects of viscous damping are neglected. If we denote the displacements of the respective masses from their equilibrium positions by $y_1(t)$, $y_2(t)$, and $y_3(t)$, respectively, the differential equations of motion are of the form

$$\left.\begin{aligned} m_1 \frac{d^2y_1}{dt^2} &= k_2(y_2 - y_1) - k_1 y_1 = -(k_1 + k_2)y_1 + k_2 y_2, \\ m_2 \frac{d^2y_2}{dt^2} &= k_3(y_3 - y_2) - k_2(y_2 - y_1) = k_2 y_1 - (k_2 + k_3)y_2 + k_3 y_3, \\ m^3 \frac{d^2y_3}{dt^2} &= -k_3(y_3 - y_2) = k_3 y_2 - k_3 y_3 \end{aligned}\right\}. \quad (293)$$

* See Reference 13.

The *natural modes* of vibration are those in which the masses oscillate in phase with a common frequency, and hence the displacements are specified by equations of the form

$$\left.\begin{aligned} y_1(t) &= x_1 \sin(\omega t + \alpha), \\ y_2(t) &= x_2 \sin(\omega t + \alpha), \\ y_3(t) &= x_3 \sin(\omega t + \alpha) \end{aligned}\right\}, \tag{294}$$

where the amplitudes x_1, x_2, and x_3 and the common circular frequency ω are to be determined. By introducing (294) into (293), and canceling the common resultant time factors, we obtain the equations

$$\left.\begin{aligned} (k_1 + k_2)x_1 \qquad - k_2x_2 \qquad &= m_1\omega^2 x_1, \\ -k_2x_1 + (k_2 + k_3)x_2 - k_3x_3 &= m_2\omega^2 x_2, \\ -k_3x_2 + k_3x_3 &= m_3\omega^2 x_3 \end{aligned}\right\}. \tag{295}$$

It should be noticed that the matrix of the coefficients in the left-hand members is *symmetric*. Also it is found that the "discriminants" Δ_m defined in Section 1.20 are of the form

$$\left.\begin{aligned} \Delta_1 &= k_1 + k_2, \\ \Delta_2 &= k_1k_2 + k_2k_3 + k_3k_1, \\ \Delta_3 &= k_1k_2k_3 \end{aligned}\right\}, \tag{296}$$

so that the matrix of coefficients is also *positive definite* when the spring constants are positive.

In the special case when $k_1 = k_2 = k_3 \equiv k$, and $m_1 = m_2 = m_3 \equiv m$, these equations reduce to equations (251) of Section 1.24 if we set

$$\kappa = \frac{1}{\lambda} = \frac{m\omega^2}{k}. \tag{297}$$

Hence the characteristic values of λ discussed in the example of that section are inversely proportional to the squares of the natural frequencies of the physical system under consideration, and the components of the characteristic vectors are in the same ratio as the three amplitudes x_1, x_2, and x_3 in a corresponding mode of vibration.

In the *fundamental mode*, corresponding to the *smallest* natural frequency, and hence to the *dominant* characteristic value of λ as defined by (297), the circular frequency is hence given by

$$\frac{m\omega_1^2}{k} = \frac{1}{5.05}: \qquad \omega_1 = 0.445\sqrt{\frac{k}{m}}.$$

Here the three masses all move in the same direction, the respective displacements from equilibrium at any instant being in the ratio 0.445:0.802:1.

By completing the analysis indicated in Section 1.24, we find that in the *second* mode there follows

$$\omega_2 = 1.247\sqrt{\frac{k}{m}}.$$

In this mode the first two masses move in the same direction, whereas the third mass moves in the opposite direction, the displacements being in the ratio $-1.247:-0.555:1$. In the *third* mode there follows

$$\omega_3 = 1.802\sqrt{\frac{k}{m}}.$$

Here the first and third masses move in the same direction, and the second mass in the opposite direction, the displacements being in the ratio $1.802:-2.247:1$.

The most general motion of the system, possible in the absence of externally applied forces, is then a superposition of the three modes just described, in which the phase angle α of equation (294) may take on different values in the individual modes.

If the three masses are unequal, equations (295) are of the form of equations (272), with g_i proportional to m_i. To illustrate the treatment of this case, we suppose that

$$k_1 = k_2 = k_3 \equiv k, \qquad m_1 = m_2 \equiv m, \qquad m_3 = 2m,$$

so that equations (295) become

$$\left.\begin{aligned} 2x_1 - x_2 \qquad &= \frac{m\omega^2}{k}x_1, \\ -x_1 + 2x_2 - x_3 &= \frac{m\omega^2}{k}x_2, \\ - x_2 + x_3 &= 2\frac{m\omega^2}{k}x_3 \end{aligned}\right\}.$$

In this case we may write

$$g_1 = 1, \qquad g_2 = 1, \qquad g_3 = 2.$$

In order to determine the *fundamental* mode *directly*, we may first *invert* these equations in the form

$$\left.\begin{aligned} x_1 + x_2 + 2x_3 &= \lambda x_1, \\ x_1 + 2x_2 + 4x_3 &= \lambda x_2, \\ x_1 + 2x_2 + 6x_3 &= \lambda x_3 \end{aligned}\right\},$$

where

$$\lambda = \frac{k}{m\omega^2}.$$

Except for refined successive estimates of $\lambda_3 = k/(m\omega_1^2)$, the calculation proceeds exactly as before. The results of successive steps are tabulated below to three significant figures:

	$\mathbf{x}^{(1)}$	$\mathbf{y}^{(1)}$	$\mathbf{x}^{(2)}$	$\mathbf{y}^{(2)}$	$\mathbf{x}^{(3)}$	$\mathbf{y}^{(3)}$	$\mathbf{x}^{(4)}$	$\mathbf{y}^{(4)}$	$\mathbf{x}^{(5)}$
x_1	1	4	0.444	3.22	0.402	3.15	0.399	3.15	0.399
x_2	1	7	0.777	6.00	0.750	5.90	0.747	5.89	0.747
x_3	1	9	1	8.00	1	7.90	1	7.89	1

Thus, after four cycles, the modal column $\{0.399, 0.747, 1\}$ is repeated. The dominant characteristic value of λ is seen to be 7.89, so that the fundamental circular frequency of the physical system is

$$\omega_1 = 0.356\sqrt{\frac{k}{m}}.$$

If only this value were of interest, and accurate values of the corresponding mode components were not required, the use of either (282a), in the form

$$\lambda(x_1^2 + x_2^2 + 2x_3^2) \approx (x_1y_1 + x_2y_2 + 2x_3y_3),$$

or (282b), in the form

$$\lambda(x_1y_1 + x_2y_2 + 2x_3y_3) \approx (y_1^2 + y_2^2 + 2y_3^2),$$

would yield the above result for the dominant value of λ after only two cycles.

If the remaining modes are required, the orthogonality relation (283) becomes

$$0.399x_1 + 0.747x_2 + 2.000x_3 = 0,$$

and permits the reduction of the order of the system to two.

1.28. Function space. In this section, we develop certain analogies between vector space and so-called "function space" and point out certain essential difficulties involved in the treatment of the latter.

If, in three-dimensional space, we consider any two nonzero vectors $\mathbf{u}$ and $\mathbf{v}$ which are not scalar multiples of each other, we see that the totality of all vectors of the form $c_1\mathbf{u} + c_2\mathbf{v}$ comprises a double infinity of vectors, namely, all vectors in that space which are parallel to the plane of $\mathbf{u}$ and $\mathbf{v}$. If $\mathbf{w}$ is any third nonzero vector which is not parallel to the plane of $\mathbf{u}$ and $\mathbf{v}$, then *all* vectors in that space are comprised in the representation $c_1\mathbf{u} + c_2\mathbf{v} + c_3\mathbf{w}$. More generally, in the language of linear vector spaces, we say that "any n linearly independent vectors form a basis in n-dimensional space."

Similarly, if we consider two functions $f(x)$ and $g(x)$, defined over an interval (a, b) and not multiples of each other over that interval, those functions which are of the form $c_1 f(x) + c_2 g(x)$ comprise a double infinity of functions. The question then arises as to the possibility of choosing a more comprehensive set of basic functions such that *any* function, satisfying appropriate regularity conditions, can be expressed as a linear combination of those functions over a certain given interval, that is, as to the possibility of choosing a "basis" in a "function space" associated with that interval. Certainly any such set of functions must have infinitely many members; that is, such a "function space" comprises infinitely many dimensions. Also, as in a vector space, we would expect the choice to be by no means a *unique* one.

In a *vector* space of n dimensions, great advantage is attained by choosing as a basis a set of n mutually *orthogonal* nonzero vectors, that is, a set of nonzero vectors such that the *scalar product* of any two distinct vectors in the set is zero. This fact suggests that we introduce an analogous definition of the scalar product of two *functions*, relative to the interval under consideration.

For real-valued functions, to which attention will be restricted here, it is found that a particularly useful definition is of the form

$$(f, g) = \int_a^b fg\, dx. \tag{298}$$

This definition is a natural generalization of the vector definition

$$(\mathbf{u}, \mathbf{v}) = \sum_{k=1}^{n} u_k v_k$$

when the dimension of the space and the number of components involved become infinitely large.

Thus (assuming here and henceforth that the functions involved are such that the integrals involved *exist*) we are led to say that *two functions $f(x)$ and $g(x)$ are orthogonal over an interval (a, b) if the integral $\int_a^b fg\, dx$ vanishes.*

In particular, when $f = g$ we may think of the number $\int_a^b f^2\, dx$ as the "square of the length" of $f(x)$ in the function space associated with the interval (a, b). It is more conventional to speak of this quantity as the square of the *norm* of f, and to write

$$\operatorname{norm} f \equiv \|f\| \equiv (f, f)^{1/2} = \left(\int_a^b f^2\, dx\right)^{1/2}. \tag{299}$$

A function whose norm is *unity* is said to be *normalized*, and is seen to be analogous to a *unit vector* in vector space.*

* In some references, the norm of f is defined as the quantity (f, f) itself.

We notice that if the norm of f is *zero*, then the integral of the non-negative function f^2 over the interval (a, b) must vanish. This means that $f(x)$ cannot differ from zero over any range of positive length in (a, b). In particular, if f is *continuous* everywhere in (a, b), and has a zero norm over that interval, then f must *vanish everywhere* in (a, b). However, it is clear that if $f(x)$ were zero everywhere except at a finite number of points in (a, b), the integral of f^2 over that interval would still vanish. It is convenient to speak of a function $f(x)$ for which $\int_a^b f^2\,dx = 0$ as a *trivial function*, and to say that such a function vanishes "almost everywhere" in (a, b).*

A set of n functions may be said to be *linearly independent* in (a, b) if *no* linear combination of those functions, with at least one nonvanishing coefficient, is a trivial function over that interval. Given any such set of real functions $f_k(x)$, for each of which the integral $\int_a^b f_k^2\,dx$ exists and is nonzero, we can determine a set of n new functions $\phi_k(x)$, each of which is a linear combination of certain of the f's, such that the ϕ's are mutually orthogonal and normalized in (a, b). The procedure is completely analogous to the Gram-Schmidt procedure of Section 1.13. We call such a set of functions an *orthonormal* set. Any two functions of the set then have the property that

$$(\phi_i, \phi_j) \equiv \int_a^b \phi_i \phi_j\,dx = \delta_{ij}, \tag{300}$$

where δ_{ij} is the Kronecker delta of equation (37).

Now for any (sufficiently regular) function $f(x)$ defined in (a, b) we may evaluate the scalar product of that function with each function ϕ_k:

$$c_k = (f, \phi_k) = \int_a^b f\phi_k\,dx. \tag{301}$$

The functions ϕ_k are analogous to a set of n mutually orthogonal unit vectors in space, and we may think of the numbers $c_1, c_2, \cdots, c_n$ as the scalar components of $f(x)$ relative to those functions. We refer to these numbers as the *Fourier constants* of $f(x)$ relative to the functions $\phi_k(x)$ in (a, b).

There then exists an n-fold infinity of functions which can be generated as linear combinations of the n ϕ's. If for any such function $F(x)$ we write

$$F(x) = \sum_{k=1}^{n} a_k \phi_k(x) \qquad (a < x < b), \tag{302}$$

* For a more satisfactory definition of this term it is desirable to consider an extension of the usual concepts of integration. The points at which a trivial function differs from zero may be *infinite* in number, so long as they are not "densely" distributed.

each coefficient a_r can be determined by forming the scalar product of ϕ_r with both members of (302), and using (300) to obtain the result

$$a_r = (F, \phi_r). \tag{303}$$

Thus the coefficient a_k in (302) is the scalar component of $F(x)$ relative to $\phi_k(x)$.

For a more general real function $f(x)$, we may assume an *approximation* of the form

$$f(x) \approx \sum_{k=1}^{n} a_k \phi_k(x) \qquad (a < x < b), \tag{304}$$

and determine the coefficients a_k in such a way that the norm of the difference between the two members of (304) over (a, b) is as small as possible, and hence also such that

$$\Delta_n \equiv \left\| f(x) - \sum_{k=1}^{n} a_k \phi_k(x) \right\|^2 = \int_a^b \left[f(x) - \sum_{k=1}^{n} a_k \phi_k(x) \right]^2 dx = \text{min.} \tag{305}$$

The approximation to be obtained, over the interval (a, b), is thus the best possible in the "least squares" sense.

If we think of a function $f(x)$ as a "vector" in function space, extending from an origin O to a "point" P in that space (see Problems 126–131), we can interpret (305) as choosing, from all points which can be attained by vectors of the form $\sum_{k=1}^{n} a_k \phi_k(x)$, that point whose "distance" from P is as small as possible.

Equation (305) is equivalent to the requirement that the expression

$$\Delta_n \equiv \int_a^b f^2 \, dx - 2 \sum_{k=1}^{n} a_k \int_a^b f \phi_k \, dx + \int_a^b \left[\sum_{k=1}^{n} a_k \phi_k(x) \right]^2 dx$$

take on a minimum value. But since the functions ϕ_k are orthonormal it follows that only squared terms in the last integrand have nonzero integrals. Hence, with the notation of (301), we obtain the result

$$\Delta_n = \int_a^b f^2 \, dx - 2 \sum_{k=1}^{n} a_k c_k + \sum_{k=1}^{n} a_k^2,$$

which can be put in the more convenient form

$$\Delta_n = \int_a^b f^2 \, dx - \sum_{k=1}^{n} c_k^2 + \sum_{k=1}^{n} (c_k - a_k)^2. \tag{306}$$

From this result it is clear that, since f and the c's are fixed, Δ_n takes a minimum value when the coefficients a_k are chosen such that

$$a_k = c_k. \tag{307}$$

Thus it follows that *the best approximation* (304) *in the least-squares sense is obtained when* a_k *is taken as the Fourier constant of* $f(x)$ *relative to* $\phi_k(x)$ *over* (a, b).

The squared norm of the deviation between $f(x)$ and its *best* n-term approximation of the form (304) is then obtained, by introducing (307) into (306), in the form

$$\left\| f(x) - \sum_{k=1}^{n} c_k \phi_k(x) \right\|_{\min}^{2} = \int_a^b f^2\,dx - \sum_{k=1}^{n} c_k^{\,2}. \tag{308}$$

From the definition (305) it follows that (308) cannot be *negative*; that is, we must have

$$\int_a^b f^2\,dx - \sum_{k=1}^{n} c_k^{\,2} \geqq 0. \tag{309}$$

This relation is known as *Bessel's inequality*.

Suppose now that the dimension n of the orthonormal set $\phi_1, \phi_2, \cdots, \phi_n$ is increased without limit. The positive series in (309) must increase with n (unless the corresponding c's vanish) so that the norm of the error involved *decreases*, but since the sum cannot become greater than the fixed number $\int_a^b f^2\,dx$, we conclude that the series $\sum_1^\infty c_k^{\,2}$ always *converges* to some positive number not greater than $\int_a^b f^2\,dx$. However, there is no assurance that the limit to which this series converges will actually *equal* the value of this integral, so that the right-hand member of (308) then will tend to zero as n increases. That is, it is *not* sufficient merely to have a set of *infinitely many* mutually orthogonal functions.

In illustration, we may recall that the functions $\cos(k\pi x/l)$ $(k = 1, 2, 3, \cdots)$ constitute an orthogonal set of functions over the interval $(0, l)$; that is, we have the relation

$$\int_0^l \cos\frac{r\pi x}{l}\cos\frac{s\pi x}{l}\,dx = 0 \qquad (r \neq s).$$

The norm of each function over $(0, l)$ is $\sqrt{l/2}$, so that the functions

$$\phi_k(x) = \sqrt{\frac{2}{l}}\cos\frac{k\pi x}{l} \qquad (k = 1, 2, \cdots) \tag{310}$$

form an infinite orthonormal set over $(0, l)$. However, for the simple function $f(x) = 1$, the relevant Fourier constants are all *zeros*, since here

$$c_k = \sqrt{\frac{2}{l}}\int_0^l 1 \cdot \cos\frac{k\pi x}{l}\,dx = 0 \qquad (k = 1, 2, \cdots).$$

Hence, in this case the right-hand member of (308) is constantly equal to l, regardless of the value of n.

In a vector space of n dimensions, if we construct a set of n mutually orthogonal unit vectors, then the possibility of expressing any other vector as a linear combination of these vectors is a consequence of the fact that no other vector can be linearly independent of them; that is, there exists no vector in that space, other than the zero vector, which is simultaneously orthogonal to these n vectors. However, in function space (of infinitely many dimensions) the difficulty consists of the fact that a function may simultaneously be orthogonal to an *infinite number* of mutually orthogonal functions. Thus, in the above case, the function $f(x) = 1$ is orthogonal to *all* the functions in the set (310) over the interval $(0, l)$. However, it can be shown that any function which has this property differs from a constant by a *trivial* function, so that for the extended set

$$\sqrt{\frac{1}{l}},\ \sqrt{\frac{2}{l}}\cos\frac{\pi x}{l},\ \sqrt{\frac{2}{l}}\cos\frac{2\pi x}{l},\ \cdots,\ \sqrt{\frac{2}{l}}\cos\frac{n\pi x}{l},\ \cdots \tag{311}$$

there is no nontrivial function whose Fourier constants *all* vanish. Such a set of orthogonal functions is said to be *complete.*

It is easily verified that the set (311) is also orthogonal (but not normalized) over the larger interval $(-l, l)$. However, it is obvious that *any odd function* of x [for which $f(-x) = -f(x)$] will possess zero Fourier constants relative to this set, over that interval. To complete the set, it is found to be sufficient to add the functions

$$\sqrt{\frac{2}{l}}\sin\frac{\pi x}{l},\ \sqrt{\frac{2}{l}}\sin\frac{2\pi x}{l},\ \cdots,\ \sqrt{\frac{2}{l}}\sin\frac{n\pi x}{l},\ \cdots. \tag{312}$$

Either of the sets (311) and (312) is complete over $(0, l)$, while the combination of the two sets is complete over $(-l, l)$ or, as a matter of fact, over *any* interval of length $2l$. These results are consequences of the known theory of *Fourier series.*

It can be shown that if the set of functions $\phi_1, \phi_2, \cdots, \phi_n, \cdots$ is *complete* in (a, b), then the right-hand member of (308) *does* indeed tend to zero for any function $f(x)$ which is of integrable square over that interval, so that

$$\sum_{k=1}^{\infty} c_k^2 = \int_a^b f^2\,dx \tag{313}$$

in that case. The proof of this relation, known as *Parseval's equality*, is involved. Furthermore, it is difficult in practice to *establish* the completeness of a given infinite orthogonal set of functions. For this reason, no attempt is made here to pursue the general theory.

However, it is important to realize that one further difficulty exists. Even though we prove that the right-hand member of (308) tends to zero as n increases, so that

$$\lim_{n\to\infty} \int_a^b \left[f(x) - \sum_{k=1}^{n} c_k \phi_k(x) \right]^2 dx = 0, \tag{314}$$

we cannot then conclude that the integrand tends to zero everywhere in (a, b). That is, there may be no *specific* value of x in (a, b) for which we are then certain that the statement

$$f(x) = \lim_{n\to\infty} \sum_{k=1}^{n} c_k \phi_k(x)$$

is true. We know only that the mean square error in (a, b) tends to zero, and we say accordingly that if (314) is true then the series *converges in the mean* to $f(x)$.

In this case it is conventional to write

$$f(x) = \underset{n\to\infty}{\text{l.i.m.}} \sum_{k=1}^{n} c_k \phi_k(x),$$

where "l.i.m." is to be read "limit in the mean" and is to be carefully distinguished from the abbreviation "lim" which corresponds to the more usual limiting process.

However, if $f(x)$ is *continuous* throughout the interval (a, b), and if we can prove that the series $\sum_{1}^{\infty} c_k \phi_k(x)$ also represents a continuous function over that interval,* then the difference between these two functions is a continuous function with zero norm, and hence is indeed zero *everywhere* in (a, b), so that the series then converges to $f(x)$ in the true sense at each point of (a, b). Unfortunately, these conditions frequently are not fulfilled in practical cases.

While the knowledge that a series represents a function which differs from $f(x)$ in (a, b) by a trivial function is often all that is required (for such purposes as *integration*), it is nevertheless frequently desirable to determine whether or not the series actually represents $f(x)$ *at a given point*. The treatment of problems of this type is again beyond the scope of the present work.

The problems just discussed have been satisfactorily solved, in the mathematical literature, for a very large class of sets of orthogonal functions which frequently arise in practice. Certain known results are summarized, for convenient reference, in the following section.

It should be pointed out first that, in analogy with the corresponding situation in vector space, it is often desirable to modify somewhat the definition of the *norm* of a function. In particular, if $f(x)$ is a *complex*

* This will be the case, in particular, if the functions ϕ_k are continuous and if the series converges *uniformly* in the interval (a, b).

function of a real variable x, of the form $u(x) + i\,v(x)$, the norm of f is usually defined to be the nonnegative *real* quantity

$$\|f\|_H = (\bar{f}, f)^{1/2} = \left(\int_a^b \bar{f} f\,dx\right)^{1/2}, \tag{315}$$

where a bar indicates that the complex conjugate is to be taken. We speak of (315) as the *Hermitian norm* of f. The Hermitian scalar product of two complex functions f and g is then defined to be one of the two different quantities $(\bar{f}, g)$ and $(f, \bar{g})$, these two quantities being complex conjugates. In particular, f and g are said to be *orthogonal in the Hermitian sense* if

$$(\bar{f}, g) = (f, \bar{g}) = 0. \tag{316}$$

In problems analogous to those discussed in Section 1.25, but involving *differential* equations, sets of real functions are often generated for which the members $\phi_1, \phi_2, \cdots, \phi_n, \cdots$ are not orthogonal in the sense of (300), but for which a relation holds of the form

$$\int_a^b r(x)\phi_i(x)\phi_j(x)\,dx = 0 \qquad (i \neq j), \tag{317}$$

where $r(x)$ is real, nontrivial, and *nonnegative* in (a, b).

We may define the left-hand member of (317) to be the *generalized* or *weighted* scalar product of ϕ_i and ϕ_j, relative to the *weighting function* $r(x)$. Finally, the norm of any function f relative to the function r is defined to be

$$\|f\|_r \equiv \left(\int_a^b r f^2\,dx\right)^{1/2}, \tag{318}$$

and a function with *unit* norm, so defined, is said to be *normalized* relative to $r(x)$. The weighted scalar product of f and g is conveniently indicated by the notation

$$(f, g)_r \equiv \int_a^b r f g\,dx. \tag{319}$$

1.29. Sturm-Liouville problems. In this section we summarize briefly certain known results concerning sets of orthogonal functions generated by certain types of boundary-value problems involving *linear ordinary differential equations*.

A problem which consists of a homogeneous linear differential equation of the form

$$\frac{d}{dx}\left(p\,\frac{dy}{dx}\right) + qy + \lambda r y = 0, \tag{320}$$

together with *homogeneous boundary conditions* of a rather general type, prescribed at the end points of an interval (a, b), generally possesses a nontrivial solution only if the parameter λ is assigned one of a certain set of permissible values. For such a value of λ, say $\lambda = \lambda_k$, the conditions of the problem are satisfied by an expression of the form $y = C\phi_k(x)$ where C is an arbitrary constant. The permissible values of λ are known as its *characteristic values* (or "eigenvalues") and the corresponding functions $\phi_k(x)$, which then satisfy the conditions of the problem when $\lambda = \lambda_k$, are known as the *characteristic functions* (or "eigenfunctions").

In most cases occurring in practice, the functions $p(x)$ and $r(x)$ are positive in the interval (a, b), except possibly at one or both of the end points.

If we define a *linear differential operator* of second order by the formal equation

$$\mathscr{L} = \frac{d}{dx}\left(p\frac{d}{dx}\right) + q, \tag{321}$$

the differential equation (320) takes the operational form

$$\mathscr{L}y + \lambda r y = 0, \tag{322}$$

and is seen to be analogous to equation (255) of Section 1.25.

We show next that, when suitable homogeneous boundary conditions are prescribed at the ends of an interval (a, b), the characteristic functions of the resulting problem have properties analogous to those discussed in Section 1.25. For this purpose, let $\phi_i(x)$ and $\phi_j(x)$ be two characteristic functions, satisfying the conditions of the problem in correspondence with *distinct* characteristic numbers λ_i and λ_j. We then have the relations

$$\frac{d}{dx}\left(p\frac{d\phi_i}{dx}\right) + q\phi_i + \lambda_i r\phi_i = 0 \tag{323a}$$

and

$$\frac{d}{dx}\left(p\frac{d\phi_j}{dx}\right) + q\phi_j + \lambda_j r\phi_j = 0. \tag{323b}$$

If we multiply (323a) by ϕ_j and (323b) by ϕ_i, and subtract the resultant equations from each other, there follows

$$\begin{aligned}(\lambda_j - \lambda_i)r\phi_i\phi_j &= \phi_j\frac{d}{dx}\left(p\frac{d\phi_i}{dx}\right) - \phi_i\frac{d}{dx}\left(p\frac{d\phi_j}{dx}\right) \\ &= \frac{d}{dx}\left[p\left(\phi_j\frac{d\phi_i}{dx} - \phi_i\frac{d\phi_j}{dx}\right)\right],\end{aligned} \tag{324}$$

and the result of integrating both members of (324) over the interval (a, b) takes the form

$$(\lambda_j - \lambda_i)\int_a^b r\phi_i\phi_j\,dx = \left[p\left(\phi_j\frac{d\phi_i}{dx} - \phi_i\frac{d\phi_j}{dx}\right)\right]_a^b. \tag{325}$$

Thus, since we have assumed that $\lambda_j \neq \lambda_i$, we conclude that if the specified boundary conditions require the right-hand member of (325) to vanish, then *the characteristic functions* ϕ_i *and* ϕ_j *are orthogonal relative to the weighting function* $r(x)$:

$$(\phi_i, \phi_j)_r \equiv \int_a^b r\phi_i\phi_j \, dx = 0 \qquad (\lambda_i \neq \lambda_j). \tag{326}$$

Appropriate boundary conditions which may be seen to give rise to this situation include the following:

1. At each end of the interval we may require that either y or dy/dx or a linear combination $\alpha y + \beta \, dy/dx$ *vanish*.

2. If it happens that $p(x)$ vanishes at $x = a$ or at $x = b$, we may require instead merely that y and dy/dx remain finite at that point, and impose one of the conditions 1 at the other point.

3. If it happens that $p(b) = p(a)$, we may require merely that $y(b) = y(a)$ and $y'(b) = y'(a)$.

In most practical cases [in particular, if p, q, and r are *regular** and both p and r *positive* throughout (a, b)], when the interval (a, b) is of *finite length* it is found that in each of the listed cases there exists an *infinite* set of distinct characteristic numbers $\lambda_1, \lambda_2, \cdots, \lambda_n, \cdots$. If also the function $q(x)$ is *nonpositive* in (a, b), and if

$$[p\phi_i\phi_i']_a^b \leqq 0, \tag{327}$$

the λ's are all *nonnegative*. Furthermore, except in the case of the periodicity condition 3, to each characteristic number there corresponds *one and only one* characteristic function, an arbitrary multiple of which satisfies all the specified conditions when λ is assigned the appropriate characteristic value. In case 3, *two* linearly independent characteristic functions generally correspond to each characteristic number. Such pairs of functions then can be orthogonalized, if this is desirable, by the Gram-Schmidt procedure.

A problem of the general type just considered is known as a *Sturm-Liouville problem.*

The importance of such problems stems from the known fact that the sets of orthogonal functions generated by these problems generally are *complete*, in the sense of the preceding section, and further, that a positive statement can be made in such cases concerning actual *convergence* of the series representation of a sufficiently well-behaved function $f(x)$ to the value of the function *at all points where* $f(x)$ *is continuous.*

In actual practice, it is often inconvenient to *normalize* the characteristic functions (so that their norm relative to r is unity). In such cases, the

* A function $f(x)$ is said to be *regular* at $x = x_0$ if it can be represented by a convergent power series over an interval including x_0.

coefficients in a series representation

$$f(x) = \sum_{k=1}^{\infty} C_k \phi_k(x) \qquad (a < x < b) \tag{328}$$

are given by the formula

$$C_i \int_a^b r\phi_i^2 \, dx = \int_a^b rf\phi_i \, dx \tag{329a}$$

or, symbolically,

$$C_i \, \|\phi_i\|_r^2 = (f, \phi_i)_r. \tag{329b}$$

This result is obtained formally by multiplying both sides of (328) by the product $r\phi_i$, integrating the results term-by-term over (a, b), and taking into account the orthogonality of the characteristic functions relative to the weighting function $r(x)$. We notice that (329b) reduces to the obvious generalization of (301) when $\|\phi_i\|_r = 1$. The theorem to which reference was made above can then be stated as follows:

Let the functions $p(x)$, $q(x)$, and $r(x)$ in (320) *be regular in the finite interval (a, b), let $p(x)$ and $r(x)$ be positive in that interval, including the end points, and let the homogeneous end conditions be such that* (327) *is satisfied. Then, if $f(x)$ is bounded and piecewise differentiable in (a, b), the series* (328) *converges to $f(x)$ at all points inside that interval where $f(x)$ is continuous,* and to the mean value $\frac{1}{2}[f(x+) + f(x-)]$ at any point where a finite jump occurs.*

While the stated conclusions follow also under even milder restrictions on $f(x)$, the conditions given here are satisfied by most functions arising in practice.

To illustrate this result, we may consider the differential equation

$$\frac{d^2y}{dx^2} + \lambda y = 0, \tag{330}$$

which is the special case of (320) in which $p(x) = r(x) = 1$ and $q(x) = 0$. If we consider the interval $(0, l)$, and impose the boundary conditions

$$y(0) = 0, \qquad y(l) = 0, \tag{331}$$

it is easily verified that the characteristic values of λ, for which this problem possesses a solution other than the trivial solution $y(x) \equiv 0$, are of the form $\lambda_k = k^2\pi^2/l^2$, where k is any positive integer, and that the corresponding

* The convergence is absolute and uniform in any *interior* subinterval which does not include a point of discontinuity as an interior or end point.

characteristic functions are given by

$$\phi_k(x) = \sin \frac{k\pi x}{l} \qquad (k = 1, 2, \cdots).$$

Thus we obtain in this way a derivation of the *Fourier sine-series* representation

$$f(x) = \sum_{k=1}^{\infty} C_k \sin \frac{k\pi x}{l} \qquad (0 < x < l), \tag{332}$$

where, with $r(x) = 1$, equation (329) determines the coefficients in the form

$$C_k = \frac{2}{l} \int_0^l f(x) \sin \frac{k\pi x}{l}\, dx. \tag{333}$$

In a similar way, the conditions $y'(0) = y'(l) = 0$ associated with (330) give rise to the Fourier *cosine-series* representation, while the periodicity conditions $y(-l) = y(l)$ and $y'(-l) = y'(l)$, relevant to the interval $(-l, l)$, lead to the *general* Fourier series representation over that interval, involving both sines and cosines of period $2l$.

By considering other appropriate special forms of (320), expansions in terms of *Bessel functions*, *Legendre polynomials*, and so forth, may be established. In certain of these cases the coefficient functions p, q, and r do not satisfy the requirements specified in the preceding theorem, but it has been found that the conclusions of the theorem are still valid.

Elementary discussions of such developments may be found in Reference 9 (Chapter 5). For more detailed treatments of these topics, References 2 and 12 are suggested.

In those cases when the interval (a, b) is of *infinite* length, or when other conditions of the stated theorem are violated, it frequently happens that the characteristic values of λ are no longer *discretely* distributed, but that all values of λ in some *continuous* range are characteristic values. In such cases, the superposition of characteristic functions is accomplished by *integration*, rather than summation. In particular, for the problems discussed relative to equation (330), it is found that *all positive values of* λ are characteristic values when the fundamental interval is of infinite length, and one is led to the *Fourier integral* representations. In certain other exceptional cases the characteristic values again may be discretely distributed, or there may be both *continuously* distributed and *discretely* distributed characteristic values of λ.

Finally, we remark that the preceding discussion can be generalized to apply to characteristic functions of boundary-value problems governed by certain linear ordinary differential equations of higher order, as well as to characteristic functions of two or more variables associated with certain *partial* differential equations. Similar problems are related to linear *integral equations* and to linear *difference equations*.

REFERENCES

1. Birkhoff, G., and S. Maclane: *A Survey of Modern Algebra*, The Macmillan Company, New York, 1950.
2. Courant, R., and D. Hilbert: *Methods of Mathematical Physics*, Interscience Publishers, Inc., New York, 1953.
3. Crout, P. D.: "A Short Method for Evaluating Determinants and Solving Systems of Linear Equations with Real or Complex Coefficients," *Trans. AIEE*, Vol. 60, pp. 1235–1241 (1941).
4. Dwyer, Paul S.: *Linear Computations*, John Wiley & Sons, Inc., New York, 1951.
5. Faddeeva, V. N. (Translated from Russian by Curtis D. Benster): *Computational Methods of Linear Algebra*, Dover Publications, Inc., New York, 1959.
6. Forsythe, George E.: "Solving Linear Algebraic Equations Can Be Interesting," *Bull. Amer. Math. Soc.*, Vol. 59, pp. 299–329 (1953).
7. Frazer, R. A., W. J. Duncan, and A. R. Collar: *Elementary Matrices*, Cambridge University Press, London, 1960.
8. Halmos, P. R.: *Finite Dimensional Vector Spaces*, D. Van Nostrand Company, Inc., Princeton, N.J., 1958.
9. Hildebrand, F. B.: *Advanced Calculus for Applications*, Prentice-Hall, Inc., Englewood Cliffs, N.J., 1962.
10. Hoffman, Kenneth, and Ray Kunze: *Linear Algebra*, Prentice-Hall, Inc., Englewood Cliffs, N.J., 1961.
11. Perlis, Sam: *Theory of Matrices*, Addison-Wesley Publishing Company, Inc., Reading, Mass., 1952.
12. Titchmarsh, E. C.: *Eigenfunction Expansions*, The Clarendon Press, Oxford, 1946.
13. Turnbull, H. W., and A. C. Aitken: *An Introduction to the Theory of Canonical Matrices*, Blackie and Son, Ltd., London, 1932.
14. Zurmühl, R.: *Matrizen*, Springer-Verlag, Berlin, 1950.

PROBLEMS

Sections 1.1, 1.2.

1. Illustrate the use of the Gauss-Jordan reduction in obtaining the general solution of each of the following sets of equations:

$$\text{(a)}\quad \begin{aligned} x_1 + 2x_2 + 2x_3 &= 1, \\ 2x_1 + 2x_2 + 3x_3 &= 3, \\ x_1 - x_2 + 3x_3 &= 5. \end{aligned} \qquad \text{(b)}\quad \begin{aligned} 2x_1 \qquad\quad + x_3 &= 4, \\ x_1 - 2x_2 + 2x_3 &= 7, \\ 3x_1 + 2x_2 \qquad\quad &= 1. \end{aligned}$$

(c)
$$\begin{aligned} 2x_1 - x_2 \qquad\qquad &= 6,\\ -x_1 + 3x_2 - 2x_3 \qquad &= 1,\\ -2x_2 + 4x_3 - 3x_4 &= -2,\\ -3x_3 + 5x_4 &= 1. \end{aligned}$$

Section 1.3.

2. Evaluate the following matrix products:

(a) $\begin{bmatrix} 1 & 2 \\ 1 & -1 \end{bmatrix}\begin{bmatrix} 1 & 0 & 1 \\ 1 & -1 & 1 \end{bmatrix}$. (b) $\begin{bmatrix} 1 & 2 \\ 3 & 6 \end{bmatrix}\begin{bmatrix} 6 & -2 \\ -3 & 1 \end{bmatrix}$.

(c) $[a_1 \quad a_2 \quad \cdots \quad a_n]\begin{Bmatrix} b_1 \\ b_2 \\ \cdot \\ \cdot \\ \cdot \\ b_n \end{Bmatrix}$. (d) $\begin{Bmatrix} b_1 \\ b_2 \\ \cdot \\ \cdot \\ \cdot \\ b_n \end{Bmatrix}[a_1 \quad a_2 \quad \cdots \quad a_n]$.

(e) $\begin{bmatrix} c_1 & 0 \\ 0 & c_2 \end{bmatrix}\begin{bmatrix} a_{11} & a_{12} \\ a_{21} & a_{22} \end{bmatrix}$. (f) $\begin{bmatrix} a_{11} & a_{12} \\ a_{21} & a_{22} \end{bmatrix}\begin{bmatrix} c_1 & 0 \\ 0 & c_2 \end{bmatrix}$.

3. If the product **A**(**B C**) is defined, show that it is of the form

$$[a_{ir}]([b_{rs}][c_{sj}]) = \left[\sum_r \sum_s a_{ir}b_{rs}c_{sj}\right]$$

and deduce that then **A**(**B C**) = (**A B**)**C**.

4. If **A B** = **C**, show that the columns of **C** are linear combinations of the columns of **A** and that the rows of **C** are linear combinations of the rows of **B**. [Show, for example, that

$$(\text{col. } i \text{ of } \mathbf{C}) = b_{1i}(\text{col. 1 of } \mathbf{A}) + b_{2i}(\text{col. 2 of } \mathbf{A}) + \cdots.]$$

5. If **A** and **B** are $n \times n$ matrices, when is it true that

$$(\mathbf{A} + \mathbf{B})(\mathbf{A} - \mathbf{B}) = \mathbf{A}^2 - \mathbf{B}^2?$$

Give an example in which this relation does *not* hold.

6. Find the most general 2×2 matrix **A** such that $\mathbf{A}^2 = \mathbf{0}$.

7. Prove that, if two square matrices of order three are both symmetrically partitioned as in the text on page 9, then these matrices may be correctly multiplied by treating the submatrices as single elements.

8. It is required to determine values of the function

$$\Phi(x) = \int_a^b K(x, \xi)f(\xi)\,d\xi,$$

at the n points $x_1, x_2, \cdots, x_n$. Show that, if in each case the integral is approximated by the use of Simpson's rule, as a linear combination of the ordinates at N equally spaced points $\xi_1 = a, \xi_2, \cdots, \xi_{N-1}, \xi_N = b$, where N is odd, the calculations can be arranged in the matrix form

$$\begin{Bmatrix} \Phi_1 \\ \Phi_2 \\ \cdot \\ \cdot \\ \cdot \\ \Phi_n \end{Bmatrix} \approx \frac{b-a}{3N-3} \begin{bmatrix} K_{11} & K_{12} & \cdots & K_{1N} \\ K_{21} & K_{22} & \cdots & K_{2N} \\ K_{31} & K_{32} & \cdots & K_{3N} \\ \cdots & \cdots & \cdots & \cdots \\ K_{n1} & K_{n2} & \cdots & K_{nN} \end{bmatrix} \begin{Bmatrix} f_1 \\ 4f_2 \\ 2f_3 \\ \cdot \\ \cdot \\ \cdot \\ f_N \end{Bmatrix},$$

where $\Phi_i \equiv \Phi(x_i)$, $K_{ij} \equiv K(x_i, \xi_j)$, and $f_j \equiv f(\xi_j)$.

9. Apply the procedure of Problem 8 to the approximate evaluation of the integral

$$\Phi(x) = \int_0^1 \sqrt{x^2 + \xi^2} \sin \pi\xi \, d\xi,$$

for $x = 0, \frac{1}{4}, \frac{1}{2}, \frac{3}{4}$, and 1, with $N = 5$. Retain three significant figures in the calculation.

Section 1.4.

10. Prove, by direct expansion or otherwise, that

$$|\mathbf{A}\,\mathbf{B}| = |\mathbf{A}|\,|\mathbf{B}|$$

when **A** and **B** are square matrices of order two.

11. Determine those values λ for which the following set of equations may possess a nontrivial solution:

$$\begin{aligned} 3x_1 + x_2 - \lambda x_3 &= 0, \\ 4x_1 - 2x_2 - 3x_3 &= 0, \\ 2\lambda x_1 + 4x_2 + \lambda x_3 &= 0. \end{aligned}$$

For each permissible value of λ, determine the most general solution.

12. Show that the equation of the straight line $ax + by + c = 0$ which passes through the points (x_1, y_1) and (x_2, y_2) can be written in the form

$$\begin{vmatrix} x & y & 1 \\ x_1 & y_1 & 1 \\ x_2 & y_2 & 1 \end{vmatrix} = 0.$$

13. Express the requirement, that four points (x_i, y_i) $(i = 1, 2, 3, 4)$ lie simultaneously on a conic of the form $ax^2 + bxy + cy^2 + d = 0$, in terms of the vanishing of a determinant.

Section 1.5.

14. If $\mathbf{A}$ and $\mathbf{B}$ commute, prove that $\mathbf{A}^T$ and $\mathbf{B}^T$ also commute.

15. A *symmetric* matrix $\mathbf{A} = [a_{ij}]$ is a square matrix for which $a_{ji} = a_{ij}$.

(a) Show that $\mathbf{A}^T = \mathbf{A}$ if and only if $\mathbf{A}$ is symmetric.

(b) Let $\mathbf{A}$ and $\mathbf{B}$ represent symmetric matrices of order n. Prove that $\mathbf{A}\,\mathbf{B}$ is also symmetric if and only if $\mathbf{A}$ and $\mathbf{B}$ are commutative.

16. Verify that, if $\mathbf{A}$ and $\mathbf{B}$ are square matrices of order two, there follows $\text{Adj}\,(\mathbf{A}\,\mathbf{B}) = (\text{Adj}\,\mathbf{B})(\text{Adj}\,\mathbf{A})$.

17. Let $\mathbf{A}$ and $\mathbf{B}$ represent diagonal matrices of order n.

(a) Prove that $\mathbf{A}\,\mathbf{B}$ is also a diagonal matrix.

(b) Prove that $\mathbf{B}\,\mathbf{A} = \mathbf{A}\,\mathbf{B}$.

18. Evaluate each of the following sums:

(a) $\sum_{k=1}^{n} \delta_{ik}\delta_{kj}$. (b) $\sum_{k=1}^{n} d_k\delta_{ik}\delta_{kj}$,

(c) $\sum_{k=1}^{n}\sum_{l=1}^{n} d_k\delta_{ik}\delta_{kl}\delta_{lj}$.

Section 1.6.

19. (a) If $\mathbf{A}\,\mathbf{B} = \mathbf{0}$ and $\mathbf{A}$ is nonsingular, prove that $\mathbf{B} = \mathbf{0}$.

(b) If $\mathbf{A}\,\mathbf{B} = \mathbf{0}$ and $\mathbf{B}$ is nonsingular, prove that $\mathbf{A} = \mathbf{0}$.

20. If $\mathbf{A}\,\mathbf{B} = \mathbf{A}\,\mathbf{C}$, where $\mathbf{A}$ is a square matrix, when does it necessarily follow that $\mathbf{B} = \mathbf{C}$? Give an example in which this conclusion does *not* follow.

21. Determine the elements of $\mathbf{A}^T$, $\text{Adj}\,\mathbf{A}$, and $\mathbf{A}^{-1}$ when

$$\mathbf{A} = \begin{bmatrix} 1 & 2 & 1 \\ 2 & 1 & 0 \\ -1 & 0 & 1 \end{bmatrix}.$$

22. Determine the elements of the matrix $\mathbf{M}$ such that $\mathbf{A}\,\mathbf{M}\,\mathbf{B} = \mathbf{C}$ when

$$\mathbf{A} = \begin{bmatrix} 2 & 1 & 1 \\ 1 & 1 & 0 \\ 0 & 0 & 1 \end{bmatrix}, \quad \mathbf{B} = \begin{bmatrix} 3 & 1 \\ 1 & 1 \end{bmatrix}, \quad \mathbf{C} = \begin{bmatrix} 1 & 1 \\ 2 & 2 \\ 1 & 1 \end{bmatrix}.$$

23. If $\mathbf{D} = [d_i\delta_{ij}]$ is a nonsingular diagonal matrix, prove that its inverse is given by

$$\mathbf{D}^{-1} = \left[\frac{1}{d_i}\delta_{ij}\right].$$

24. (a) Prove that $\mathbf{A}(\text{Adj}\,\mathbf{A}) = \mathbf{0}$ if $\mathbf{A}$ is singular, and illustrate by an example.

(b) Prove that $|\text{Adj}\,\mathbf{A}| = |\mathbf{A}|^{n-1}$ ($\mathbf{A}$ of order n) and illustrate by an example.

25. If $|\mathbf{A}| \neq 0$, prove that $|\mathbf{A}^{-1}| = |\mathbf{A}|^{-1}$.

26. If **A** and **B** commute and **B** is nonsingular, prove that **A** and $\mathbf{B}^{-1}$ also commute.

27. For any matrix **A**, a matrix **P** is said to be a *left inverse* of **A** if $\mathbf{P\,A} = \mathbf{I}$ and a matrix **Q** is said to be a *right inverse* of **A** if $\mathbf{A\,Q} = \mathbf{I}$. Show that the matrix

$$\mathbf{A} = \begin{bmatrix} 1 & 1 \\ 1 & 2 \\ 1 & 1 \end{bmatrix}$$

has a two-parameter set of left inverses but no right inverse, whereas $\mathbf{A}^T$ has a two-parameter set of right inverses but no left inverse. [It is true (see Problem 31) that **A** cannot have *both* right and left inverses unless **A** is nonsingular.]

Section 1.7.

28. If **A** is an $m \times n$ matrix and if rank $\mathbf{A} \geqq m$, prove that rank $\mathbf{A} = m$.

29. Use the result of Problem 4 to prove that

$$\text{rank}\,(\mathbf{A\,B}) \leqq \min\,(\text{rank}\,\mathbf{A}, \text{rank}\,\mathbf{B}).$$

30. (a) If $a_{ij} = r_i s_j$, prove that $\mathbf{A} = [a_{ij}]$ is of rank one or zero.

(b) If $\mathbf{A} = [a_{ij}]$ is of rank one, prove that a_{ij} can be written as $r_i s_j$. [Such a matrix is called a *dyad*.]

31. Suppose that **A** is an $m \times n$ matrix. Prove that if **A** has a left inverse **P** (see Problem 27) then the rank of **A** is n, and also that if **A** has a right inverse **Q** then the rank of **A** is m. [Use the result of Problem 29, noticing that **P** and **Q** are $n \times m$.] Hence deduce that *if* **A** *has both a left inverse and a right inverse then* **A** *must be nonsingular.*

Section 1.8.

32. If **A** is an $m \times n$ matrix, show that each of the three elementary operations on *rows* of **A** can be accomplished by *pre*multiplying **A** by a matrix **P**, where **P** is formed by performing that operation on corresponding rows of the *unit* matrix **I** of order m. In each case, show also that **P** is nonsingular.

33. If **A** is an $m \times n$ matrix, show that each of the elementary operations on *columns* of **A** can be accomplished by *post*multiplying **A** by a matrix **Q**, where **Q** is formed by performing that operation on corresponding columns of the unit matrix **I** of order n. In each case, show also that **Q** is nonsingular.

34. Show that any nonsingular matrix can be reduced to the unit matrix of the same order by use of only elementary *row* operations and also by use of only elementary *column* operations. [Consider the process used in the Gauss-Jordan reduction.] Thus deduce that *if* $\mathbf{B} = \mathbf{P\,A\,Q}$, *where* **P** *and* **Q** *are nonsingular, then* **B** *can be obtained from* **A** *by use of elementary row and column operations.*

Section 1.9.

35. (a) By investigating ranks of relevant matrices, show that the following set of equations possesses a one-parameter family of solutions:

$$\begin{aligned} 2x_1 - x_2 - x_3 &= 2, \\ x_1 + 2x_2 + x_3 &= 2, \\ 4x_1 - 7x_2 - 5x_3 &= 2. \end{aligned}$$

(b) Determine the general solution.

36. (a) Show that the set

$$\begin{aligned} 2x_1 - 2x_2 + x_3 &= \lambda x_1, \\ 2x_1 - 3x_2 + 2x_3 &= \lambda x_2, \\ -x_1 + 2x_2 \qquad &= \lambda x_3 \end{aligned}$$

can possess a nontrivial solution only if $\lambda = 1$ or $\lambda = -3$.

(b) Obtain the general solution in each case.

37. The matrix

$$\begin{bmatrix} 0 & a & 1 & b \\ a & 0 & b & 1 \\ a & a & 2 & 2 \end{bmatrix}$$

is the augmented matrix of a system of linear algebraic equations. Determine for what fixed values of a and b (if any) the system possesses the following:

(a) A unique solution.
(b) A one-parameter solution.
(c) A two-parameter solution.
(d) No solution.

Section 1.10.

38. Determine the dimension of the vector space generated by each of the following sets of vectors:

(a) $\{1, 1, 0\}, \{1, 0, 1\}, \{0, 1, 1\}$.
(b) $\{1, 0, 0\}, \{0, 1, 0\}, \{0, 0, 1\}, \{1, 1, 1\}$.
(c) $\{1, 1, 1\}, \{1, 0, 1\}, \{1, 2, 1\}$.

39. Determine whether the vector $\{6, 1, -6, 2\}$ is in the vector space generated by the vectors $\{1, 1, -1, 1\}$, $\{-1, 0, 1, 1\}$, and $\{1, -1, -1, 0\}$.

40. (a) Determine the angle θ between the vectors

$$\mathbf{u} = \{1, 1, 1, 1\}, \qquad \mathbf{v} = \{1, 0, 0, 1\}.$$

(b) Determine the Hermitian angle θ_H between the complex vectors

$$\mathbf{u} = \{0, i, 1\}, \qquad \mathbf{v} = \{i, 1 + i, 1\}.$$

41. (a) If $\mathbf{A}$ is an $m \times n$ matrix and $\mathbf{Q}$ is an $n \times s$ matrix, and if the elements of the jth column of $\mathbf{Q}$ are considered to comprise the elements of a vector $\mathbf{v}_j$, show that the jth column of $\mathbf{A\,Q}$ is the vector $\mathbf{A\,v}_j$.

(b) If $\mathbf{A}$ is an $m \times n$ matrix and $\mathbf{P}$ is an $r \times m$ matrix, and if the elements of the ith row of $\mathbf{P}$ are considered to comprise the elements of a transposed vector $\mathbf{u}_i^T$, show that the ith row of $\mathbf{P\,A}$ is the transposed vector $\mathbf{u}_i^T\,\mathbf{A}$.

(c) With the notation of parts (a) and (b), show that the typical element b_{ij} of the product $\mathbf{P\,A\,Q}$ is the scalar $\mathbf{u}_i^T\,\mathbf{A}\,\mathbf{v}_j$.

42. (a) Prove that if the Gramian of two real vectors $\mathbf{u}_1$ and $\mathbf{u}_2$ vanishes, then $\mathbf{u}_1$ and $\mathbf{u}_2$ are linearly dependent. [Notice that, if $G = 0$, then the equations

$$c_1\mathbf{u}_1^T\,\mathbf{u}_1 + c_2\mathbf{u}_1^T\,\mathbf{u}_2 = 0 \qquad \text{and} \qquad c_1\mathbf{u}_2^T\,\mathbf{u}_1 + c_2\mathbf{u}_2^T\,\mathbf{u}_2 = 0$$

possess a nontrivial solution. Multiply the first equation by c_1, the second by c_2, add, and interpret the result.]

(b) Generalize the result of part (a) to the case of n vectors.

43. If $\mathbf{u}$ and $\mathbf{v}$ are real vectors, use the fact that the quantity

$$(\mathbf{u} + \lambda\,\mathbf{v})^T\,(\mathbf{u} + \lambda\,\mathbf{v}) = (\mathbf{u}, \mathbf{u}) + 2\lambda(\mathbf{u}, \mathbf{v}) + \lambda^2(\mathbf{v}, \mathbf{v})$$

is *nonnegative* for all real values of λ to deduce the *Schwarz inequality*:

$$|(\mathbf{u}, \mathbf{v})| \leqq (\mathbf{u}, \mathbf{u})^{1/2}\,(\mathbf{v}, \mathbf{v})^{1/2}.$$

44. Deal as in Problem 43 with the nonnegative real quantity

$$(\bar{\mathbf{u}} + \lambda\,\bar{\mathbf{v}})^T\,(\mathbf{u} + \lambda\,\mathbf{v}),$$

where $\mathbf{u}$ and $\mathbf{v}$ are complex vectors and λ is real, to deduce the generalized form of the *Schwarz inequality*:

$$\tfrac{1}{2}\,|(\bar{\mathbf{u}}, \mathbf{v}) + (\mathbf{u}, \bar{\mathbf{v}})| \leqq (\bar{\mathbf{u}}, \mathbf{u})^{1/2}\,(\bar{\mathbf{v}}, \mathbf{v})^{1/2}.$$

45. Establish the *parallelogram law*,

$$l^2(\mathbf{u} + \mathbf{v}) + l^2(\mathbf{u} - \mathbf{v}) = 2l^2(\mathbf{u}) + 2l^2(\mathbf{v}),$$

where $\mathbf{u}$ and $\mathbf{v}$ are real vectors, and interpret the result geometrically when $\mathbf{u}$ and $\mathbf{v}$ are three-dimensional vectors. Also show that this result remains true when $\mathbf{u}$ and $\mathbf{v}$ are complex if l is replaced by l_H.

Section 1.11.

46. Show that the set of equations

$$\begin{aligned} x_1 + x_2 + x_3 &= 3, \\ x_1 + x_2 - x_3 &= 1, \\ 3x_1 + 3x_2 - 5x_3 &= 1 \end{aligned}$$

possesses a one-parameter family of solutions, and verify directly that the vector $\mathbf{c}$ whose elements comprise the right-hand members is orthogonal to all vector solutions of the transposed homogeneous set of equations.

47. (a) Prove that if the set $\mathbf{A}\,\mathbf{x} = \mathbf{0}$ possesses an r-parameter set of nontrivial solutions, then the same is true of the transposed set $\mathbf{A}^T\,\mathbf{x}' = \mathbf{0}$, and conversely, when $\mathbf{A}$ is square.

(b) Interpret the result at the end of Section 1.11 in the case when the transposed set $\mathbf{A}^T\,\mathbf{x}' = \mathbf{0}$ possesses no nontrivial solution.

48. Prove that no nonzero vector $\mathbf{v}$ can be in both the solution space of the system $\mathbf{A}\,\mathbf{x} = \mathbf{0}$ and in the row space of $\mathbf{A}$. [Assume that

$$\mathbf{v} = k_1\boldsymbol{\alpha}_1 + \cdots + k_r\boldsymbol{\alpha}_r = C_1\mathbf{u}_1 + \cdots + C_{n-r}\mathbf{u}_{n-r},$$

with the notation of Section 1.11, and form the scalar product of each $\mathbf{u}_i$ into the two equal right-hand members of this relation.]

Section 1.12.

49. Show that the problem

$$x_1 - 2x_2 = \lambda x_1,$$
$$x_1 - x_2 = \lambda x_2$$

does not possess real nontrivial solutions for any values of λ.

50. (a) Determine the characteristic numbers (λ_1, λ_2) and corresponding unit characteristic vectors $(\mathbf{e}_1, \mathbf{e}_2)$ of the matrix

$$\mathbf{A} = \begin{bmatrix} 5 & 2 \\ 2 & 2 \end{bmatrix}.$$

(b) Verify that $\mathbf{e}_1$ and $\mathbf{e}_2$ are orthogonal.

(c) If $\mathbf{v} = \{1, 1\}$, determine α_1 and α_2 so that

$$\mathbf{v} = \alpha_1\mathbf{e}_1 + \alpha_2\mathbf{e}_2.$$

(d) Use the results of part (a), together with equation (105), to obtain the solution of the following set of equations:

$$5x_1 + 2x_2 = \lambda x_1 + 2,$$
$$2x_1 + 2x_2 = \lambda x_2 + 1.$$

Consider the exceptional cases separately.

51. (a) Suppose that the n characteristic vectors of the real symmetric matrix $\mathbf{A}$ are *not* normalized (reduced to unit length). If they are denoted by $\mathbf{u}_1, \mathbf{u}_2, \cdots, \mathbf{u}_n$, show that (105) must be replaced by the equation

$$\mathbf{x} = \sum_{k=1}^{n} \frac{(\mathbf{u}_k, \mathbf{c})}{\lambda_k - \lambda} \frac{\mathbf{u}_k}{(\mathbf{u}_k, \mathbf{u}_k)}.$$

(b) Verify this result in the case of Problem 50(d).

52. Suppose that a sequence of approximations $\mathbf{x}^{(1)}, \mathbf{x}^{(2)}, \cdots, \mathbf{x}^{(r)}, \cdots$ to the vector solution $\mathbf{x}$ of the equation

$$\mathbf{x} = \mathbf{M}\,\mathbf{x} + \mathbf{c}$$

is generated by the recurrence formula

$$\mathbf{x}^{(r+1)} = \mathbf{M}\,\mathbf{x}^{(r)} + \mathbf{c} \qquad (r = 0, 1, 2, \cdots),$$

where $\mathbf{x}^{(0)}$ is an initial approximation, and assume that $\mathbf{M}$ is an $n \times n$ real, symmetric matrix. Let the characteristic numbers of $\mathbf{M}$ be denoted by $\lambda_1, \cdots, \lambda_n$, with corresponding mutually orthogonal characteristic vectors $\mathbf{u}_1, \cdots, \mathbf{u}_n$.

(a) If $\boldsymbol{\epsilon}^{(r)} \equiv \mathbf{x} - \mathbf{x}^{(r)}$, show that

$$\boldsymbol{\epsilon}^{(r+1)} = \mathbf{M}\,\boldsymbol{\epsilon}^{(r)} \qquad (r = 0, 1, 2, \cdots).$$

(b) Noticing that the initial error vector $\boldsymbol{\epsilon}^{(0)}$ can be expressed in the form

$$\boldsymbol{\epsilon}^{(0)} = \sum_{k=1}^{n} \alpha_k \mathbf{u}_k,$$

for *some* values of $\alpha_1, \cdots, \alpha_n$, show that there follows

$$\boldsymbol{\epsilon}^{(r)} = \sum_{k=1}^{n} \alpha_k \lambda_k{}^r \mathbf{u}_k.$$

(c) Deduce that $\mathbf{x}^{(r)}$ converges to $\mathbf{x}$ as $r \to \infty$, regardless of the form of the initial approximation $\mathbf{x}^{(0)}$, if and only if $|\lambda_k| < 1$ $(k = 1, 2, \cdots, n)$. [This result can also be established, by a less simple argument, when $\mathbf{M}$ is not necessarily symmetric.]

53. To illustrate the result of Problem 52, suppose that the system

$$\begin{aligned} x_1 - \alpha x_2 \qquad &= c_1, \\ -\alpha x_1 + x_2 - \alpha x_3 &= c_2, \\ - \alpha x_2 + x_3 &= c_3 \end{aligned}$$

is solved iteratively by use of the relations

$$\begin{aligned} x_1{}^{(r+1)} &= \alpha x_2{}^{(r)} + c_1, \\ x_2{}^{(r+1)} &= \alpha(x_1{}^{(r)} + x_3{}^{(r)}) + c_2, \\ x_3{}^{(r+1)} &= \alpha x_2{}^{(r)} + c_3. \end{aligned}$$

Show that convergence is guaranteed if and only if $|\alpha| < \sqrt{2}/2$.

Section 1.13.

54. Construct a set of three mutually orthogonal unit vectors which are linear combinations of the vectors $\{1, 0, 2, 2\}$, $\{1, 1, 0, 1\}$, and $\{1, 1, 0, 0\}$.

55. Prove that the vector $\mathbf{v} = \{2, 1, 2, 0\}$ is in the space generated by the three vectors defined in Problem 54 and express $\mathbf{v}$ as a linear combination of the selected vectors $\mathbf{e}_1$, $\mathbf{e}_2$, and $\mathbf{e}_3$.

Sections 1.14, 1.15.

56. If A is a (homogeneous) quadratic form in $x_1, x_2, \cdots, x_n$, prove that

$$A = \frac{1}{2}\sum_{k=1}^{n} x_k \frac{\partial A}{\partial x_k}.$$

57. Construct an orthonormal modal matrix $\mathbf{Q}$ corresponding to the matrix

$$\mathbf{A} = \begin{bmatrix} 1 & 0 & 0 \\ 0 & 3 & -1 \\ 0 & -1 & 3 \end{bmatrix},$$

and verify that $\mathbf{Q}^T \mathbf{A} \mathbf{Q} = [\lambda_i \delta_{ij}]$. (Notice the footnote on page 31.)

58. Reduce the quadratic form $A = x_1^2 + 3x_2^2 + 3x_3^2 - 2x_2x_3$ to a canonical form by making an appropriate change in variables, $\mathbf{x} = \mathbf{Q}\,\mathbf{x}'$, where $\mathbf{Q}$ is an orthogonal matrix.

59. Let $\mathbf{M}$ represent a modal matrix of a real symmetric matrix $\mathbf{A}$, the modal columns of which are orthogonal, but not necessarily reduced to unit length. If the characteristic vectors whose elements comprise successive columns of $\mathbf{M}$ are denoted by $\mathbf{u}_1, \mathbf{u}_2, \cdots, \mathbf{u}_n$, show that

$$\mathbf{M}^T \mathbf{M} = \begin{bmatrix} l_1^2 & 0 & \cdots & 0 \\ 0 & l_2^2 & \cdots & 0 \\ \cdots & \cdots & \cdots & \cdots \\ 0 & 0 & \cdots & l_n^2 \end{bmatrix}$$

and

$$\mathbf{M}^T \mathbf{A} \mathbf{M} = \begin{bmatrix} \lambda_1 l_1^2 & 0 & \cdots & 0 \\ 0 & \lambda_2 l_2^2 & \cdots & 0 \\ \cdots & \cdots & \cdots & \cdots \\ 0 & 0 & \cdots & \lambda_n l_n^2 \end{bmatrix},$$

where $l_i^2 = \mathbf{u}_i^T \mathbf{u}_i$. Hence deduce also that the form $A = \mathbf{x}^T \mathbf{A}\,\mathbf{x}$ is reduced to the form

$$A = \lambda_1 l_1^2 x_1'^2 + \lambda_2 l_2^2 x_2'^2 + \cdots + \lambda_n l_n^2 x_n'^2$$

by the change in variables $\mathbf{x} = \mathbf{M}\,\mathbf{x}'$. [Notice that this form reduces to the canonical form (127) if the vectors $\mathbf{u}_i$ are normalized, so that $\mathbf{M}$ is an orthogonal matrix.]

60. (a) Prove that the product of two orthogonal matrices is also an orthogonal matrix.

(b) Prove that the inverse of an orthogonal matrix is also an orthogonal matrix.

Section 1.16.

61. Let $\mathbf{A} = \begin{bmatrix} 1 & 2 & 3 \\ 4 & 5 & 6 \end{bmatrix}$. Determine nonsingular matrices $\mathbf{P}$ and $\mathbf{Q}$ such that $\mathbf{P}\,\mathbf{A}\,\mathbf{Q} = \mathbf{B}$, where $\mathbf{B}$ is obtained by interchanging the two rows of $\mathbf{A}$ and then adding twice the first column to the third column. (See also Problems 32 and 33.)

Section 1.17.

62. (a) Determine the characteristic numbers (λ_1, λ_2) and corresponding Hermitian unit characteristic vectors $(\mathbf{e}_1, \mathbf{e}_2)$ of the problem

$$9x_1 + (2 + 2i)x_2 = \lambda x_1,$$
$$(2 - 2i)x_1 + 2x_2 = \lambda x_2,$$

where $i^2 = -1$, and verify that $\mathbf{e}_1$ and $\mathbf{e}_2$ are orthogonal in the Hermitian sense.

(b) If $\mathbf{H}$ is the coefficient matrix of the system of part (a) and $\mathbf{U}$ is the orthonormal modal matrix made up of $\mathbf{e}_1$ and $\mathbf{e}_2$, verify that

$$\bar{\mathbf{U}}^T \mathbf{H} \mathbf{U} = [\lambda_i \delta_{ij}].$$

(c) If $\mathbf{v} = \{1 + i, 1\}$, determine α_1 and α_2 such that

$$\mathbf{v} = \alpha_1 \mathbf{e}_1 + \alpha_2 \mathbf{e}_2,$$

where $\mathbf{e}_1$ and $\mathbf{e}_2$ are the vectors determined in part (a).

63. Describe the modification of the Gram-Schmidt orthogonalization procedure of Section 1.13 which applies when orthogonality and unit length are defined in the Hermitian sense.

64. Prove that an orthonormal modal matrix $\mathbf{U}$ of a Hermitian matrix $\mathbf{H}$ is a unitary matrix [i.e., that equation (149) is satisfied].

65. Prove that (148) holds when $\mathbf{H}$ is Hermitian, $\mathbf{U}$ is a corresponding orthonormal modal matrix, and λ_k is the kth characteristic number of $\mathbf{H}$.

66. (a) Show that, if a matrix is both unitary and Hermitian, it must satisfy the equation $\mathbf{U}^2 = \mathbf{I}$.

(b) Prove that any matrix of order two, of this type, is either the positive or negative unit matrix, or else is of the form

$$\mathbf{U} = \begin{bmatrix} a & r\,e^{i\alpha} \\ r\,e^{-i\alpha} & -a \end{bmatrix},$$

where a, r, and α are real and $a^2 + r^2 = 1$.

67. A matrix $\mathbf{N}$ is called a *normal matrix* if it commutes with its conjugate transpose, so that

$$\mathbf{N}\,\bar{\mathbf{N}}^T = \bar{\mathbf{N}}^T\,\mathbf{N}.$$

(Notice that hence, in particular, any *real* matrix $\mathbf{A}$ for which $\mathbf{A}^T = \pm\mathbf{A}$ or any *complex* matrix $\mathbf{B}$ for which $\bar{\mathbf{B}}^T = \pm\mathbf{B}$ is normal.)

(a) Prove that if $\mathbf{N}$ is normal, then $l_H(\mathbf{N}\,\mathbf{v}) = l_H(\bar{\mathbf{N}}^T\,\mathbf{v})$ and hence deduce that *if* $\mathbf{N}$ *is normal, then the relations*

$$\mathbf{N}\,\mathbf{v} = \mathbf{0}, \qquad \bar{\mathbf{N}}^T\,\mathbf{v} = \mathbf{0}$$

imply each other.

(b) Prove that $\mathbf{N} - \lambda\,\mathbf{I}$ is normal if $\mathbf{N}$ is normal.

(c) By replacing $\mathbf{N}$ by $\mathbf{N} - \lambda_k \mathbf{I}$ and $\mathbf{v}$ by $\mathbf{u}_k$ in the result of (a), prove that $\mathbf{N}\,\mathbf{u}_k = \lambda_k \mathbf{u}_k$ implies $\bar{\mathbf{N}}^T\,\mathbf{u}_k = \bar{\lambda}_k \mathbf{u}_k$, so that *if* $\mathbf{u}_k$ *is a characteristic vector of a normal matrix* $\mathbf{N}$, *corresponding to* λ_k, *then* $\mathbf{u}_k$ *also is a characteristic vector of* $\bar{\mathbf{N}}^T$, *corresponding to* $\bar{\lambda}_k$.

(d) Prove that *two characteristic vectors of a normal matrix* $\mathbf{N}$, *corresponding to distinct characteristic numbers, are orthogonal in the Hermitian sense.* [Use the relations $\mathbf{N}\,\mathbf{u}_1 = \lambda_1 \mathbf{u}_1$ and $\bar{\mathbf{N}}^T\,\mathbf{u}_2 = \bar{\lambda}_2 \mathbf{u}_2$.]

68. Assuming that a normal matrix possesses s linearly independent characteristic vectors in correspondence with a characteristic number of multiplicity s (see Problem 71), prove that *if* $\mathbf{N}$ *is normal, then an associated orthonormal modal matrix* $\mathbf{U}$ *is unitary and has the property that*

$$\bar{\mathbf{U}}^T\,\mathbf{N}\,\mathbf{U} = [\lambda_i \delta_{ij}],$$

where λ_i *is the ith characteristic number of* $\mathbf{N}$.

Section 1.18.

69. Show that if the first two columns of an orthogonal matrix $\mathbf{Q}$ comprise the elements of two unit characteristic vectors of a real symmetric matrix $\mathbf{A}$, then $\mathbf{Q}^{-1}\,\mathbf{A}\,\mathbf{Q}$ is of the form

$$\begin{bmatrix} \lambda_1 & 0 & 0 & \cdots & 0 \\ 0 & \lambda_2 & 0 & \cdots & 0 \\ 0 & 0 & \alpha_{33} & \cdots & \alpha_{3n} \\ \cdots & \cdots & \cdots & \cdots & \cdots \\ 0 & 0 & \alpha_{3n} & \cdots & \alpha_{nn} \end{bmatrix},$$

where λ_1 and λ_2 are the characteristic numbers corresponding respectively to the two characteristic vectors.

70. Modify the argument of Section 1.18 to deal with a *Hermitian* matrix $\mathbf{H}$. (Notice that $\bar{\mathbf{U}}^T\,\mathbf{H}\,\mathbf{U}$ is also Hermitian.)

71. Modify the argument of Section 1.18 to deal with a general *normal* matrix $\mathbf{N}$ (see Problems 67 and 68). [Use the fact that $\bar{\mathbf{N}}^T\,\mathbf{e}_1 = \bar{\lambda}_1 \mathbf{e}_1$ to show that, when $\mathbf{e}_1$ is the first column of $\mathbf{U}$, both $\bar{\mathbf{U}}^T\,\mathbf{N}\,\mathbf{U}$ and its conjugate transpose will have zeros following the first element throughout the first column.]

Section 1.19.

72. Determine whether the real form

$$A = x_1^2 + 2x_2^2 + x_3^2 - 2x_1x_2 + 2x_2x_3$$

is positive definite, by examining the characteristic numbers of the associated matrix.

73. Determine a real change in variables which reduces the forms

$$A = 3x_1^2 - 2x_1x_2 + 3x_2^2, \qquad B = 2x_1^2 + 2x_2^2$$

simultaneously to the canonical forms

$$A = \lambda_1\alpha_1^2 + \lambda_2\alpha_2^2, \qquad B = \alpha_1^2 + \alpha_2^2,$$

by using the methods of Section 1.19.

Section 1.20.

74. Find the *sum* and *product* of all characteristic numbers of the matrix

$$\mathbf{A} = \begin{bmatrix} 2 & 1 & -1 & 0 \\ 1 & 3 & 4 & 2 \\ -1 & 4 & 1 & 2 \\ 0 & 2 & 2 & 1 \end{bmatrix}.$$

75. Determine whether the matrix **A** of Problem 74 is positive definite.

76. A real symmetric matrix **A** is said to be *negative definite* if its associated quadratic form $\mathbf{x}^T \mathbf{A} \mathbf{x}$ is nonpositive for all real **x**, and is zero only when $\mathbf{x} = \mathbf{0}$. State conditions under which this situation exists, (a) in terms of the characteristic numbers of **A**, and (b) in terms of the discriminants of **A**. (Notice that **A** is negative definite if and only if $-\mathbf{A}$ is positive definite.)

77. Determine for what values of c, if any, it is true that

$$x^2 + y^2 + z^2 \geqq 2c(xy + yz + zx)$$

for all real values of x, y, and z, with equality holding only when $x = y = z = 0$.

78. Prove that if **P** is a nonsingular real matrix then $\mathbf{A} = \mathbf{P}^T \mathbf{P}$ is positive definite.

79. Prove that if **A** is real, symmetric, and positive definite then there exists a nonsingular real matrix **P** such that $\mathbf{A} = \mathbf{P}^T \mathbf{P}$. [First show that **A** can be reduced to a form $\mathbf{R}^T \mathbf{D} \mathbf{R}$, where $\mathbf{D} = \text{diag}(\lambda_1, \cdots, \lambda_n)$ and $|\mathbf{R}| \neq 0$, and then show that **D** can be rewritten appropriately as $\mathbf{G}^T \mathbf{G}$.]

Section 1.21.

80. A geometrical vector $\mathscr{X}$ is represented by the numerical vector $\mathbf{x} = \{1, 1, 1\}$ in terms of components along unit vectors $\mathbf{i}_1$, $\mathbf{i}_2$, and $\mathbf{i}_3$ coinciding with the axes of a rectangular $x_1x_2x_3$ coordinate system. If new axes are chosen in such a way that the new unit vectors are related to the original ones by the equations

$$\mathbf{i}_1' = \frac{\mathbf{i}_1 + \mathbf{i}_2}{\sqrt{2}}, \qquad \mathbf{i}_2' = \frac{\mathbf{i}_1 - \mathbf{i}_2}{\sqrt{2}}, \qquad \mathbf{i}_3' = \mathbf{i}_3,$$

determine the representation $\mathbf{x}'$ of $\mathscr{X}$ in terms of components of $\mathscr{X}$ along the new axes. Show also that the new coordinate system is also rectangular.

81. A numerical vector $\mathbf{y}$, representing $\mathscr{Y}$, is related to the numerical vector $\mathbf{x}$ of Problem 80 by the equation $\mathbf{y} = \mathbf{A}\,\mathbf{x}$, where

$$\mathbf{A} = \begin{bmatrix} 1 & 1 & 1 \\ 0 & 1 & 1 \\ 0 & 0 & 1 \end{bmatrix}.$$

Determine the components of the representation $\mathbf{y}'$ in the new system, *first*, by determining the components of $\mathbf{y}$ and transforming them directly, and *second*, by using equation (199) in connection with the result of Problem 80.

82. Prove that if the new unit vectors of (190) are mutually orthogonal, then the matrix (196) is an orthogonal matrix.

83. (a) Show that an orthogonal matrix of order two is necessarily of one of the following two types:

$$\mathbf{Q}^{(+)} = \begin{bmatrix} \cos\alpha & \sin\alpha \\ -\sin\alpha & \cos\alpha \end{bmatrix}, \qquad \mathbf{Q}^{(-)} = \begin{bmatrix} \cos\alpha & \sin\alpha \\ \sin\alpha & -\cos\alpha \end{bmatrix}.$$

[Notice that $|\mathbf{Q}^{(+)}| = +1$, and $|\mathbf{Q}^{(-)}| = -1$.]

(b) If $\mathbf{x}$ and $\mathbf{x}'$ are considered as two distinct vectors referred to the same axes, and are related by the equation $\mathbf{x} = \mathbf{Q}\,\mathbf{x}'$, verify that $\mathbf{x}$ is rotated into $\mathbf{x}'$ through the angle α by a positive (counterclockwise) rotation if $\mathbf{Q} = \mathbf{Q}^{(+)}$.

(c) If $\mathbf{x}$ and $\mathbf{x}'$ are considered as comprising the components of representations of the same geometrical vector, referred to original and rotated axes, respectively, verify that the coordinate transformation $\mathbf{x} = \mathbf{Q}^{(+)}\mathbf{x}'$ corresponds to a negative rotation of the original axes, through the angle α.

(d) If $\mathbf{Q} = \mathbf{Q}^{(-)}$ in parts (b) and (c), verify that the transformations then each involve a *reversed* rotation combined with a suitable *reflection.*

Section 1.22.

84. If $\mathbf{A}$ is a symmetric $n \times n$ matrix with n distinct characteristic numbers λ_i, show that any polynomial $P(\mathbf{A})$ can be expressed in the form

$$P(\mathbf{A}) = c_1\mathbf{A}^{n-1} + c_2\mathbf{A}^{n-2} + \cdots + c_{n-1}\mathbf{A} + c_n\mathbf{I}$$

where the c's are determined by the n simultaneous linear equations

$$P(\lambda_i) = \sum_{k=1}^{n} c_k\lambda_i^{n-k} \qquad (i = 1, 2, \cdots, n).$$

[In some cases this direct determination is more convenient than the use of Sylvester's formula (226).]

85. Let $\mathbf{A} = \begin{bmatrix} 2 & 1 \\ 1 & 2 \end{bmatrix}$ and $\mathbf{B} = \mathbf{A}^5 - 3\mathbf{A}^4 + 2\mathbf{A} - \mathbf{I}$.

(a) Determine the characteristic numbers and corresponding characteristic vectors of **B**.

(b) Determine whether **B** is positive definite.

(c) Determine the elements of $\mathbf{A}^{100}$.

86. Suppose that **A** is real and symmetric of order two, with a repeated characteristic number $\lambda_1 = \lambda_2$.

(a) Obtain from (230) the evaluation

$$e^{\mathbf{A}} = e^{\lambda_1}\,\mathbf{A} - (\lambda_1 - 1)e^{\lambda_1}\,\mathbf{I}.$$

(b) Prove that **A** must in this case be a scalar matrix $\mathbf{A} = k\,\mathbf{I}$, and show that the evaluation of part (a) reduces to

$$e^{k\mathbf{I}} = e^k\,\mathbf{I}.$$

87. Suppose that the elements of a matrix $\mathbf{A}(t) = [a_{ij}(t)]$ are differentiable functions of a variable t.

(a) From the definition

$$\frac{d\mathbf{A}(t)}{dt} = \lim_{\Delta t \to 0} \frac{\mathbf{A}(t + \Delta t) - \mathbf{A}(t)}{\Delta t} \equiv \lim_{\Delta t \to 0} \frac{\Delta \mathbf{A}}{\Delta t},$$

prove that $d\mathbf{A}(t)/dt = [da_{ij}/dt]$.

(b) Prove that

$$\frac{d}{dt}(\mathbf{A}\,\mathbf{B}) = \frac{d\mathbf{A}}{dt}\mathbf{B} + \mathbf{A}\frac{d\mathbf{B}}{dt}.$$

(c) Specialize the result of part (b) in the case when $\mathbf{B} = \mathbf{A}$, and give an example to show that $d\mathbf{A}^2/dt \neq 2\mathbf{A}\,d\mathbf{A}/dt$ in general.

88. (a) If **A** is a real symmetric matrix, verify that the differential equation

$$\frac{d\mathbf{x}}{dt} = \mathbf{A}\,\mathbf{x}$$

is satisfied by $\mathbf{x} = e^{t\mathbf{A}}\,\mathbf{c}$, where **c** is a constant vector.

(b) Use this result and an appropriate modification of equation (230) to solve the system

$$\frac{dx_1}{dt} = x_1 + 2x_2,$$

$$\frac{dx_2}{dt} = 2x_1 + x_2,$$

subject to the initial conditions $\{x_1(0), x_2(0)\} = \{c_1, c_2\}$.

89. (a) Show that if **A** is a symmetric matrix of order n, with distinct characteristic numbers, then

$$\mathbf{A}^N = \sum_{k=1}^{n} \lambda_k^N\, Z_k(\mathbf{A}),$$

where Z_k is defined by (227) and N is a positive integer.

(b) Let λ_n be the dominant characteristic number (i.e., the characteristic number with largest absolute value). Noticing that for sufficiently large N the nth term in the preceding sum will then predominate, show that if $\mathbf{x}$ is an arbitrary vector there follows

$$\mathbf{A}^N \mathbf{x} \approx \lambda_n{}^N \mathbf{v}, \qquad \mathbf{A}^{N+1} \mathbf{x} \approx \lambda_n{}^{N+1} \mathbf{v},$$

where $\mathbf{v} = [Z_n(\mathbf{A})]\mathbf{x}$, when N is large, unless it happens that $\mathbf{v} = \mathbf{0}$.

(c) Hence deduce that, in general, if an arbitrary vector $\mathbf{x}$ is premultiplied repeatedly by a symmetric matrix $\mathbf{A}$, the vector obtained after $N + 1$ such multiplications is approximately λ_n times that obtained after N multiplications, where λ_n is the dominant characteristic number of $\mathbf{A}$, and hence also that the vectors obtained after successive multiplications tend, in general, to become multiples of the characteristic vector associated with λ_n.

(d) Show that the exceptional case, in which the vector $\mathbf{x}$ is such that $[Z_n(\mathbf{A})]\mathbf{x} = \mathbf{0}$, will occur if $\mathbf{x}$ happens to be a characteristic vector of $\mathbf{A}$, corresponding to a characteristic number $\lambda_k \neq \lambda_n$, or if $\mathbf{x}$ is a linear combination of such vectors.

[A more complete treatment of this procedure, from a somewhat different point of view, is given in Section 1.23.]

Section 1.23.

90. Determine the dominant characteristic number and the corresponding characteristic vector for the system

$$\begin{aligned} x_1 + x_2 + x_3 &= \lambda x_1, \\ x_1 + 3x_2 + 3x_3 &= \lambda x_2, \\ x_1 + 3x_2 + 6x_3 &= \lambda x_4. \end{aligned}$$

(Retain slide-rule accuracy.)

91. Show that the iterative method does not converge to a characteristic vector if $\mathbf{A} = \begin{bmatrix} 1 & 1 \\ -1 & -1 \end{bmatrix}$, regardless of the initial approximation. Explain.

92. Investigate the application of the iterative method in the case of the matrix $\mathbf{A} = \begin{bmatrix} 1 & 1 \\ -2 & -1 \end{bmatrix}$. Explain.

Section 1.24.

93. Determine the two largest characteristic numbers, and corresponding characteristic vectors, of the system

$$\begin{aligned} x_1 + x_2 + x_3 + x_4 &= \lambda x_1, \\ x_1 + 2x_2 + 2x_3 + 2x_4 &= \lambda x_2, \\ x_1 + 2x_2 + 3x_3 + 3x_4 &= \lambda x_3, \\ x_1 + 2x_2 + 3x_3 + 4x_4 &= \lambda x_4. \end{aligned}$$

(Retain slide-rule accuracy.)

94. Determine all characteristic numbers, and the corresponding characteristic vectors, of the system

$$\begin{aligned} x_1 - x_2 \qquad\qquad &= \lambda x_1, \\ -x_1 + 2x_2 - x_3 \qquad &= \lambda x_2, \\ -x_2 + 2x_3 - x_4 &= \lambda x_3, \\ -x_3 + x_4 &= \lambda x_4. \end{aligned}$$

(Retain slide-rule accuracy.)

95. Determine the smallest characteristic number for the system (241), without using the inversion (251), by making use of the procedure described in connection with (239) and (240). (Retain slide-rule accuracy.)

96. Suppose that the iterative method of Section 1.23 fails to converge for a real symmetric matrix **A**, so that λ_n and $-\lambda_n$ are both dominant characteristic numbers. Take $\lambda_n > 0$, and write $\lambda_{n-1} = -\lambda_n$.

Show that, if r is sufficiently large, the input in the rth cycle is given approximately by

$$\mathbf{x}^{(r)} \approx \mathbf{v}_n + \mathbf{v}_{n-1},$$

where $\mathbf{v}_n$ and $\mathbf{v}_{n-1}$ are constant multiples of the unit characteristic vectors relevant to λ_n and $\lambda_{n-1} \equiv -\lambda_n$, respectively, whereas the output is then given approximately by

$$\mathbf{y}^{(r)} \approx \lambda_n(\mathbf{v}_n - \mathbf{v}_{n-1}).$$

Show further that *if this output is taken as the input for the next cycle*, so that

$$\mathbf{x}^{(r+1)} = \mathbf{y}^{(r)},$$

there follows also

$$\mathbf{y}^{(r+1)} \approx \lambda_n^2(\mathbf{v}_n + \mathbf{v}_{n-1}),$$

so that λ_n can then be determined approximately by the relation

$$\mathbf{y}^{(r+1)} \approx \lambda_n^2 \mathbf{x}^{(r)},$$

after which approximations to $\mathbf{v}_n$ and $\mathbf{v}_{n-1}$ are given by

$$\mathbf{v}_n \approx \frac{1}{2}\left[\mathbf{x}^{(r)} + \frac{1}{\lambda_n}\mathbf{y}^{(r)}\right], \qquad \mathbf{v}_{n-1} \approx \frac{1}{2}\left[\mathbf{x}^{(r)} - \frac{1}{\lambda_n}\mathbf{y}^{(r)}\right],$$

when r is sufficiently large.

97. Illustrate the technique developed in Problem 96 in the case of the symmetric matrix $\mathbf{A} = \begin{bmatrix} -4 & 3 \\ 3 & 4 \end{bmatrix}$.

Section 1.25.

98. Prove that the characteristic numbers of the problem $\mathbf{A}\,\mathbf{x} = \lambda\,\mathbf{B}\,\mathbf{x}$ are real when **A** and **B** are real and symmetric, and either **A** or **B** is positive definite.

99. Determine the characteristic numbers and vectors of the problem $\mathbf{A}\,\mathbf{x} = \lambda\,\mathbf{B}\,\mathbf{x}$, where

$$\mathbf{A} = \begin{bmatrix} 5 & 2 \\ 2 & 3 \end{bmatrix}, \qquad \mathbf{B} = \begin{bmatrix} 1 & 0 \\ 0 & 2 \end{bmatrix},$$

and verify that the characteristic vectors are orthogonal relative to both **A** and **B**.

100. Construct an orthonormal modal matrix associated with Problem 99, where the normalization is relative to **B**.

101. Use the results of Problems 99 and 100 to determine a change in variables which reduces the quadratic forms

$$A = 5x_1^2 + 4x_1x_2 + 3x_2^2, \qquad B = x_1^2 + 2x_2^2$$

simultaneously to the canonical forms

$$A = \lambda_1\alpha_1^2 + \lambda_2\alpha_2^2, \qquad B = \alpha_1^2 + \alpha_2^2.$$

102. Show that the condition (282a) is equivalent to the condition

$$(\mathbf{x}^{(r)}, \mathbf{x}^{(r)})_{\mathbf{A}} \approx \lambda_n(\mathbf{x}^{(r)}, \mathbf{x}^{(r)})_{\mathbf{B}}$$

and that (282b) is equivalent to the condition

$$(\mathbf{y}^{(r)}, \mathbf{y}^{(r)})_{\mathbf{B}} \approx \lambda_n(\mathbf{x}^{(r)}, \mathbf{x}^{(r)})_{\mathbf{A}}.$$

103. Prove that any set of nonzero real vectors which are mutually orthogonal relative to a real positive definite matrix **B** is a linearly independent set. [Assume the contrary and deduce a contradiction.]

104. Consider the equation $\mathbf{A}\,\mathbf{x} = \lambda\,\mathbf{B}\,\mathbf{x}$, where

$$\mathbf{A} = \begin{bmatrix} 2 & -1 & 0 \\ -1 & 2 & -1 \\ 0 & -1 & 2 \end{bmatrix}, \qquad \mathbf{B} = \begin{bmatrix} 10 & 1 & 0 \\ 1 & 10 & 1 \\ 0 & 1 & 10 \end{bmatrix}.$$

(a) Determine the largest characteristic number λ_3 by an iterative process, after rewriting the equation in the form $\mathbf{B}^{-1}\,\mathbf{A}\,\mathbf{x} = \lambda\,\mathbf{x}$.

(b) Determine λ_3 by an iterative process without rewriting the equation in the form suggested in part (a).

(Retain slide-rule accuracy.)

105. (a) Determine the smallest characteristic number λ_1 for the equation of Problem 104 by an iterative process, after rewriting the equation in the form $\mathbf{A}^{-1}\,\mathbf{B}\,\mathbf{x} = \lambda^{-1}\,\mathbf{x}$.

(b) Determine λ_1 by an iterative process without rewriting the equation in the form suggested in part (a).

(Retain slide-rule accuracy.)

Section 1.26.

106. If $\mathbf{A} = \begin{bmatrix} 2 & 1 \\ -2 & -1 \end{bmatrix}$, determine the characteristic vectors $\mathbf{u}_1$ and $\mathbf{u}_2$ of the problem $\mathbf{A}\,\mathbf{x} = \lambda\,\mathbf{x}$, and the characteristic vectors $\mathbf{u}_1'$ and $\mathbf{u}_2'$ of the associated problem $\mathbf{A}^T\,\mathbf{x}' = \lambda\,\mathbf{x}'$, and verify the validity of equation (289).

107. (a) Suppose that $\mathbf{A}$ posesses n distinct characteristic numbers $\lambda_1, \cdots, \lambda_n$, with corresponding characteristic vectors $\mathbf{u}_1, \cdots, \mathbf{u}_n$, and denote corresponding characteristic vectors of $\mathbf{A}^T$ by $\mathbf{u}_1', \cdots, \mathbf{u}_n'$. Obtain the solution of the problem $\mathbf{A}\,\mathbf{x} - \lambda\,\mathbf{x} = \mathbf{c}$ in the form

$$\mathbf{x} = \sum_{k=1}^{n} \frac{(\mathbf{u}_k', \mathbf{c})}{\lambda_k - \lambda} \frac{\mathbf{u}_k}{(\mathbf{u}_k', \mathbf{u}_k)},$$

when $\lambda \neq \lambda_1, \cdots, \lambda_n$. [Compare Problem 51.]

(b) Discuss the situation when λ assumes a characteristic value λ_p. (Notice also that this case is described by the result of replacing $\mathbf{A}$ by $\mathbf{A} - \lambda_p\mathbf{I}$ in the statement at the end of Section 1.11.)

108. With the terminology of Problem 107, use the result at the end of Section 1.9 to show that the elements of $\mathbf{u}_r$ are proportional to the cofactors of respective elements in any *row* of the matrix

$$\mathbf{A} - \lambda_r\mathbf{I} = \begin{bmatrix} a_{11} - \lambda_r & a_{12} & \cdots & a_{1n} \\ a_{21} & a_{22} - \lambda_r & \cdots & a_{2n} \\ \cdots & \cdots & \cdots & \cdots \\ a_{n1} & a_{n2} & \cdots & a_{nn} - \lambda_r \end{bmatrix},$$

whereas the elements of $\mathbf{u}_r'$ are proportional to the cofactors of respective elements in any *column* of that matrix, if not all the relevant cofactors vanish. Verify this conclusion in the example of Problem 106.

109. Let $\mathbf{M}$ represent a modal matrix of *any* square matrix $\mathbf{A}$ of order n with n linearly independent characteristic vectors $\mathbf{u}_1, \cdots, \mathbf{u}_n$ corresponding to the respective characteristic numbers $\lambda_1, \cdots, \lambda_n$, where the $\mathbf{u}$'s which are used to construct $\mathbf{M}$ need be neither orthogonal nor of unit length and the λ's need be neither real nor distinct.

(a) By appropriately modifying the argument of equations (119)–(122) of Section 1.14, prove that

$$\mathbf{M}^{-1}\,\mathbf{A}\,\mathbf{M} = \mathbf{D},$$

where $\mathbf{D}$ is the diagonal matrix

$$\mathbf{D} = [\lambda_i \delta_{ij}],$$

so that $\mathbf{A}$ is thus diagonalized by a *similarity* transformation.

(b) Deduce that any such matrix $\mathbf{A}$ can be determined from its characteristic numbers and vectors by use of the relation

$$\mathbf{A} = \mathbf{M}\,\mathbf{D}\,\mathbf{M}^{-1}$$

and show also that

$$\mathbf{A}^r = \mathbf{M}\,\mathbf{D}^r\,\mathbf{M}^{-1}$$

when r is any nonnegative integer.

110. Consider the nonsymmetric matrix

$$\mathbf{A} = \begin{bmatrix} 2 & 1 \\ -1 & 0 \end{bmatrix}.$$

(a) Show that $\mathbf{A}$ has a double characteristic number $\lambda_1 = \lambda_2 = 1$, corresponding to a single characteristic vector $\mathbf{u}_1$.

(b) Determine a matrix $\mathbf{P}$ such that $\mathbf{P}^{-1}\mathbf{A}\mathbf{P}$ is in the Jordan canonical form,

$$\mathbf{P}^{-1}\mathbf{A}\mathbf{P} = \begin{bmatrix} 1 & 1 \\ 0 & 1 \end{bmatrix},$$

by taking the vector $\mathbf{u}_1$ as the first column of $\mathbf{P}$ and determining the unknown elements of the second column of $\mathbf{P}$ in such a way that the upper right element of $\mathbf{P}^{-1}\mathbf{A}\mathbf{P}$ is unity. Notice that infinitely many such $\mathbf{P}$'s are obtained.

111. Prove the following theorem: *If* $\mathbf{A}$ *is a real matrix with nonnegative elements and if the sum of the elements in each column* (or *in each row*) *of* $\mathbf{A}$ *is less than one, then the characteristic numbers of* $\mathbf{A}$ *are smaller than one in absolute value.* [Let $\mathbf{u} = \{u_1, \cdots, u_n\}$ be a characteristic vector corresponding to λ. Deduce from the relation $\lambda u_i = \sum_k a_{ik} u_k$ that $|\lambda|\,|u_i| \leq \sum_k a_{ik}\,|u_k|$, and sum over i. Note that λ is also a characteristic number of $\mathbf{A}^T$.]

Section 1.27.

Determine the natural frequencies and natural modes of vibration of the mechanical system of Figure 1.1 in the following cases:

112. Assume $k_1 = 2k$, $k_2 = k_3 = k$; $m_1 = m_2 = m_3 = m$.

113. Assume $k_1 = 2k$, $k_2 = k_3 = k$; $m_1 = m_2 = m$, $m_3 = 2m$.

114. Assume $k_1 = 0$, $k_2 = k_3 = k$; $m_1 = m_2 = m_3 = m$.

115. Assume $k_1 = 0$, $k_2 = k_3 = k$; $m_1 = m_2 = m$, $m_3 = 2m$.

[In most physical problems of this type the fundamental mode (corresponding to the *smallest* natural frequency) is such that the initial approximation $\{1, 1, 1, \cdots, 1\}$ is a convenient one. In the highest natural mode, the successive masses generally tend to oscillate with opposite phases, so that the initial approximation $\{1, -1, 1, \cdots, \pm 1\}$ usually leads to more rapid convergence. In Problems 114 and 115, the system of masses and springs is unattached to a support, and the characteristic number associated with $\omega = 0$ corresponds to motion of the system as a rigid body.]

Minimal Properties of Characteristic Numbers.

116. Let $\mathbf{A}$ denote a real symmetric matrix of order n, with characteristic numbers $\lambda_1, \cdots, \lambda_n$, arranged in increasing *algebraic* order ($\lambda_1 \leqq \lambda_2 \leqq \cdots \leqq \lambda_n$), and corresponding normalized and orthogonalized characteristic vectors $\mathbf{e}_1, \cdots, \mathbf{e}_n$.

(a) If $\mathbf{x}$ is an arbitrary real vector with n components, and hence expressible in the form

$$\mathbf{x} = c_1\mathbf{e}_1 + c_2\mathbf{e}_2 + \cdots + c_n\mathbf{e}_n = \sum_{k=1}^{n} c_k\mathbf{e}_k,$$

establish the relations

$$\mathbf{x}^T\,\mathbf{x} = c_1^2 + c_2^2 + \cdots + c_n^2 = \sum_{k=1}^{n} c_k^2,$$

$$\mathbf{A}\,\mathbf{x} = \lambda_1 c_1\mathbf{e}_1 + \lambda_2 c_2\mathbf{e}_2 + \cdots + \lambda_n c_n\mathbf{e}_n = \sum_{k=1}^{n} \lambda_k c_k\mathbf{e}_k,$$

and

$$\mathbf{x}^T\,\mathbf{A}\,\mathbf{x} = \lambda_1 c_1^2 + \lambda_2 c_2^2 + \cdots + \lambda_n c_n^2 = \sum_{k=1}^{n} \lambda_k c_k^2.$$

(b) Deduce that

$$\frac{\mathbf{x}^T\,\mathbf{A}\,\mathbf{x}}{\mathbf{x}^T\,\mathbf{x}} = \frac{\lambda_1 c_1^2 + \lambda_2 c_2^2 + \cdots + \lambda_n c_n^2}{c_1^2 + c_2^2 + \cdots + c_n^2},$$

and hence also that

$$\left|\frac{\mathbf{x}^T\,\mathbf{A}\,\mathbf{x}}{\mathbf{x}^T\,\mathbf{x}}\right| \leqq |\lambda_i|_{\max}.$$

(c) Prove that

$$\lambda_n - \frac{\mathbf{x}^T\,\mathbf{A}\,\mathbf{x}}{\mathbf{x}^T\,\mathbf{x}} = \frac{\sum_{k=1}^{n} (\lambda_n - \lambda_k)c_k^2}{\sum_{k=1}^{n} c_k^2} \geqq 0,$$

for *any* real vector $\mathbf{x}$,

(d) If $\mathbf{x}$ is orthogonal to the characteristic vectors $\mathbf{e}_{r+1}, \mathbf{e}_{r+2}, \cdots, \mathbf{e}_n$, show that

$$\lambda_r - \frac{\mathbf{x}^T\,\mathbf{A}\,\mathbf{x}}{\mathbf{x}^T\,\mathbf{x}} = \frac{\sum_{k=1}^{r} (\lambda_r - \lambda_k)c_k^2}{\sum_{k=1}^{r} c_k^2} \geqq 0.$$

(e) Show that, if $\mathbf{x} = \mathbf{e}_i$, there follows

$$\frac{\mathbf{x}^T\,\mathbf{A}\,\mathbf{x}}{\mathbf{x}^T\,\mathbf{x}} = \frac{\mathbf{e}_i^T\,\mathbf{A}\,\mathbf{e}_i}{\mathbf{e}_i^T\,\mathbf{e}_i} = \lambda_i \qquad (i = 1, 2, \cdots, n).$$

117. Let $\mathbf{A}$ be a real symmetric matrix, with characteristic numbers $\lambda_1 \leqq \lambda_2 \leqq \cdots \leqq \lambda_n$ and corresponding normalized and orthogonalized characteristic vectors $\mathbf{e}_1, \mathbf{e}_2, \cdots, \mathbf{e}_n$. Deduce the following results from the results of Problem 116.

(a) The number λ_n is the maximum value of $(\mathbf{x}^T\,\mathbf{A}\,\mathbf{x})/(\mathbf{x}^T\,\mathbf{x})$ for all real vectors $\mathbf{x}$, and this maximum value is taken on when $\mathbf{x}$ is identified with a characteristic vector associated with λ_n.

(b) The number λ_r is the maximum value of $(\mathbf{x}^T\,\mathbf{A}\,\mathbf{x})/(\mathbf{x}^T\,\mathbf{x})$ for all real vectors $\mathbf{x}$ which are simultaneously orthogonal to the characteristic vectors associated with $\lambda_{r+1}, \lambda_{r+2}, \cdots, \lambda_n$, and this maximum value is taken on when $\mathbf{x}$ is identified with a characteristic vector associated with λ_r.

(c) The number λ_n is the maximum value of $\mathbf{e}^T \mathbf{A} \mathbf{e}$ for all real *unit* vectors $\mathbf{e}$, the number λ_r is the maximum value for all unit vectors simultaneously orthogonal to $\mathbf{e}_{r+1}, \mathbf{e}_{r+2}, \cdots, \mathbf{e}_n$, and these successive maxima are taken on when $\mathbf{e}$ is identified with the relevant unit characteristic vector.

118. Suppose that Problems 116 and 117 are modified in such a way that $\lambda_1 \leqq \lambda_2 \leqq \cdots \leqq \lambda_n$ are the characteristic numbers of the problem $\mathbf{A}\,\mathbf{x} = \lambda\,\mathbf{B}\,\mathbf{x}$, where $\mathbf{A}$ and $\mathbf{B}$ are real and symmetric, and also $\mathbf{B}$ is positive definite, and $\mathbf{e}_1, \mathbf{e}_2, \cdots, \mathbf{e}_n$ comprise an orthonormal set of corresponding characteristic vectors, the orthogonality and normality being relative to the matrix $\mathbf{B}$. Show that the results of those Problems again apply if $\mathbf{x}^T\,\mathbf{x}$ is replaced by $\mathbf{x}^T\,\mathbf{B}\,\mathbf{x}$ throughout, and if "unit vectors" are of unit length relative to $\mathbf{B}$.

119. Suppose that the characteristic numbers of a real symmetric matrix $\mathbf{A}$ are arranged in order of increasing *absolute value.*

(a) Deduce from the result of Problem 116(b) that the use of equation (250a), in connection with iterative approximation to characteristic quantities, leads to approximations to λ_n which are not greater than λ_n in absolute value.

(b) Show that the use of equation (250b) amounts to approximating λ_n by a ratio of the form

$$\frac{\lambda_1{}^2 c_1{}^2 + \lambda_2{}^2 c_2{}^2 + \cdots + \lambda_n{}^2 c_n{}^2}{\lambda_1 c_1{}^2 + \lambda_2 c_2{}^2 + \cdots + \lambda_n c_n{}^2},$$

and deduce that such an approximation is conservative if all characteristic numbers λ_i are *positive.*

120. (a) If $\mathbf{A}$ is a real symmetric matrix, all of whose elements are *nonnegative* ($a_{ij} \geqq 0$), deduce from preceding results that the characteristic number of largest magnitude is *positive* (although its negative *may* then *also* be a characteristic number), and that all components of the corresponding characteristic vector $\mathbf{e}_n$ are of the same sign, and hence may be taken to be all nonnegative. [Consider the nature of $\mathbf{e}^T\,\mathbf{A}\,\mathbf{e}$.]

(b) Show that the result of part (a) is also true of the dominant characteristic quantities for the problem $\mathbf{A}\,\mathbf{x} = \lambda\,\mathbf{B}\,\mathbf{x}$ if $\mathbf{B}$ is real, symmetric, and positive definite, $\mathbf{A}$ is real and symmetric, and $a_{ij} \geqq 0$.

[The result of part (a) is also true for *any* real square matrix $\mathbf{A}$ with nonnegative elements.]

121. If $\mathbf{A}$ is a real symmetric matrix with characteristic numbers λ_i and corresponding characteristic vectors $\mathbf{e}_i$, show that there follows

$$\mathbf{e}_i{}^T(\mathbf{A}\,\mathbf{x} - \lambda_i \mathbf{x}) = 0 \qquad (i = 1, 2, \cdots, n),$$

for any real vector $\mathbf{x}$. Hence, with the notation $\mathbf{y} = \mathbf{A}\,\mathbf{x}$ for the "transform" of $\mathbf{x}$, deduce that

$$\mathbf{e}_i{}^T(\mathbf{y} - \lambda_i \mathbf{x}) = 0 \qquad (i = 1, 2. \cdots, n),$$

for any real vector $\mathbf{x}$.

122. Suppose that $\mathbf{A}$ is a real symmetric matrix with *no negative elements.*

(a) Deduce from the results of Problems 120(a) and 121 that the components of the vector $\mathbf{y} - \lambda_n \mathbf{x}$, where $\mathbf{y} \equiv \mathbf{A}\,\mathbf{x}$, then cannot all be of the same sign (unless they all vanish, so that $\mathbf{x}$ is a multiple of $\mathbf{e}_n$).

(b) Deduce that, *in this case*, if the input $\mathbf{x}$ of the iterative method of Section 1.23 posesses only nonegative elements, then the dominant characteristic number λ_n is not larger than the largest ratio y_i/x_i of corresponding elements of the output and input vectors, and not less than the smallest such ratio:

$$\min_i \frac{y_i}{x_i} \leq \lambda_n \leq \max_i \frac{y_i}{x_i}.$$

[This result also holds for *any* real square matrix *with no negative elements.*]

123. Prove that the statement of Problem 122(b) is true also for the application of the iterative method to the determination of the dominant characteristic number of the problem $\mathbf{A}\,\mathbf{x} = \lambda\,\mathbf{B}\,\mathbf{x}$ if $\mathbf{B}$ is real, symmetric, and positive definite whereas $\mathbf{A}$ is real and symmetric and composed only of nonnegative elements, and if also $\mathbf{A}$ and $\mathbf{B}$ are *commutative.* [Use Problem 120(b) and a generalization of Problem 121, with $\mathbf{y} = \mathbf{B}^{-1}\,\mathbf{A}\,\mathbf{x} = \mathbf{A}\,\mathbf{B}^{-1}\,\mathbf{x}$.]

124. Suppose that $\mathbf{A}$ is a real, symmetric, positive definite matrix such that, whereas all diagonal elements are positive, all elements off the diagonal are either negative or zero. [See, for example, the matrix of coefficients in (295).]

(a) Show that, if α is any positive constant *larger than the largest diagonal element of* $\mathbf{A}$, then the matrix $\mathbf{M} \equiv \alpha\,\mathbf{I} - \mathbf{A}$ is a symmetric matrix, all of whose elements are nonnegative.

(b) Show that the characteristic numbers μ_i of $\mathbf{M}$ are given by $\mu_i = \alpha - \lambda_i$, where λ_i are the characteristic numbers of $\mathbf{A}$, and that the characteristic vector of $\mathbf{M}$ associated with μ_i is that of $\mathbf{A}$ associated with λ_i.

(c) Use the result of Problem 120(a) to show that the largest μ_i is positive. Deduce that the dominant μ_i is $\mu_1 = \alpha - \lambda_1$, where λ_1 is the smallest characteristic number of $\mathbf{A}$, and that all components of the corresponding characteristic vector have the same sign. Hence show that the *smallest* characteristic number of a matrix $\mathbf{A}$ of the type under consideration is not larger than the largest diagonal element of $\mathbf{A}$, and that all components of the associated characteristic vector may be taken as nonnegative. [Notice that the matrix $\mathbf{M}$ can be obtained more easily than the matrix $\mathbf{A}^{-1}$, for the purpose of determining λ_1 and the corresponding characteristic vector by matrix iteration.]

125. Generalize the results and procedures of Problem 124 to the case of the problem $\mathbf{A}\,\mathbf{x} = \lambda\,\mathbf{B}\,\mathbf{x}$ where $\mathbf{A}$ is a matrix of the type described in that problem, and $\mathbf{B}$ is a positive definite matrix with no negative elements. [Here $\mathbf{M} = \alpha\,\mathbf{B} - \mathbf{A}$, where $\alpha > (b_{ii}/a_{ii})_{\max}$.]

Section 1.28.

126. Prove that the relation

$$\|f(x) + g(x)\|^2 = \|f(x)\|^2 + \|g(x)\|^2$$

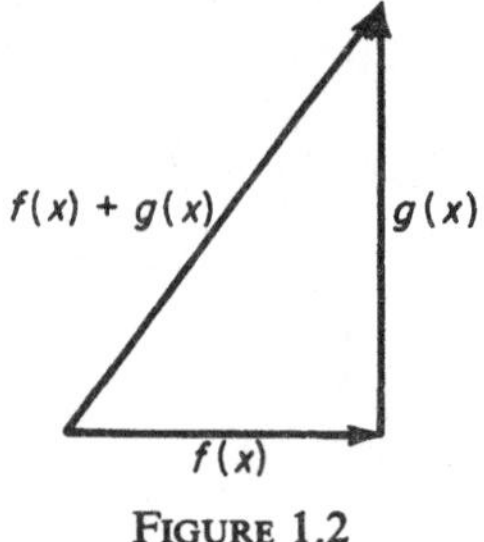

FIGURE 1.2

is true, over a prescribed interval (a, b), if and only if f and g are orthogonal over (a, b). Notice that if we think of $\|f(x)\|^2 = \int_a^b f^2\,dx$ as the "square of the length of $f(x)$" in the function space relevant to (a, b), then this result is analogous to the Pythagorean theorem (see Figure 1.2).

127. By noticing that, if all integrals are evaluated over an interval $(a. b)$, the quantity

$$\int [f(x) + \lambda g(x)]^2\,dx = \int f^2\,dx + 2\lambda \int fg\,dx + \lambda^2 \int g^2\,dx$$

is necessarily nonnegative for any real value of λ, deduce that

$$\left(\int fg\,dx\right)^2 \leqq \left(\int f^2\,dx\right)\left(\int g^2\,dx\right)$$

and hence

$$\left|\int fg\,dx\right| \leqq \left(\sqrt{\int f^2\,dx}\right)\left(\sqrt{\int g^2\,dx}\right).$$

This relation is known as the *Schwarz inequality*. Show also that equality holds if and only if $g(x)$ differs from a constant multiple of $f(x)$ by a trivial function.

Deduce that if we define the "angle between $f(x)$ and $g(x)$" in the function space relevant to (a, b) by the equation

$$\cos \theta[f, g] = \frac{\int_a^b fg\,dx}{\sqrt{\int_a^b f^2\,dx}\sqrt{\int_a^b g^2\,dx}} = \frac{(f, g)}{\|f\| \cdot \|g\|}$$

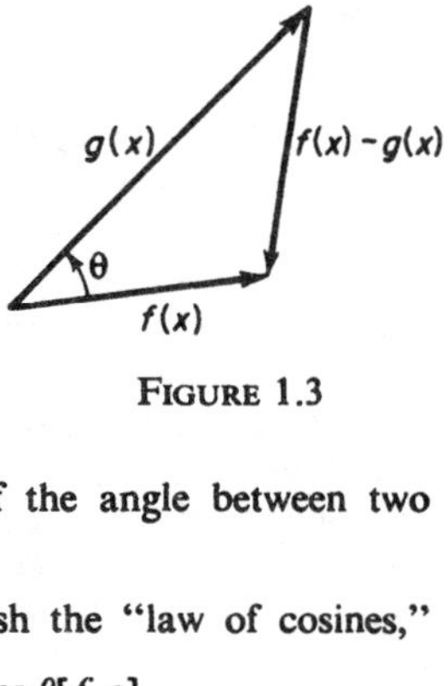

FIGURE 1.3

then θ is a *real* angle. Notice that this definition is completely analogous to the geometrical definition of the angle between two vectors (see Figure 1.3).

128. With the terminology of Problem 127, establish the "law of cosines,"

$$\|f - g\|^2 = \|f\|^2 + \|g\|^2 - 2\,\|f\| \cdot \|g\| \cos \theta[f, g],$$

in function space.

129. Verify the truth of the identity

$$\int (f - g)^2\,dx = \begin{cases} \left(\sqrt{\int f^2\,dx} - \sqrt{\int g^2\,dx}\right)^2 + 2\left[\sqrt{\int f^2\,dx}\,\sqrt{\int g^2\,dx} - \int fg\,dx\right] \\ \left(\sqrt{\int f^2\,dx} + \sqrt{\int g^2\,dx}\right)^2 - 2\left[\sqrt{\int f^2\,dx}\,\sqrt{\int g^2\,dx} + \int fg\,dx\right] \end{cases}$$

where each integral is evaluated over the interval (a, b). Use the Schwarz inequality (Problem 127) to show that each quantity in square brackets is nonnegative, and hence deduce the relation

$$\|f\| - \|g\| \leqq \|f - g\| \leqq \|f\| + \|g\|,$$

where neither equality can hold unless $g(x)$ differs from a constant multiple of $f(x)$ by a trivial function.

To what geometrical relation is this function-theoretical result analogous?

130. As a special case of the Schwarz inequality (Problem 127), deduce that

$$\frac{1}{b-a}\int_a^b f\,dx \leq \sqrt{\frac{1}{b-a}\int_a^b f^2\,dx}.$$

[The left-hand member is the mean value of $f(x)$ over (a, b), the right-hand member the so-called *root mean square* (rms) value.]

131. Establish the validity of the following statement: "The rms value of the sum of two functions over a given interval is not greater than the sum of the separate rms values, and not less than their difference."

132. Let $f_1(x)$, $f_2(x)$, $\cdots$, $f_n(x)$, $\cdots$ comprise an infinite set of functions defined over (a, b). Describe a procedure for forming from this set a set of linear combinations $\phi_1(x), \phi_2(x), \cdots, \phi_n(x), \cdots$ which is orthonormal over (a, b). [See Section 1.13.]

133. Show that the functions $f_0(x) = x$ and $f_k(x) = \sin \mu_k x$ $(k = 1, 2, \cdots)$ comprise an orthogonal set over $(0, 1)$ if the constants μ_k are the positive roots of the transcendental equation $\tan \mu_k = \mu_k$.

134. Show that the complex functions $f_k(x) = e^{ikx}$, where k takes on *all* integral values, comprise a set which is orthogonal in the Hermitian sense over any real interval $(a, a + 2\pi)$. Determine the normalizing factors.

Section 1.29.

135. Show that the functions defined in Problem 133 are the characteristic functions of the following Sturm-Liouville problem:

$$\frac{d^2y}{dx^2} + \mu^2 y = 0; \qquad y(0) = 0, \quad y(1) = y'(1).$$

136. Determine the coefficients in the expansion

$$1 = A_0 x + \sum_{k=1}^{\infty} A_k \sin \mu_k x \quad (0 < x < 1),$$

where $\tan \mu_k = \mu_k$.

137. (a) If a function $F(x)$ possesses the expansion

$$F(x) = A_0 x + \sum_{k=1}^{\infty} A_k \sin \mu_k x \qquad (0 < x < 1),$$

where $\tan \mu_k = \mu_k$, obtain the solution of the problem

$$\frac{d^2y}{dx^2} + \lambda y = F(x); \qquad y(0) = 0, \quad y(1) = y'(1),$$

in the form

$$y(x) = \frac{A_0}{\lambda} x + \sum_{k=1}^{\infty} \frac{A_k}{\lambda - \mu_k^2} \sin \mu_k x \qquad (0 < x < 1),$$

when $\lambda \neq 0, \mu_1^2, \mu_2^2, \cdots$.

(b) Use the result of Problem 136 to obtain this solution in the special case when $F(x) = 1$.

138. Suppose that y is a nontrivial solution of the equation

$$\mathscr{L}y + \lambda ry = 0 \qquad (a < x < b)$$

where $\mathscr{L}$ is defined by equation (321), with p, q, and r real.

(a) By multiplying the equal members of this equation by the complex conjugate function $\bar{y}$, integrating the resultant relation over (a, b), and transforming the result by integration by parts, show that

$$\lambda \int_a^b ry\bar{y}\, dx = \int_a^b py'\bar{y}'\, dx - \int_a^b qy\bar{y}\, dx - [p\bar{y}y']_a^b.$$

(b) If $p(x)$ and $r(x)$ are *positive* and $q(x)$ *nonpositive* over (a, b) and if y satisfies the end conditions

$$\alpha_1 y(a) + \beta_1 y'(a) = 0, \qquad \alpha_2 y(b) + \beta_2 y'(b) = 0,$$

show that λ is real and positive when also α_1 and β_1 are of the *same* sign (or $\alpha_1\beta_1 = 0$) and α_2 and β_2 are of *opposite* sign (or $\alpha_2\beta_2 = 0$). [Thus, under these conditions, the characteristic values of λ are real and positive and the corresponding characteristic functions can be taken to be real (by rejecting a permissible imaginary multiplicative constant).]

CHAPTER TWO

Calculus of Variations and Applications

2.1. Maxima and minima. Applications of the *calculus of variations* are concerned chiefly with the determination of maxima and minima of certain expressions involving unknown functions. Certain techniques involved are analogous to procedures in the differential calculus, which are briefly reviewed in this section.

An important problem in the differential calculus is that of determining maximum and minimum values of a function $y = f(x)$ for values of x in a certain interval (a, b). If in that interval $f(x)$ has a continuous derivative, it is recalled that a *necessary* condition for the existence of a maximum or minimum at a point x_0 *inside* (a, b) is that $dy/dx = 0$ at x_0. A *sufficient* condition that y be a maximum (or a minimum) at x_0, relative to values at neighboring points, is that, in addition, $d^2y/dx^2 < 0$ (or $d^2y/dx^2 > 0$) at that point.

If z is a function of two independent variables, say $z = f(x, y)$, in a region $\mathscr{R}$, and if the partial derivatives $\partial z/\partial x$ and $\partial z/\partial y$ exist and are continuous throughout $\mathscr{R}$, then *necessary* conditions that z possess a relative maximum or minimum at an interior point (x_0, y_0) of $\mathscr{R}$ are that $\partial z/\partial x = 0$ and $\partial z/\partial y = 0$ simultaneously at (x_0, y_0). These two requirements are equivalent to the single requirement that

$$dz = \frac{\partial z}{\partial x}\,dx + \frac{\partial z}{\partial y}\,dy = 0$$

at a point (x_0, y_0), for arbitrary values of both dx and dy. *Sufficient* conditions for either a maximum or a minimum involve certain inequalities among the second partial derivatives (see Problem 1).

More generally, a *necessary* condition that a continuously differentiable function $f(x_1, x_2, \cdots, x_n)$ of n variables $x_1, x_2, \cdots, x_n$ have a relative maximum or minimum value at an interior point of a region is that

$$df = \frac{\partial f}{\partial x_1} dx_1 + \frac{\partial f}{\partial x_2} dx_2 + \cdots + \frac{\partial f}{\partial x_n} dx_n = 0 \tag{1}$$

at that point, for all *permissible* values of the differentials $dx_1, \cdots, dx_n$.

At a point satisfying (1) the function f is said to be *stationary*.

If the n variables are all independent, the n differentials can be assigned arbitrarily, and it follows that (1) then is equivalent to the n conditions

$$\frac{\partial f}{\partial x_1} = \frac{\partial f}{\partial x_2} = \cdots = \frac{\partial f}{\partial x_n} = 0. \tag{2}$$

Sufficient conditions that values of the variables satisfying (1) or (2) actually determine maxima (or minima) involve certain inequalities among the higher partial derivatives (see Problem 1).

Suppose, however, that the n variables are *not* independent, but are related by, say, N conditions each of the form

$$\phi_k(x_1, \cdots, x_n) = 0.$$

Then, at least theoretically, these N equations generally can be solved to express N of the variables in terms of the $n - N$ remaining variables, and hence to express f and df in terms of $n - N$ *independent* variables and their differentials. Alternatively, N linear relations among the n *differentials* can be obtained by differentiation. These conditions permit the expression of N of the *differentials* as linear combinations of the differentials of the $n - N$ independent variables. If (1) is expressed in terms of these differentials, their coefficients must then vanish, giving $n - N$ conditions for stationary values of f which supplement the N constraint conditions.

A procedure which is often still more convenient in this case consists of the introduction of the so-called *Lagrange multipliers*. To illustrate their use, we consider here the problem of obtaining stationary values of $f(x, y, z)$,

$$df = f_x \, dx + f_y \, dy + f_z \, dz = 0, \tag{3}$$

subject to the two constraints

$$\phi_1(x, y, z) = 0, \tag{4a}$$

$$\phi_2(x, y, z) = 0. \tag{4b}$$

Since the three variables x, y, z must satisfy the two auxiliary conditions (4a,b), only one variable can be considered as independent. Equations (4a,b) imply the differential relations

$$\phi_{1x} \, dx + \phi_{1y} \, dy + \phi_{1z} \, dz = 0, \tag{5a}$$

$$\phi_{2x} \, dx + \phi_{2y} \, dy + \phi_{2z} \, dz = 0. \tag{5b}$$

The procedure outlined above would consist of first solving (5a,b) for, say, dx and dy in terms of dz (if this is possible) and of introducing the results into (3), to give a result of the form

$$df \equiv (\cdots)\, dz = 0.$$

Since dz can be assigned arbitrarily, the vanishing of the indicated expression in parentheses in this form is the desired condition that f be stationary when (4a,b) are satisfied.

As an alternative procedure, we first multiply (5a) and (5b) respectively by the quantities λ_1 and λ_2, to be specified presently, and add the results to (3). Since the right-hand members are all zeros, there follows

$$(f_x + \lambda_1\phi_{1x} + \lambda_2\phi_{2x})\, dx + (f_y + \lambda_1\phi_{1y} + \lambda_2\phi_{2y})\, dy + (f_z + \lambda_1\phi_{1z} + \lambda_2\phi_{2z})\, dz = 0, \qquad (6)$$

for arbitrary values of λ_1 *and* λ_2. Now let λ_1 and λ_2 be determined so that two of the parentheses in (6) vanish.* Then the differential multiplying the remaining parenthesis can be arbitrarily assigned, and hence that parenthesis must also vanish. Thus we must have

$$\left.\begin{aligned} \frac{\partial f}{\partial x} + \lambda_1 \frac{\partial \phi_1}{\partial x} + \lambda_2 \frac{\partial \phi_2}{\partial x} = 0, \\ \frac{\partial f}{\partial y} + \lambda_1 \frac{\partial \phi_1}{\partial y} + \lambda_2 \frac{\partial \phi_2}{\partial y} = 0, \\ \frac{\partial f}{\partial z} + \lambda_1 \frac{\partial \phi_1}{\partial z} + \lambda_2 \frac{\partial \phi_2}{\partial z} = 0 \end{aligned}\right\}. \qquad (7a,b,c)$$

Equations (7a,b,c) and (4a,b) comprise five equations determining x, y, z and λ_1, λ_2. The quantities λ_1 and λ_2 are known as *Lagrange multipliers*. Their introduction frequently simplifies the relevant algebra in problems of the type just considered. In many applications they are found to have physical significance as well. We notice that *the conditions* (7) *are the conditions that* $f + \lambda_1\phi_1 + \lambda_2\phi_2$ *be stationary when no constraints are present.*

The procedure outlined is applicable without modification to the general case of n variables and $N < n$ constraints.

In illustration of the method, we attempt to determine the point on the curve of intersection of the surfaces

$$z = xy + 5, \qquad x + y + z = 1 \qquad (8a,b)$$

which is nearest the origin. Thus, we must minimize the quantity

$$f = x^2 + y^2 + z^2$$

* It can be shown that if this were not possible, then the functions ϕ_1 and ϕ_2 would be *functionally dependent*, so that the two constraints (4a) and (4b) would be either equivalent or incompatible.

subject to the two constraints (8a,b). With

$$\phi_1 = z - xy - 5, \qquad \phi_2 = x + y + z - 1,$$

equations (7a,b,c) take the form

$$\left.\begin{aligned} 2x - \lambda_1 y + \lambda_2 &= 0, \\ 2y - \lambda_1 x + \lambda_2 &= 0, \\ 2z + \lambda_1 \quad + \lambda_2 &= 0 \end{aligned}\right\}. \tag{9a,b,c}$$

The elimination of λ_1 and λ_2 from equations (9a,b,c) yields the two alternatives

$$x + y - z + 1 = 0 \qquad \textit{or} \qquad x = y. \tag{10a,b}$$

The simultaneous solution of (8a,b) and (10a) leads to the coordinates of the two points $(2, -2, 1)$ and $(-2, 2, 1)$, which are each three units distant from the origin, whereas the equations (8a,b) and (10b) have *no* real common solution. Geometrical considerations indicate that there is indeed at least one point nearest the origin; since the two points obtained are *necessarily* the only possible ones, they must accordingly be the points required.

As an illustration closely related to certain topics in Chapter 1, we may seek those points on a central quadric surface

$$\phi \equiv a_{11}x^2 + a_{22}y^2 + a_{33}z^2 + 2a_{12}xy + 2a_{23}yz + 2a_{13}xz = \text{constant}$$

for which distance from the origin is maximum or minimum relative to neighboring points. We are thus to render the form

$$f \equiv x^2 + y^2 + z^2$$

stationary, subject to the constraint $\phi =$ constant. Here, if we denote the Lagrange multiplier by $-1/\lambda$, the requirement that $\phi - \lambda f$ be stationary leads to the conditions

$$\left.\begin{aligned} a_{11}x + a_{12}y + a_{13}z &= \lambda x, \\ a_{12}x + a_{22}y + a_{23}z &= \lambda y, \\ a_{13}x + a_{23}y + a_{33}z &= \lambda z \end{aligned}\right\}.$$

This set of equations comprises a characteristic-value problem of the type discussed in Section 1.12. Each "characteristic value" of λ, for which a nontrivial solution exists, leads to the three coordinates of one or more *points* (x, y, z), determined within a common arbitrary multiplicative factor which is available for the satisfaction of the equation of the surface. Section 1.21 shows that it is always possible to rotate the coordinate axes in such a way that each new axis coincides with the direction from the origin to such a point, and that the equation of the surface, referred to the new axes, then involves only squares of the new coordinates. That is, the new axes (which coincide with the "characteristic vectors" of the problem) are the *principal axes* of the quadric surface. The characteristic values of λ are inversely proportional to the squares of the semiaxes. Repeated roots of the characteristic equation correspond to surfaces of revolution, in which cases the new axes can be so chosen in infinitely many ways, while zero roots correspond to surfaces which extend infinitely far from the origin.

The basic problem in the *calculus of variations* is to determine a *function* such that a certain definite integral involving that function and certain of its derivatives takes on a maximum or minimum value. The elementary part of the theory is concerned with a *necessary* condition (generally in the form of a differential equation with boundary conditions) which the required function must satisfy. To show mathematically that the function obtained actually maximizes (or minimizes) the integral is much more difficult than in the corresponding problems of differential calculus. *Sufficient* conditions are developed in more advanced works. In physically motivated problems, such additional considerations frequently may be avoided.

As an example of a problem of this sort, we notice that in order to determine the surface of revolution, obtained by rotating about the x axis a curve passing through two given points (x_1, y_1) and (x_2, y_2), which has minimum surface area, we must determine the function $y(x)$ which specifies the curve to be revolved, in such a way that the integral

$$I = 2\pi \int_{x_1}^{x_2} y(1 + y'^2)^{1/2}\, dx$$

is a minimum, and also so that $y(x_1) = y_1$ and $y(x_2) = y_2$. Here it is assumed that y_1 and y_2 are nonnegative.

In most cases it is to be required that the function and the derivatives explicitly involved be continuous in the region of definition.

2.2. The simplest case. We now consider the problem of determining a continuously differentiable function $y(x)$ for which the integral

$$I = \int_{x_1}^{x_2} F(x, y, y')\, dx \tag{11}$$

takes on a maximum or minimum value,* and which satisfies the prescribed end conditions

$$y(x_1) = y_1, \qquad y(x_2) = y_2.$$

To fix ideas, we may suppose that I is to be *minimized*.

Suppose that $y(x)$ is the actual minimizing function, and choose *any* continuously differentiable function $\eta(x)$ which *vanishes* at the end points

* We suppose that F has continuous second partial derivatives with respect to its three arguments.

$x = x_1$ and $x = x_2$. Then for any constant ϵ the function $y(x) + \epsilon\eta(x)$ will satisfy the end conditions (Figure 2.1). The integral

$$I(\epsilon) = \int_{x_1}^{x_2} F(x, y + \epsilon\eta, y' + \epsilon\eta')\,dx, \tag{12}$$

obtained by replacing y by $y + \epsilon\eta$ in (11), is then a function of ϵ, once y and η are assigned, which takes on its minimum value when $\epsilon = 0$. But this is possible only if

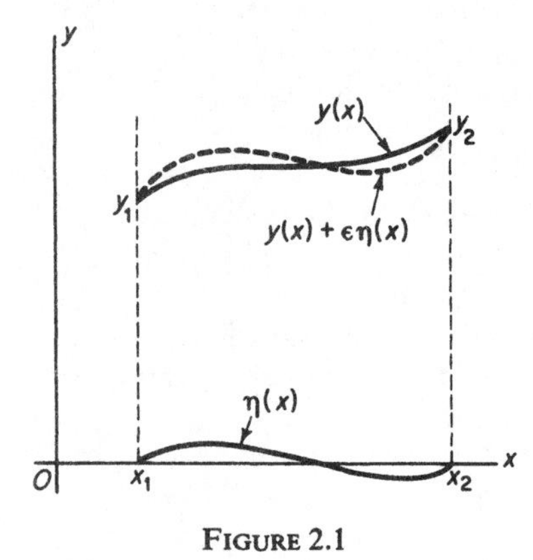

FIGURE 2.1

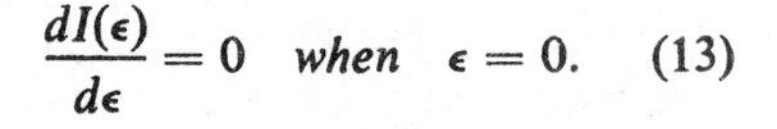

$$\frac{dI(\epsilon)}{d\epsilon} = 0 \quad when \quad \epsilon = 0. \tag{13}$$

If we denote the integrand in (12) by $\tilde{F}$,

$$\tilde{F} = F(x, y + \epsilon\eta, y' + \epsilon\eta'),$$

and notice that

$$\frac{d\tilde{F}}{d\epsilon} = \frac{\partial \tilde{F}}{\partial y}\eta + \frac{\partial \tilde{F}}{\partial y'}\eta',$$

we obtain from (12) the result

$$\frac{dI(\epsilon)}{d\epsilon} = \int_{x_1}^{x_2} \left(\frac{\partial \tilde{F}}{\partial y}\eta + \frac{\partial \tilde{F}}{\partial y'}\frac{d\eta}{dx}\right) dx,$$

by differentiating under the integral sign. Finally, since $\tilde{F} \to F$ when $\epsilon \to 0$, and the same is true of the partial derivatives, the necessary condition (13) takes the form

$$\int_{x_1}^{x_2} \left(\frac{\partial F}{\partial y}\eta + \frac{\partial F}{\partial y'}\frac{d\eta}{dx}\right) dx = 0. \tag{14}$$

The next step in the development consists of integrating the second term by parts, to transform (14) to the condition

$$\int_{x_1}^{x_2} \left[\frac{\partial F}{\partial y}\eta - \frac{d}{dx}\left(\frac{\partial F}{\partial y'}\right)\eta\right] dx + \left[\frac{\partial F}{\partial y'}\eta(x)\right]_{x_1}^{x_2} = 0. \tag{15}$$

But since $\eta(x)$ vanishes at the end points, by assumption, the integrated terms vanish and (15) becomes

$$\int_{x_1}^{x_2} \left[\frac{\partial F}{\partial y} - \frac{d}{dx}\left(\frac{\partial F}{\partial y'}\right)\right]\eta\,dx = 0. \tag{16}$$

Finally, since $\eta(x)$ is arbitrary, we conclude that its coefficient in (16) must vanish identically over (x_1, x_2). For if this were not so we could choose a continuously differentiable function $\eta(x)$, which vanishes at the ends of the interval, in such a way that the (continuous) integrand in (16) is positive whenever it is not zero,* and a contradiction would be obtained.

The end result is that if $y(x)$ minimizes (or maximizes) the integral (11), it must satisfy the *Euler equation*

$$\frac{d}{dx}\left(\frac{\partial F}{\partial y'}\right) - \frac{\partial F}{\partial y} = 0. \tag{17a}$$

Here the partial derivatives $\partial F/\partial y$ and $\partial F/\partial y'$ have been formed by treating x, y, and y' as independent variables. Since $\partial F/\partial y'$ is, in general, a function of x explicitly and also implicitly through y and $y' = dy/dx$, the first term in (17a) can be written in the expanded form

$$\frac{\partial}{\partial x}\left(\frac{\partial F}{\partial y'}\right) + \frac{\partial}{\partial y}\left(\frac{\partial F}{\partial y'}\right)\frac{dy}{dx} + \frac{\partial}{\partial y'}\left(\frac{\partial F}{\partial y'}\right)\frac{dy'}{dx}.$$

Thus (17a) is equivalent to the equation

$$F_{y'y'}\frac{d^2y}{dx^2} + F_{y'y}\frac{dy}{dx} + (F_{y'x} - F_y) = 0. \tag{17b}$$

This equation is of second order in y unless $F_{y'y'} = \partial^2 F/\partial y'^2 \equiv 0$, so that in general two constants are available for the satisfaction of the end conditions.

It is useful to notice that (17b) is equivalent to the form

$$\frac{1}{y'}\left[\frac{d}{dx}\left(F - \frac{\partial F}{\partial y'}\frac{dy}{dx}\right) - \frac{\partial F}{\partial x}\right] = 0, \tag{17c}$$

as can be verified by expansion (see also Problem 7). From this result it follows that if F does not involve x *explicitly* a first integral of Euler's equation is

$$F - y'\frac{\partial F}{\partial y'} = C \quad \textit{if} \quad \frac{\partial F}{\partial x} \equiv 0, \tag{18a}$$

while (17a) shows that if F does not involve y explicitly a first integral is

$$\frac{\partial F}{\partial y'} = C \quad \textit{if} \quad \frac{\partial F}{\partial y} \equiv 0. \tag{18b}$$

Solutions of Euler's equation are known as *extremals* of the problem considered. In general, they comprise a two-parameter family of functions in the case just treated.

An extremal which satisfies the appropriate end conditions is often called a *stationary function* of the variational problem, and is said to make the relevant integral *stationary*, whether or not it also makes the integral *maximum* or *minimum*, relative to all slightly varied admissible functions.

* This fact, which is intuitively plausible, can be proved analytically.

Thus, by definition, the integral I of equation (11) will be said to be *stationary* when $y(x)$ is so determined that equation (15) holds for every permissible function $\eta(x)$.*

2.3. Illustrative examples. In Section 2.1 it was pointed out that to find the minimal surface of revolution passing through two given points it is necessary to minimize the integral

$$\frac{I}{2\pi} = \int_{x_1}^{x_2} y(1 + y'^2)^{1/2}\, dx. \tag{19}$$

With $F = y(1 + y'^2)^{1/2}$, the Euler equation (17a) becomes

$$\frac{d}{dx}\left[\frac{yy'}{(1 + y'^2)^{1/2}}\right] - (1 + y'^2)^{1/2} = 0$$

or, after a reduction or use of (17b),

$$yy'' - y'^2 - 1 = 0. \tag{20}$$

Following the usual procedure for solving equations of this type, we set

$$y' = p, \qquad y'' = \frac{dp}{dx} = p\frac{dp}{dy},$$

so that (20) becomes

$$py\frac{dp}{dy} = p^2 + 1.$$

This equation is separable, and is integrated to give

$$y = c_1(1 + p^2)^{1/2} \equiv c_1\left[1 + \left(\frac{dy}{dx}\right)^2\right]^{1/2},$$

as would be obtained more directly by use of (18a), since here F does not explicitly involve x. There follows

$$\frac{dy}{dx} = \left(\frac{y^2}{c_1^{\,2}} - 1\right)^{1/2},$$

and hence finally

$$y = c_1 \cosh\left(\frac{x}{c_1} + c_2\right). \tag{21}$$

Thus, as is well known, the required minimal surface (if it exists) must be obtained by revolving a catenary. It then remains to be seen whether the arbitrary constants c_1 and c_2 can indeed be so chosen that the curve (21) passes through any two assigned points in the upper half plane.

* The usage of the terms "extremal" and "stationary function" varies within the literature.

The determination of these constants is found to involve the solution of a transcendental equation which possesses two, one, or no solutions, depending upon the prescribed values $y(x_1)$ and $y(x_2)$. In particular, it is found that all curves representing (21) and passing through *one* prescribed point $P_1(x_1, y_1)$ are tangent to a certain curve $\mathscr{C}_1$ which passes through the point $(x_1, 0)$, and that no such curve *crosses* $\mathscr{C}_1$. When the second prescribed point $P_2(x_2, y_2)$ is separated from P_1 by $\mathscr{C}_1$, there is accordingly *no* curve representing (21) passing through *both* points, and hence *no* admissible minimal surface of revolution. When P_1 and P_2 are on the same side of $\mathscr{C}_1$, there are *two* "stationary curves," the shorter of which generates a minimal surface. Finally, when P_2 is *on* $\mathscr{C}_1$, there is *one* stationary curve but the surface of revolution which it generates is *not* minimal.

The situations in which there is no admissible minimizing curve are those in which smaller and smaller areas are generated by rotating curves which more and more nearly approach a broken line consisting of segments from (x_1, y_1) to $(x_1, 0)$ to $(x_2, 0)$ to (x_2, y_2), and in which that *unattainable* limiting area is smaller than the area generated by any admissible curve.

Physically, the problem can be interpreted as that of determining the shape of a *soap film* connecting parallel circular wire hoops of radii y_1 and y_2, perpendicular to the x axis, with centers at $(x_1, 0)$ and $(x_2, 0)$. The exceptional cases arise when the separation $x_2 - x_1$ is increased to or beyond the point where the film no longer can *join* the hoops, but breaks into two parts, each then spanning a hoop in its plane.

The classical "elementary" application of the calculus of variations consists of *proving* mathematically that the shortest distance between two points in a plane is a straight line. If the points, in the xy plane, are (x_1, y_1) and (x_2, y_2) and if the equation of the minimizing curve is $y = y(x)$, we are then to minimize

$$I = \int_{x_1}^{x_2} (1 + y'^2)^{1/2}\, dx.$$

Since here $F = (1 + y'^2)^{1/2}$ does not involve either x or y explicitly, either of the forms (18a, b) can be used to give a first integral of Euler's equation directly. However, here it is easier to use the form (17b) to deduce that $y'' = 0$ and consequently $y = c_1 x + c_2$. From this result we can conclude that *if a minimizing curve exists and if it can be specified by an equation of the form $y = y(x)$, then that curve necessarily must be a straight line.* It is clear that the case in which $x_1 = x_2$ is exceptional, and must be treated separately.

In the preceding examples no proof was given that the stationary function obtained actually possesses the required minimizing property. Such considerations comprise most of the less elementary theory of the calculus of variations. In a great number of physically motivated problems it is intuitively clear that a minimizing function does indeed *exist*. Then if the present methods and their extensions show that only the particular function obtained could possibly be the minimizing function, the problem can be considered as solved for practical purposes. If several alternatives (stationary functions) are determined, direct calculation will show which one actually leads to the smaller value of the quantity to be minimized.

In many practical situations the stationary functions are of importance whether or not they maximize or minimize the relevant integral. This fact is illustrated, for example, in Section 2.9.

2.4. Natural boundary conditions and transition conditions. When the value of the unknown function $y(x)$ is *not* preassigned at one or both of the end points $x = x_1, x_2$ the difference $\epsilon\eta(x)$ between the true function $y(x)$ and the varied function $y(x) + \epsilon\eta(x)$ need not vanish there. However, the left-hand member of (15) must vanish when $y(x)$ is identified with the minimizing (or maximizing) function, for *all* permissible variations $\epsilon\eta(x)$. Thus it must vanish, *in particular*, for all variations which *are* zero at both ends. For all such η's the second term in (15) is zero and equation (16) again follows, and yields the Euler equation (17) as before.

Hence the first term in (15) must be zero for *all* permissible η's, since the *coefficient* of η in the integrand must be zero. Thus it follows that the second term in (15) must itself vanish,

$$\left[\frac{\partial F}{\partial y'}\,\eta(x)\right]_{x=x_2} - \left[\frac{\partial F}{\partial y'}\,\eta(x)\right]_{x=x_1} = 0, \tag{22}$$

for all permissible values of $\eta(x_2)$ and $\eta(x_1)$. If $y(x)$ is not preassigned at either end point, then $\eta(x_1)$ and $\eta(x_2)$ are both completely arbitrary and we conclude that their coefficients in (22) must each vanish, yielding the conditions

$$\left[\frac{\partial F}{\partial y'}\right]_{x=x_1} = 0, \qquad \left[\frac{\partial F}{\partial y'}\right]_{x=x_2} = 0 \tag{23a,b}$$

which must be satisfied instead.

The requirements that (23a) hold when $y(x_1)$ is not given, and that (23b) hold when $y(x_2)$ is not given, are often called the *natural boundary conditions* of the problem.* If, for example, $y(x_1)$ were preassigned as y_1 whereas $y(x_2)$ were not given in advance, then the relevant end conditions would be $y(x_1) = y_1$ and $(\partial F/\partial y')_{x=x_2} = 0$, and (23a) would *not* apply.

In some situations the integrand F in (11) is such that one or both of the terms $\partial F/\partial y$ and $d(\partial F/\partial y')/dx$ are discontinuous at one or more points inside the interval (x_1, x_2), but the conditions assumed in the preceding section are satisfied in the subintervals separated by these points. To illustrate the treatment of such cases, we suppose here that there is only *one* point of

* In some references, only the conditions (23a,b) themselves are called the "natural boundary conditions."

discontinuity, at $x = c$. Then the integral (11) must be expressed as the sum of integrals over $(x_1, c-)$ and $(c+, x_2)$ before the steps leading to (15) are taken, and equation (15) is replaced by the relation

$$\int_{x_1}^{c-}\left[\frac{\partial F}{\partial y} - \frac{d}{dx}\left(\frac{\partial F}{\partial y'}\right)\right]\eta\, dx + \int_{c+}^{x_2}\left[\frac{\partial F}{\partial y} - \frac{d}{dx}\left(\frac{\partial F}{\partial y'}\right)\right]\eta\, dx$$
$$+ \left[\frac{\partial F}{\partial y'}\eta(x)\right]_{x_1}^{c-} + \left[\frac{\partial F}{\partial y'}\eta(x)\right]_{c+}^{x_2} = 0. \qquad (24)$$

If we require that the minimizing (or maximizing) function $y(x)$ be *continuous* at $x = c$, and accordingly require that all admissible functions $y(x) + \epsilon\eta(x)$ have the same property, it follows that

$$\eta(c+) = \eta(c-) = \eta(c),$$

so that (24) can be written in the form

$$\int_{x_1}^{c-}\left[\frac{\partial F}{\partial y} - \frac{d}{dx}\left(\frac{\partial F}{\partial y'}\right)\right]\eta\, dx + \int_{c+}^{x_2}\left[\frac{\partial F}{\partial y} - \frac{d}{dx}\left(\frac{\partial F}{\partial y'}\right)\right]\eta\, dx + \left[\frac{\partial F}{\partial y'}\right]_{x_2}\eta(x_2)$$
$$- \left[\frac{\partial F}{\partial y'}\right]_{x_1}\eta(x_1) - \left\{\left[\frac{\partial F}{\partial y'}\right]_{c+} - \left[\frac{\partial F}{\partial y'}\right]_{c-}\right\}\eta(c) = 0. \qquad (25)$$

Hence we may deduce that the Euler equation (17) must hold in each of the subintervals (x_1, c) and (c, x_2), that $\partial F/\partial y'$ must vanish at any end point $x = x_1$ or $x = x_2$ where y is not prescribed, as before, and also that the *natural transition conditions*

$$y(c+) = y(c-), \qquad \lim_{x\to c+}\frac{\partial F}{\partial y'} = \lim_{x\to c-}\frac{\partial F}{\partial y'} \qquad (26\text{a,b})$$

must be satisfied at the point $x = c$. Whereas (26a) represents the requirement that y itself be continuous at $x = c$, the condition (26b) may demand that the *derivative* y' be *discontinuous* at that point.

To illustrate the preceding considerations, we consider the determination of stationary functions associated with the integral

$$I = \int_0^1 (Ty'^2 - \rho\omega^2 y^2)\, dx, \qquad (27)$$

where T, ρ, and ω are given constants or functions of x. The Euler equation is

$$\frac{d}{dx}\left(T\frac{dy}{dx}\right) + \rho\omega^2 y = 0, \qquad (28)$$

regardless of what (if anything) is prescribed in advance at the ends of the interval.

Thus, in particular, when T, ρ, and ω are *positive constants*, the extremals are of the form

$$y = c_1 \cos \alpha x + c_2 \sin \alpha x \qquad \left(\alpha^2 = \frac{\rho\omega^2}{T}\right), \tag{29}$$

where c_1 and c_2 are constants. When the conditions

$$y(0) = 0, \qquad y(1) = 1$$

are prescribed, there follows $c_1 = 0$; $c_2 \sin \alpha = 1$ and hence

$$y = \frac{\sin \alpha x}{\sin \alpha} \qquad (\alpha \neq \pi, 2\pi, \cdots). \tag{30}$$

When the condition

$$y(1) = 1$$

is prescribed, but $y(0)$ is *not* preassigned, the appropriate condition at $x = 0$ follows from (23b), with $F = Ty'^2 - \rho\omega^2 y^2$, in the form

$$Ty'(0) = 0$$

and hence

$$y = \frac{\cos \alpha x}{\cos \alpha} \qquad \left(\alpha \neq \frac{\pi}{2}, \frac{3\pi}{2}, \cdots\right). \tag{31}$$

When neither $y(0)$ nor $y(1)$ is prescribed, the conditions (23a,b) require

$$Ty'(0) = Ty'(1) = 0$$

and hence

$$y = \begin{cases} 0 & (\alpha \neq \pi, 2\pi, \cdots), \\ c_1 \cos \alpha x & (\alpha = \pi, 2\pi, \cdots), \end{cases} \tag{32}$$

where c_1 is arbitrary. In the exceptional cases noted in (30) and (31), no stationary function exists. The limiting cases in which $\alpha = 0$ must be treated separately, since (29) is incomplete in that case.

If $T = T_1$ and $\rho = \rho_1$ when $0 \leqq x < c$ whereas $T = T_2$ and $\rho = \rho_2$ when $c < x \leqq 1$, where T_1, T_2, ρ_1, ρ_2, and ω are positive constants, and if the conditions

$$y(0) = 0, \qquad y(1) = 1$$

are prescribed, there follows

$$y = \begin{cases} c_1 \cos \alpha_1 x + c_2 \sin \alpha_1 x & (0 \leqq x < c), \\ d_1 \cos \alpha_2 x + d_2 \sin \alpha_2 x & (c < x \leqq 1), \end{cases}$$

where $\alpha_i^2 = \rho_i \omega^2 / T_i$. The natural transition conditions (26a,b) also give

$$\lim_{x \to c-} y(x) = \lim_{x \to c+} y(x), \qquad T_1 \lim_{x \to c-} y'(x) = T_2 \lim_{x \to c+} y'(x).$$

Thus we have four conditions which are to be satisfied by the four constants of integration, and a stationary function is determined provided that α does not take on one of a certain infinite set of exceptional values (which correspond to the vanishing of the determinant of a certain coefficient matrix).

2.5. The variational notation. We next introduce the notation of "variations" in order to establish more clearly the analogy between the calculus of variations and the differential calculus.

Suppose that we consider a set $\mathscr{S}$ of functions satisfying certain conditions. For example, we might define $\mathscr{S}$ to be the set of all functions of a single variable x which possess a continuous first derivative at all points in an interval $a \leqq x \leqq b$. Then any quantity which takes on a specific numerical value corresponding to each function in $\mathscr{S}$ is said to be a *functional* on the set $\mathscr{S}$.

In illustration, we may speak of the quantities

$$I_1 = \int_a^b y(x)\,dx, \qquad I_2 = \int_a^b \{y(x)y''(x) - [y'(x)]^2\}\,dx$$

as functionals, since corresponding to any function $y(x)$ for which the indicated operations are defined each quantity has a definite numerical value.

With the above definition, it is proper also to speak of such quantities as $f[y(x)]$ and $g[x, y(x), y'(x), \cdots, y^{(n)}(x)]$ as functionals in those cases when the variable x is considered as fixed in a given discussion and the function $y(x)$ is varied.

In Section 2.2, we considered an integrand of the form

$$F = F(x, y, y')$$

which *for a fixed value of* x depends upon the *function* $y(x)$ and its derivative. We then changed the function $y(x)$, to be determined, into a new function $y(x) + \epsilon\eta(x)$. The change $\epsilon\eta(x)$ in $y(x)$ is called the *variation* of y and is conventionally denoted by δy,

$$\delta y \equiv \epsilon\eta(x). \tag{33}$$

Corresponding to this change in $y(x)$, for a fixed value of x, the functional F changes by an amount ΔF, where

$$\Delta F = F(x, y + \epsilon\eta, y' + \epsilon\eta') - F(x, y, y'). \tag{34}$$

If the right-hand member is expanded in powers of ϵ, there follows

$$\Delta F = \frac{\partial F}{\partial y}\epsilon\eta + \frac{\partial F}{\partial y'}\epsilon\eta' + (\text{terms involving higher powers of } \epsilon). \tag{35}$$

In analogy with the definition of the *differential*, the first two terms in the right-hand member of (35) are *defined* to be the *variation of F*,

$$\delta F = \frac{\partial F}{\partial y}\epsilon\eta + \frac{\partial F}{\partial y'}\epsilon\eta'. \tag{36}$$

In the special case when $F = y$, this definition is properly consistent with (33). Further, when $F = y'$ it yields the additional relation

$$\delta y' = \epsilon \eta', \tag{37}$$

so that (36) can be rewritten in the form

$$\delta F = \frac{\partial F}{\partial y}\,\delta y + \frac{\partial F}{\partial y'}\,\delta y'. \tag{38}$$

For a complete analogy with the definition of the differential, we would perhaps have anticipated the definition

$$\delta F = \frac{\partial F}{\partial x}\,\delta x + \frac{\partial F}{\partial y}\,\delta y + \frac{\partial F}{\partial y'}\,\delta y'.$$

But here x *is not varied*, so that we have

$$\delta x \equiv 0, \tag{39}$$

and hence the analogy is indeed complete.

We notice that the *differential* of a function is a first-order approximation to the change in that function *along a particular curve*, while the *variation* of a functional is a first-order approximation to the change *from curve to curve*.

It is easily verified directly, from the definition, that the laws of variation of sums, products, ratios, powers, and so forth, are completely analogous to the corresponding laws of differentiation. Thus, for example, there follows

$$\delta(F_1 F_2) = F_1\,\delta F_2 + F_2\,\delta F_1,$$
$$\delta\left(\frac{F_1}{F_2}\right) = \frac{F_2\,\delta F_1 - F_1\,\delta F_2}{F_2{}^2},$$

and so forth.

Analogous definitions are introduced in the more general case. Thus, for example, if x and y are independent variables, and u and v are dependent variables, we may consider a functional

$$F = F(x, y, u, v, u_x, u_y, v_x, v_y).$$

We now vary both u and v, holding x and y fixed, into new functions $u + \epsilon\xi$ and $v + \epsilon\eta$, and define the variations of u and v as follows:

$$\delta u = \epsilon\xi(x, y), \qquad \delta v = \epsilon\eta(x, y). \tag{40}$$

The change in F is then found (by expansion in powers of ϵ) to be

$$\Delta F = \frac{\partial F}{\partial u}\,\epsilon\xi + \frac{\partial F}{\partial v}\,\epsilon\eta + \frac{\partial F}{\partial u_x}\,\epsilon\xi_x + \frac{\partial F}{\partial u_y}\,\epsilon\xi_y + \frac{\partial F}{\partial v_x}\,\epsilon\eta_x + \frac{\partial F}{\partial v_y}\,\epsilon\eta_y$$
$$+ \text{(terms involving higher powers of } \epsilon). \tag{41}$$

The first-order terms are *defined* to comprise the variation of F. Hence, using (40), we have the definition

$$\delta F = \frac{\partial F}{\partial u}\,\delta u + \frac{\partial F}{\partial v}\,\delta v + \frac{\partial F}{\partial u_x}\,\delta u_x + \frac{\partial F}{\partial u_y}\,\delta u_y + \frac{\partial F}{\partial v_x}\,\delta v_x + \frac{\partial F}{\partial v_y}\,\delta v_y. \quad (42)$$

Since the independent variables x and y are held fixed, there follows also

$$\delta x = \delta y \equiv 0. \quad (43)$$

From (33) and (37) we obtain the result

$$\frac{d}{dx}(\delta y) = \epsilon\,\frac{d\eta}{dx} = \delta\,\frac{dy}{dx}.$$

Hence, *if x is the independent variable* (and, accordingly, $\delta x \equiv 0$) *the operators δ and d/dx are commutative:*

$$\frac{d}{dx}\,\delta y = \delta\,\frac{dy}{dx}. \quad (44)$$

Similarly, from (40) and (42) we can deduce that *if x and y are independent variables* (and hence $\delta x = \delta y \equiv 0$) *the operators δ and $\partial/\partial x$ or $\partial/\partial y$ are commutative:*

$$\frac{\partial}{\partial x}\,\delta u = \delta\,\frac{\partial u}{\partial x}, \qquad \frac{\partial}{\partial y}\,\delta u = \delta\,\frac{\partial u}{\partial y}. \quad (45)$$

That is, *the derivative of the variation with respect to an independent variable is the same as the variation of the derivative.*

It should be noticed that this is not generally true unless the differentiation is with respect to an independent variable. For if x and y are both functions of an independent variable t we may write

$$\frac{dy}{dx} = \frac{dy/dt}{dx/dt} = \frac{y'}{x'}$$

where a prime now denotes t differentiation. Thus we then have

$$\delta\,\frac{dy}{dx} = \delta\left(\frac{y'}{x'}\right) = \frac{x'\,\delta y' - y'\,\delta x'}{x'^2}.$$

But now δ and d/dt *are* commutative, so that

$$\delta\,\frac{dy}{dx} = \frac{\dfrac{dx}{dt}\dfrac{d}{dt}(\delta y) - \dfrac{dy}{dt}\dfrac{d}{dt}(\delta x)}{\left(\dfrac{dx}{dt}\right)^2} = \frac{\dfrac{d}{dt}(\delta y)}{\dfrac{dx}{dt}} - \frac{\dfrac{dy}{dt}\dfrac{d}{dt}(\delta x)}{\dfrac{dx}{dt}\dfrac{dx}{dt}}$$

or, finally,

$$\delta \frac{dy}{dx} = \frac{d}{dx}\,\delta y - \frac{dy}{dx}\frac{d}{dx}\,\delta x. \tag{44'}$$

If $\delta x \equiv 0$, equation (44′) reduces to (44).

The quantity δF is sometimes called the *first* variation of F, the *second* variation then being defined as the group of second-order terms in ϵ in (35) or (41). However, when the term "variation" is used alone the *first* variation is generally implied.

For a functional expressed as a definite integral,

$$I = \int_{x_1}^{x_2} F(x, y, y')\,dx, \tag{46a}$$

where x is the independent variable in F, there follows from the definition

$$\delta I = \delta \int_{x_1}^{x_2} F\,dx = \int_{x_1}^{x_2} \delta F\,dx, \tag{46b}$$

so that variation and integration between limits (which are not to be varied) are commutative processes.

We now show that *the integral*

$$I = \int_{x_1}^{x_2} F(x, y, y')\,dx \tag{47}$$

is stationary if and only if its (first) variation vanishes,

$$\delta I \equiv \delta \int_{x_1}^{x_2} F(x, y, y')\,dx = 0, \tag{48}$$

for every permissible variation δy.

According to (46) and (38), equation (48) is equivalent to

$$\delta I = \int_{x_1}^{x_2} \delta F(x, y, y')\,dx = \int_{x_1}^{x_2} \left[\frac{\partial F}{\partial y}\,\delta y + \frac{\partial F}{\partial y'}\,\delta y'\right] dx = 0.$$

If we replace $\delta y'$ by $d(\delta y)/dx$, in accordance with (44), and integrate by parts, there follows

$$\delta I = \int_{x_1}^{x_2} \left[\frac{\partial F}{\partial y} - \frac{d}{dx}\left(\frac{\partial F}{\partial y'}\right)\right] \delta y\,dx + \left[\frac{\partial F}{\partial y'}\,\delta y\right]_{x_1}^{x_2}. \tag{49}$$

But the right-hand member of (49) is proportional to the left-hand member of (15), as was to be shown.

Thus a *stationary function* for an *integral functional* is one for which the *variation* of that integral is zero, just as a *stationary point* of a *function* is one at which the *differential* of the function is zero.

The use of the variational notation leads to concise derivations and computations. This notation will be used in the remainder of this chapter; its justification in any particular case follows the lines of the preceding argument.

2.6. The more general case. We next consider the case when the integral to be maximized or minimized is of the form

$$I = \iint_{\mathscr{R}} F(x, y, u, v, u_x, u_y, v_x, v_y)\, dx\, dy. \tag{50}$$

Here x and y are independent variables, u and v are continuously differentiable functions of x and y to be determined, and the integration is carried out over a simple two-dimensional region $\mathscr{R}$ of the xy plane.

The condition

$$\delta I = 0 \tag{51}$$

then becomes

$$\delta I = \iint_{\mathscr{R}} \left[\left(\frac{\partial F}{\partial u}\,\delta u + \frac{\partial F}{\partial u_x}\,\delta u_x + \frac{\partial F}{\partial u_y}\,\delta u_y\right) + \left(\frac{\partial F}{\partial v}\,\delta v + \frac{\partial F}{\partial v_x}\,\delta v_x + \frac{\partial F}{\partial v_y}\,\delta v_y\right)\right] dx\, dy = 0. \tag{52}$$

Here the variations δu and δv are to be continuously differentiable over $\mathscr{R}$ and are to *vanish* on the boundary $\mathscr{C}$ when u and v are *prescribed* on $\mathscr{C}$, but are otherwise completely arbitrary.

In order to transform the terms involving variations of derivatives, we make use of the formulas*

$$\iint_{\mathscr{R}} \frac{\partial \phi}{\partial x}\, dx\, dy = \oint_{\mathscr{C}} \phi \cos \nu\, ds, \qquad \iint_{\mathscr{R}} \frac{\partial \phi}{\partial y}\, dx\, dy = \oint_{\mathscr{C}} \phi \sin \nu\, ds, \tag{53a,b}$$

in which ν represents the angle between the positive x axis and the outward normal at a point of the boundary $\mathscr{C}$ of $\mathscr{R}$ and s is arc length along $\mathscr{C}$.

The general procedure may be illustrated by considering the treatment of a typical term:

$$\iint_{\mathscr{R}} \frac{\partial F}{\partial u_x}\,\delta\!\left(\frac{\partial u}{\partial x}\right) dx\, dy = \iint_{\mathscr{R}} \frac{\partial F}{\partial u_x}\frac{\partial}{\partial x}(\delta u)\, dx\, dy$$
$$= \iint_{\mathscr{R}} \frac{\partial}{\partial x}\left(\frac{\partial F}{\partial u_x}\,\delta u\right) dx\, dy - \iint_{\mathscr{R}} \frac{\partial}{\partial x}\left(\frac{\partial F}{\partial u_x}\right) \delta u\, dx\, dy.$$

* These formulas are specializations of *Green's theorem* in the plane and are valid (in particular) when ϕ, $\partial\phi/\partial x$, and $\partial\phi/\partial y$ are continuous inside and on $\mathscr{C}$, and when $\mathscr{C}$ is a piecewise smooth closed curve which does not cross itself.

If the first of the last two integrals is transformed by use of (53a), there follows (see Figure 2.2)

$$\iint_{\mathscr{R}} \frac{\partial F}{\partial u_x} \delta u_x \, dx\, dy = \oint_{\mathscr{C}} \frac{\partial F}{\partial u_x} \cos \nu \, \delta u \, ds - \iint_{\mathscr{R}} \frac{\partial}{\partial x}\left(\frac{\partial F}{\partial u_x}\right) \delta u \, dx \, dy. \quad (54)$$

When the other terms in (52) are treated similarly, (52) takes the form

$$\begin{aligned}\delta I = \oint_{\mathscr{C}}\left[\left(\frac{\partial F}{\partial u_x}\cos\nu + \frac{\partial F}{\partial u_y}\sin\nu\right)\delta u + \left(\frac{\partial F}{\partial v_x}\cos\nu + \frac{\partial F}{\partial v_y}\sin\nu\right)\delta v\right] ds \\ + \iint_{\mathscr{R}}\left\{\left[\frac{\partial F}{\partial u} - \frac{\partial}{\partial x}\left(\frac{\partial F}{\partial u_x}\right) - \frac{\partial}{\partial y}\left(\frac{\partial F}{\partial u_y}\right)\right]\delta u \right. \\ \left. + \left[\frac{\partial F}{\partial v} - \frac{\partial}{\partial x}\left(\frac{\partial F}{\partial v_x}\right) - \frac{\partial}{\partial y}\left(\frac{\partial F}{\partial v_y}\right)\right]\delta v\right\} dx\, dy = 0. \quad (55)\end{aligned}$$

If the variations δu and δv are independent of each other, that is, if u and v can be varied independently, then as in an earlier argument it follows that the coefficients of δu and δv in the integrand of the double integral must *each* vanish identically in $\mathscr{R}$, giving the two Euler equations

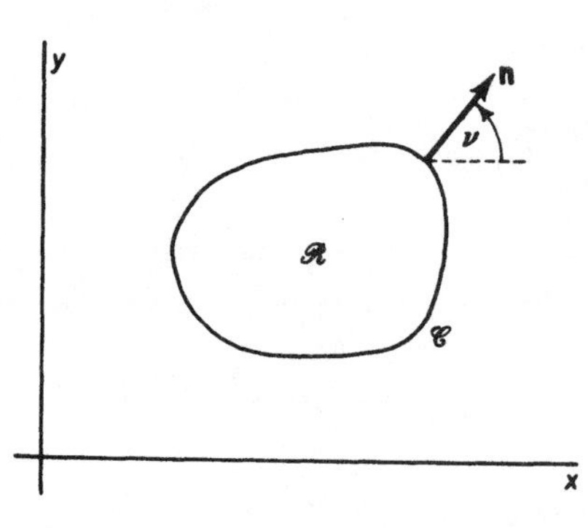

FIGURE 2.2

$$\left.\begin{aligned}\frac{\partial}{\partial x}\left(\frac{\partial F}{\partial u_x}\right) + \frac{\partial}{\partial y}\left(\frac{\partial F}{\partial u_y}\right) - \frac{\partial F}{\partial u} = 0, \\ \frac{\partial}{\partial x}\left(\frac{\partial F}{\partial v_x}\right) + \frac{\partial}{\partial y}\left(\frac{\partial F}{\partial v_y}\right) - \frac{\partial F}{\partial v} = 0\end{aligned}\right\}. \quad (56a,b)$$

Furthermore, the *natural boundary conditions* require that

$$\frac{\partial F}{\partial u_x}\cos\nu + \frac{\partial F}{\partial u_y}\sin\nu = 0 \quad on\, \mathscr{C} \quad (57a)$$

when u is not prescribed on $\mathscr{C}$, and that

$$\frac{\partial F}{\partial v_x}\cos\nu + \frac{\partial F}{\partial v_y}\sin\nu = 0 \quad on\, \mathscr{C} \quad (57b)$$

when v is not given on the boundary.

The conditions (56a,b) comprise two partial differential equations in u and v and are, in general, linear or quasi-linear of second order in u and v, as can be shown by expansion. We notice that in the differentiations with respect to u, v, u_x, u_y, v_x, and v_y all of the *eight* variables listed in (50) are treated as though they were independent. Thus $\partial F/\partial u_x$ is formed by holding x, y, u, v, u_y, v_x, and v_y constant in the expression for F. However, in the differentiations with respect to x or y only these *two* variables are treated as independent. Since, in general, $\partial F/\partial u_x$ will involve x (and y) both explicitly and implicitly, the *first term* in (56a) becomes, on expansion,

$$(F_{u_x})_x + (F_{u_x})_u \frac{\partial u}{\partial x} + (F_{u_x})_v \frac{\partial v}{\partial x} + (F_{u_x})_{u_x} \frac{\partial u_x}{\partial x} + (F_{u_x})_{u_y} \frac{\partial u_y}{\partial x} + (F_{u_x})_{v_x} \frac{\partial v_x}{\partial x} + (F_{u_x})_{v_y} \frac{\partial v_y}{\partial x}$$

or

$$F_{u_x u_x} \frac{\partial^2 u}{\partial x^2} + F_{u_x u_y} \frac{\partial^2 u}{\partial x\, \partial y} + F_{u u_x} \frac{\partial u}{\partial x} + F_{u_x v_x} \frac{\partial^2 v}{\partial x^2} + F_{u_x v_y} \frac{\partial^2 v}{\partial x\, \partial y} + F_{v u_x} \frac{\partial v}{\partial x} + F_{x u_x}.$$

Equations (56a,b) represent *necessary* conditions that I of equation (50) take on a maximum or minimum value. They are subject, of course to the appropriate boundary conditions along $\mathscr{C}$. The formulation of *sufficient* conditions again is extremely involved, and is omitted here.

Completely analogous conditions are obtained in the general case of m dependent and n independent variables. Here m Euler equations analogous to (56) are obtained, each having $n + 1$ terms. Still more generally, partial derivatives of higher order than the first may be involved in the integral to be made stationary. The extension of the present methods to such cases is straightforward.

If the integrand F involves n independent variables $x, y, \cdots$, and m dependent variables $u, v, \cdots$, together with partial derivatives, of various orders, of $u, v, \cdots$ with respect to $x, y, \cdots$, one obtains one Euler equation for each of the m dependent variables. The equation corresponding to u is then of the form

$$F_u - \left(\frac{\partial}{\partial x} F_{u_x} + \frac{\partial}{\partial y} F_{u_y} + \cdots\right) + \left(\frac{\partial^2}{\partial x^2} F_{u_{xx}} + \frac{\partial^2}{\partial x\, \partial y} F_{u_{xy}} + \frac{\partial^2}{\partial y^2} F_{u_{yy}} + \cdots\right) - \left(\frac{\partial^3}{\partial x^3} F_{u_{xxx}} + \cdots\right) + \left(\frac{\partial^4}{\partial x^4} F_{u_{xxxx}} + \cdots\right) - \cdots = 0. \quad (58)$$

As an application of the preceding results, we obtain the partial differential equation satisfied by the equation of a *minimal surface*, that is, the surface passing through a given simple closed curve $\mathscr{C}$ in space and having minimum surface area bounded by $\mathscr{C}$. If the equation of the surface is assumed to be expressible in the form $z = z(x, y)$, the area to be minimized is then given by the integral

$$S = \iint_{\mathscr{R}} (1 + z_x^2 + z_y^2)^{1/2}\, dx\, dy, \tag{59}$$

where $\mathscr{R}$ is the region in the xy plane bounded by the projection $\mathscr{C}_0$ of $\mathscr{C}$ onto the xy plane and where z is given along $\mathscr{C}_0$. With $F = (1 + z_x^2 + z_y^2)^{1/2}$, the Euler equation is

$$\frac{\partial}{\partial x}\left(\frac{\partial F}{\partial z_x}\right) + \frac{\partial}{\partial y}\left(\frac{\partial F}{\partial z_y}\right) - \frac{\partial F}{\partial z} = 0$$

or

$$\frac{\partial}{\partial x}\left[\frac{z_x}{(1 + z_x^2 + z_y^2)^{1/2}}\right] + \frac{\partial}{\partial y}\left[\frac{z_y}{(1 + z_x^2 + z_y^2)^{1/2}}\right] = 0.$$

After some reduction, this equation takes the form

$$(1 + z_y^2)z_{xx} - 2z_x z_y z_{xy} + (1 + z_x^2)z_{yy} = 0,$$

or, with the conventional abbreviations for the partial derivatives,

$$p = z_x, \qquad q = z_y, \qquad r = z_{xx}, \qquad s = z_{xy}, \qquad t = z_{yy},$$

the differential equation of minimal surfaces becomes

$$(1 + q^2)r - 2pqs + (1 + p^2)t = 0. \tag{60}$$

As a second example, we seek the function $\phi(x, y, z)$ for which the mean square value of the magnitude of the gradient over a certain region $\mathscr{R}$ of space is minimum, when ϕ is given on the boundary of $\mathscr{R}$. This problem is closely related to many physical considerations, as will be seen. A necessary condition for the determination of ϕ is then

$$\delta \iiint_{\mathscr{R}} (\phi_x^2 + \phi_y^2 + \phi_z^2)\, dx\, dy\, dz = 0. \tag{61}$$

With $F = \phi_x^2 + \phi_y^2 + \phi_z^2$, the Euler equation can be obtained by reference to (58), in the form

$$\frac{\partial}{\partial x}\left(\frac{\partial F}{\partial \phi_x}\right) + \frac{\partial}{\partial y}\left(\frac{\partial F}{\partial \phi_y}\right) + \frac{\partial}{\partial z}\left(\frac{\partial F}{\partial \phi_z}\right) = 0,$$

since F does not involve ϕ explicitly. This equation reduces to

$$\phi_{xx} + \phi_{yy} + \phi_{zz} \equiv \nabla^2\phi = 0, \tag{62}$$

so that ϕ must satisfy *Laplace's equation.*

Conversely, suppose that ϕ satisfies (62) everywhere in a region $\mathscr{R}$, and takes on prescribed values on the boundary of that region. We may then multiply both sides of (62) by any continuously differentiable variation $\delta\phi$

which vanishes on the boundary of $\mathscr{R}$, and integrate the results over $\mathscr{R}$ to obtain

$$\iiint_{\mathscr{R}} (\phi_{xx} + \phi_{yy} + \phi_{zz})\,\delta\phi\,dx\,dy\,dz = 0.$$

If the first term is transformed by integration by parts, there follows

$$\iint\left\{\int_{x_1}^{x_2} \phi_{xx}\,\delta\phi\,dx\right\} dy\,dz = \iint\left\{\Big[\phi_x\,\delta\phi\Big]_{x_1}^{x_2} - \int_{x_1}^{x_2} \phi_x\,\delta\phi_x\,dx\right\} dy\,dz$$
$$= -\iiint \phi_x\,\delta\phi_x\,dx\,dy\,dz$$
$$= -\frac{1}{2}\,\delta\iiint \phi_x^2\,dx\,dy\,dz.$$

By treating the other terms similarly, we thus recover (61) from (62) when ϕ is prescribed on the boundary of $\mathscr{R}$. Thus, in this sense, the two problems are *equivalent*.

Hence, the so-called *Dirichlet problem*, in which we seek a function which satisfies Laplace's equation in a region $\mathscr{R}$ and which takes on prescribed values along the boundary of $\mathscr{R}$, can be expressed as a variational problem. As will be shown, the variational problem (61) often can be treated conveniently by approximate methods, to yield an approximate solution to the corresponding Dirichlet problem.

2.7. Constraints and Lagrange multipliers. In certain cases in which one or more functions are to be determined by a variational procedure, the variations cannot all be arbitrarily assigned, but are governed by one or more auxiliary conditions or *constraints*. Methods analogous to those described in Section 2.1 are available for the treatment of such cases.

We illustrate the procedure first in the special case of two dependent variables u and v, and one independent variable x,

$$\delta\int_{x_1}^{x_2} F(x, u, v, u_x, v_x)\,dx = 0, \tag{63}$$

in which the constraint is of the form

$$\phi(u, v) = 0. \tag{64}$$

We suppose that u and v are prescribed at the end points, in consistency with (64). Then, by proceeding as in Section 2.6, we may transform (63) into the condition

$$\int_{x_1}^{x_2}\left\{\left[\frac{\partial F}{\partial u} - \frac{d}{dx}\left(\frac{\partial F}{\partial u_x}\right)\right]\delta u + \left[\frac{\partial F}{\partial v} - \frac{d}{dx}\left(\frac{\partial F}{\partial v_x}\right)\right]\delta v\right\} dx = 0, \tag{65}$$

Since u and v must satisfy (64), the variations δu and δv cannot *both* be assigned arbitrarily inside (x_1, x_2) so that their coefficients in (65) need not vanish separately. However, from (64) there follows $\delta\phi = 0$, or

$$\phi_u \, \delta u + \phi_v \, \delta v = 0. \tag{66}$$

If we multiply (66) by a quantity λ (a *Lagrange multiplier*), which may be a function of x, and integrate the result with respect to x over (x_1, x_2), there follows

$$\int_{x_1}^{x_2} (\lambda\phi_u \, \delta u + \lambda\phi_v \, \delta v) \, dx = 0 \tag{67}$$

for any λ. The result of adding (65) and (67),

$$\int_{x_1}^{x_2} \left\{ \left[\frac{\partial F}{\partial u} - \frac{d}{dx}\left(\frac{\partial F}{\partial u_x}\right) + \lambda\phi_u \right] \delta u + \left[\frac{\partial F}{\partial v} - \frac{d}{dx}\left(\frac{\partial F}{\partial v_x}\right) + \lambda\phi_v \right] \delta v \right\} dx = 0, \tag{68}$$

must then also be true for any λ. Let λ be chosen so that, say, the coefficient of δu in (68) vanishes. Then, since the single variation δv *can* be arbitrarily assigned inside (x_1, x_2), its coefficient must also vanish.* Thus we must have

$$\left. \begin{aligned} \frac{d}{dx}\left(\frac{\partial F}{\partial u_x}\right) - \frac{\partial F}{\partial u} - \lambda\phi_u = 0, \\ \frac{d}{dx}\left(\frac{\partial F}{\partial v_x}\right) - \frac{\partial F}{\partial v} - \lambda\phi_v = 0 \end{aligned} \right\}. \tag{69a,b}$$

Equation (69a,b) and (64) comprise three equations in the three functions u, v, and λ.

If λ is eliminated between (69a) and (69b), the result

$$\phi_v \left[\frac{d}{dx}\left(\frac{\partial F}{\partial u_x}\right) - \frac{\partial F}{\partial u} \right] - \phi_u \left[\frac{d}{dx}\left(\frac{\partial F}{\partial v_x}\right) - \frac{\partial F}{\partial v} \right] = 0, \tag{69c}$$

together with (64), gives two conditions governing u and v. It may be noticed that this same relation would be obtained more directly by solving (66) for, say, δu as a multiple of δv, introducing the result into (65), and equating to zero the resultant coefficient of δv in (65). In more involved cases, the use of Lagrange multipliers is frequently more advantageous.

The extension to the more general case is perfectly straightforward.

In some cases the constraint condition is prescribed directly in the variational form

$$f \, \delta u + g \, \delta v = 0,$$

rather than in the form of (64). Whether or not a function ϕ can be found whose variation is given by the left-hand member, the preceding derivation

* If ϕ_u vanishes identically, λ is to be chosen such that the coefficient of δv vanishes Clearly, ϕ_u and ϕ_v cannot *both* vanish identically.

shows that the required necessary conditions are given by replacing ϕ_u and ϕ_v by f and g, respectively, in (69a,b) or (69c).

Also, a constraint condition may be expressed by the requirement that a certain definite *integral* involving the unknown function or functions take on a prescribed value. We illustrate a procedure which may be used in such cases by supposing that $y(x)$ is to be determined such that

$$I \equiv \int_{x_1}^{x_2} F(x, y, y')\, dx = \max \textit{ or } \min, \tag{70}$$

where y is prescribed at the end points,

$$y(x_1) = y_1, \qquad y(x_2) = y_2, \tag{71a,b}$$

and where y also is to satisfy the single constraint condition

$$J \equiv \int_{x_1}^{x_2} G(x, y, y')\, dx = k, \tag{72}$$

where k is a prescribed constant.

Here, in order to define an admissible set of varied functions which satisfy (72) as well as (71), we express the variation of y in terms of *two* parameters ϵ_1 and ϵ_2,

$$\delta y(x) = \epsilon_1 \eta_1(x) + \epsilon_2 \eta_2(x), \tag{73}$$

where $\eta_1(x)$ and $\eta_2(x)$ are continuously differentiable functions which each vanish at both $x = x_1$ and $x = x_2$. Then, with the abbreviations

$$\left.\begin{aligned} I(\epsilon_1, \epsilon_2) &\equiv \int_{x_1}^{x_2} F(x, y + \epsilon_1\eta_1 + \epsilon_2\eta_2, y' + \epsilon_1\eta_1' + \epsilon_2\eta_2')\, dx, \\ J(\epsilon_1, \epsilon_2) &\equiv \int_{x_1}^{x_2} G(x, y + \epsilon_1\eta_1 + \epsilon_2\eta_2, y' + \epsilon_1\eta_1' + \epsilon_2\eta_2')\, dx \end{aligned}\right\}, \tag{74a,b}$$

it must be true that $I(\epsilon_1, \epsilon_2)$ takes on a relative maximum or minimum value, subject to the constraint $J(\epsilon_1, \epsilon_2) = k$, when $\epsilon_1 = 0$ and $\epsilon_2 = 0$.

Hence, as was indicated in Section 2.1, we may introduce a Lagrange multiplier λ and write the relevant necessary conditions in the form

$$\frac{\partial}{\partial \epsilon_1}[I(\epsilon_1, \epsilon_2) + \lambda J(\epsilon_1, \epsilon_2)] = \frac{\partial}{\partial \epsilon_2}[I(\epsilon_1, \epsilon_2) + \lambda J(\epsilon_1, \epsilon_2)] = 0 \quad \textit{when } \epsilon_1 = \epsilon_2 = 0. \tag{75}$$

When the indicated operations are effected, and the results are appropriately integrated by parts, these conditions become

$$\int_{x_1}^{x_2} \left[\left(\frac{d}{dx}\frac{\partial F}{\partial y'} - \frac{\partial F}{\partial y}\right) + \lambda\left(\frac{d}{dx}\frac{\partial G}{\partial y'} - \frac{\partial G}{\partial y}\right)\right] \eta_1\, dx = 0 \tag{76a}$$

and

$$\int_{x_1}^{x_2}\left[\left(\frac{d}{dx}\frac{\partial F}{\partial y'}-\frac{\partial F}{\partial y}\right)+\lambda\left(\frac{d}{dx}\frac{\partial G}{\partial y'}-\frac{\partial G}{\partial y}\right)\right]\eta_2\,dx=0. \tag{76b}$$

Now, unless G is such that the coefficient of $\lambda\eta_2$ in the integrand of (76b) vanishes identically,* the function $\eta_2(x)$ certainly can be so chosen that

$$\int_{x_1}^{x_2}\left(\frac{d}{dx}\frac{\partial G}{\partial y'}-\frac{\partial G}{\partial y}\right)\eta_2\,dx\neq 0,$$

in consequence of which the constant λ then can be so determined that (76b) is satisfied. Since $\eta_1(x)$ then is arbitrary inside (x_1, x_2), its coefficient in the integrand of (76a) must vanish, giving the desired Euler equation

$$\frac{d}{dx}\left[\frac{\partial}{\partial y'}(F+\lambda G)\right]-\frac{\partial}{\partial y}(F+\lambda G)=0. \tag{77}$$

Thus the general solution of this equation will involve the constant parameter λ as well as the usual two constants of integration, in correspondence with the presence of the *three* conditions (71a,b) and (72) which are to be satisfied in this case.

The result established may be summarized as follows: *In order to maximize (or minimize) an integral $\int_a^b F\,dx$ subject to a constraint $\int_a^b G\,dx = k$, first write $H = F + \lambda G$, where λ is a constant, and maximize (or minimize) $\int_a^b H\,dx$ subject to no constraint. Carry the Lagrange multiplier λ through the calculation, and determine it, together with the constants of integration arising in the solution of Euler's equation, so that the constraint $\int_a^b G\,dx = k$ is satisfied, and so that the end conditions are satisfied.*

It may be seen that when one of the end conditions (71a,b) is not imposed, the condition $\partial H/\partial y' = 0$ must be substituted for it at the end in question.

The same procedure is applicable in the more general case, in which two or more independent and dependent variables are involved.

To illustrate the procedure, we determine the curve of length l which passes through the points (0, 0) and (1, 0) and for which the area I between the curve and the x axis is a maximum. We are thus to maximize the integral

$$I=\int_0^1 y\,dx, \tag{78a}$$

subject to the end conditions

$$y(0)=y(1)=0 \tag{78b}$$

* It can be shown that those cases in which this situation exists are trivial, and of no interest.

and to the constraint

$$J \equiv \int_0^1 (1 + y'^2)^{1/2}\,dx = l, \tag{78c}$$

where l is a prescribed constant greater than unity.

The Euler equation corresponding to the maximization of the integral of $H = y + \lambda(1 + y'^2)^{1/2}$ is then of the form

$$\lambda \frac{d}{dx}\left[\frac{y'}{(1 + y'^2)^{1/2}}\right] - 1 = 0$$

or, after integration and simplification,

$$[\lambda^2 - (x - c_1)^2]y'^2 = (x - c_1)^2.$$

By solving for y' and integrating again, we find that the extremals are of the form

$$y = \pm[\lambda^2 - (x - c_1)^2]^{1/2} + c_2, \tag{79}$$

and hence (as might have been expected) are arcs of the circles

$$(x - c_1)^2 + (y - c_2)^2 = \lambda^2. \tag{80}$$

The three constants are to be determined so that the circle passes through the end points, and so that the relevant arc length is l.

We may notice that if $l > \pi/2$, the "solution" defined by (80) will *not* determine y as a single-valued function of x. In order to avoid such exceptional situations, one may employ a *parametric representation*, expressing x and y in terms of a parameter t, rather than assuming that one of the variables x and y is a single-valued function of the other. Thus, in the present case, we could reformulate the problem in the form*

$$I \equiv \frac{1}{2}\int_{t_1}^{t_2} (x\dot{y} - y\dot{x})\,dt = \max \tag{81a}$$

where $$x(t_1) = 0, \quad x(t_2) = 1, \quad y(t_1) = 0, \quad y(t_2) = 0 \tag{81b}$$

and $$J \equiv \int_{t_1}^{t_2} (\dot{x}^2 + \dot{y}^2)^{1/2}\,dt = l. \tag{81c}$$

Here a dot is used to denote t differentiation. With

$$H = \frac{1}{2}(x\dot{y} - y\dot{x}) + \lambda(\dot{x}^2 + \dot{y}^2)^{1/2},$$

* In transforming (78a), we make use of the fact that the element of area is correctly given by $dA = \frac{1}{2}(x\,dy - y\,dx)$, whether or not y is a single-valued function of x, whereas the relation $dA = y\,dx$ may not be valid in the general case.

the *two* relevant Euler equations take the forms

$$\frac{1}{2}\dot{y} - \frac{d}{dt}\left(-\frac{1}{2}y + \frac{\lambda \dot{x}}{\sqrt{\dot{x}^2 + \dot{y}^2}}\right) = 0 \tag{82a}$$

and

$$-\frac{1}{2}\dot{x} - \frac{d}{dt}\left(\frac{1}{2}x + \frac{\lambda \dot{y}}{\sqrt{\dot{x}^2 + \dot{y}^2}}\right) = 0, \tag{82b}$$

and the relation (80) can be deduced without difficulty in this more general case.

2.8. Variable end points. In some variational problems, the boundary of the region of integration is not completely specified, but is to be determined together with the unknown function or functions.

To illustrate such situations, we suppose that $y(x)$ is to be determined such that

$$\delta I \equiv \delta \int_{x_1}^{x_2} F(x, y, y')\, dx = 0 \tag{83}$$

where x_1 is fixed and the value of $y(x_1)$ is given,

$$y(x_1) = y_1, \tag{84}$$

but where the point $(x_2, y(x_2))$ is required only to lie on a certain curve $y = g(x)$, so that

$$y(x_2) = g(x_2) \tag{85}$$

where $g(x)$ is a given function of x but x_2 is *not* preassigned.

Since here x_2 may be varied, the basic requirement (83) becomes*

$$\frac{\partial I}{\partial x_2}\delta x_2 + \int_{x_1}^{x_2}\left(\frac{\partial F}{\partial y}\delta y + \frac{\partial F}{\partial y'}\delta y'\right) dx = 0$$

or

$$[F]_{x=x_2}\delta x_2 + \left[\frac{\partial F}{\partial y'}\delta y\right]_{x_1}^{x_2} - \int_{x_1}^{x_2}\left(\frac{d}{dx}\frac{\partial F}{\partial y'} - \frac{\partial F}{\partial y}\right)\delta y\, dx = 0$$

or

$$[F]_{x=x_2}\delta x_2 + \left[\frac{\partial F}{\partial y'}\right]_{x=x_2}\delta y(x_2) - \int_{x_1}^{x_2}\left(\frac{d}{dx}\frac{\partial F}{\partial y'} - \frac{\partial F}{\partial y}\right)\delta y\, dx = 0. \tag{86}$$

In order to relate δx_2 to $\delta y(x_2)$, we must make use of the fact that the *varied* end point is to *remain* on the curve $y = g(x)$. Thus, if the true function $y(x)$ is changed to

$$y(x) + \Delta y(x) = y(x) + \delta y(x) = y(x) + \epsilon\eta(x),$$

* It is assumed here that the *integrand* F does not depend upon x_2.

and if x_2 correspondingly changes by Δx_2, the requirement that the new end point lie on the curve $y = g(x)$ is of the form

$$y(x_2 + \Delta x_2) + \epsilon\eta(x_2 + \Delta x_2) = g(x_2 + \Delta x_2). \tag{87}$$

When the equal members of (85) are subtracted from the corresponding members of (87), there follows

$$y(x_2 + \Delta x_2) - y(x_2) + \epsilon\eta(x_2 + \Delta x_2) = g(x_2 + \Delta x_2) - g(x_2).$$

Thus

$$y'(x_2)\,\Delta x_2 + \epsilon\eta(x_2) = g'(x_2)\,\Delta x_2 + (\text{higher-order terms in } \epsilon \text{ and } \Delta x_2)$$

or

$$\Delta x_2 = \frac{\epsilon\eta(x_2)}{g'(x_2) - y'(x_2)} + (\text{higher-order terms in } \epsilon),$$

and hence we deduce that

$$\delta x_2 = \frac{\delta y(x_2)}{g'(x_2) - y'(x_2)}. \tag{88}$$

The result of introducing (88) into (86) then is the condition

$$\left[\frac{F}{g' - y'} + \frac{\partial F}{\partial y'}\right]_{x=x_2} \delta y(x_2) - \int_{x_1}^{x_2} \left(\frac{d}{dx}\frac{\partial F}{\partial y'} - \frac{\partial F}{\partial y}\right) \delta y\, dx = 0, \tag{89}$$

which yields the usual Euler equation, subject to the condition (84) and to the condition

$$\left[F + (g' - y')\frac{\partial F}{\partial y'}\right]_{x=x_2} = 0. \tag{90}$$

This last condition is called a *transversality condition*.

In illustration, when $F = (1 + y'^2)^{1/2}$ and $g(x) = mx + b$, where m and b are prescribed constants, the transversality condition (90) becomes $y'(x_2) = -1/m$. This result clearly corresponds to the fact that the shortest distance from the point (x_1, y_1) to the nearest point on the line $y = mx + b$ is measured in the direction *perpendicular* to that line.

2.9. Sturm-Liouville problems. To illustrate another important class of problems, we consider next the determination of stationary values of the quantity λ defined by the *ratio*

$$\lambda = \frac{\int_a^b (py'^2 - qy^2)\, dx}{\int_a^b ry^2\, dx} \equiv \frac{I_1}{I_2} \tag{91}$$

where p, q, and r are given functions of the independent variable x. The variation of the ratio is of the form

$$\delta\lambda = \frac{I_2\,\delta I_1 - I_1\,\delta I_2}{I_2^{\,2}} = \frac{1}{I_2}(\delta I_1 - \lambda\,\delta I_2) \tag{92}$$

where

$$\delta I_1 = 2\Big[py'\,\delta y\Big]_a^b - 2\int_a^b [(py')' + qy]\,\delta y\,dx \tag{93a}$$

and

$$\delta I_2 = 2\int_a^b ry\,\delta y\,dx. \tag{93b}$$

If (93a,b) are introduced into (92), there follows

$$\delta\lambda = 2\,\frac{\Big[py'\,\delta y\Big]_a^b - \int_a^b [(py')' + qy + \lambda ry]\,\delta y\,dx}{\int_a^b ry^2\,dx}. \tag{94}$$

Thus the condition $\delta\lambda = 0$ leads to the relevant Euler equation, in the form

$$\frac{d}{dx}\left(p\frac{dy}{dx}\right) + qy + \lambda ry = 0, \tag{95}$$

and to the natural boundary conditions, which here require that py' vanish at an end where y is not preassigned:

$$y(a)\ \textit{prescribed}\quad or\quad [py']_{x=a} = 0,\qquad y(b)\ \textit{prescribed}\quad or\quad [py']_{x=b} = 0. \tag{96}$$

In particular, when the boundary conditions are *homogeneous*, of the form

$$y(a) = 0\quad or\quad y'(a) = 0,\qquad y(b) = 0\quad or\quad y'(b) = 0, \tag{97}$$

the problem is one of *characteristic values*, and is a special case of the general *Sturm-Liouville problem* (see Section 1.29). It follows that the problem of determining characteristic functions of (95), subject to (97), is equivalent to the problem of determining functions satisfying (97) which render (91) stationary.

Stationary values of λ then must be characteristic numbers of the problem. To verify this fact directly, suppose that λ_k and $\phi_k(x)$ are corresponding characteristic quantities, so that

$$(p\phi_k')' + q\phi_k + \lambda_k r\phi_k = 0. \tag{98}$$

Then if y is replaced by ϕ_k in (91), λ should reduce to λ_k. Before making the substitution, we transform I_1 by integrating the first term by parts, using (97), to rewrite (91) in the form

$$\lambda = -\frac{\int_a^b [(py')' + qy]y\,dx}{\int_a^b ry^2\,dx}. \tag{91'}$$

Now, when y is replaced by ϕ_k, there follows

$$\lambda = \frac{\int_a^b [-(p\phi_k')' - q\phi_k]\phi_k \, dx}{\int_a^b r\phi_k^2 \, dx} = \frac{\int_a^b [\lambda_k r\phi_k]\phi_k \, dx}{\int_a^b r\phi_k^2 \, dx} = \lambda_k, \tag{99}$$

as was to be shown. The equality of the two square brackets in (99) follows from (98).

If we arbitrarily impose the constraint

$$\int_a^b ry^2 \, dx = 1, \tag{100}$$

in the homogeneous case, it follows from (91) that the stationary condition takes the form

$$\delta\lambda \equiv \delta\int_a^b (py'^2 - qy^2) \, dx = 0, \tag{101}$$

where y is to satisfy (100) and the relevant end conditions. From (92) it follows also that the condition

$$\delta(I_1 - \lambda I_2) = 0, \tag{102}$$

with the provision $y \not\equiv 0$, is equivalent to either (92) or the combination of (101) and (100). In this last form, the constant λ plays the role of a Lagrange multiplier, and is to be determined together with the function y so that $I_1 - \lambda I_2$ is stationary and $y(x) \not\equiv 0$. The condition (100) is recognized as a *normalizing* condition, the weighting function $r(x)$ being that function with respect to which pairs of characteristic functions of the problem are orthogonal (see Section 1.29).

If the constraint (100) were suppressed, the problem (101) generally would determine only one stationary function, $y \equiv 0$. However, when the condition (100) is added, the problem generally has an infinite set of stationary functions, for each of which λ is stationary for small variations in y.

For a physical interpretation of these results, we may recall that in the case of small free vibrations of an elastic string of length l with fixed ends, under tension $F(x)$, the amplitude $y(x)$ satisfies the equation

$$\frac{d}{dx}\left(F\frac{dy}{dx}\right) + \omega^2 \rho y = 0 \tag{103}$$

and the end conditions

$$y(0) = 0, \qquad y(l) = 0, \tag{104a,b}$$

where $\rho(x)$ is the linear mass density and ω the circular frequency. Equation (103) is identified with (95) if we set

$$p = F, \qquad q = 0, \qquad r = \rho, \qquad \lambda = \omega^2. \tag{105}$$

Hence the vibration modes are stationary curves of the problem

$$\delta\omega^2 \equiv \delta \frac{\int_0^l Fy'^2\,dx}{\int_0^l \rho y^2\,dx} = 0, \tag{106}$$

and stationary values of the ratio are squares of the natural circular frequencies. Alternatively, from (102), the variational problem can be taken in the form

$$\delta \int_0^l (Fy'^2 - \omega^2 \rho y^2)\,dx = 0. \tag{107}$$

The statement of (106) is a special case of *Rayleigh's principle* which applies to more general elastic systems (see Reference 8). It can be shown that *the smallest stationary value of* ω^2 *is truly the minimum value of the ratio in* (106), for all continuously differentiable functions $y(x)$ which vanish when $x = 0$ and $x = l$ (see Problem 55).

Equation (107) is closely connected with *Hamilton's principle*, which is treated in the following section.

Methods for obtaining *approximations* to stationary functions of such problems are to be considered in Sections 2.19 and 2.20.

2.10. Hamilton's principle. One of the most basic and important principles of mathematical physics bears the name of Hamilton.* From it can be deduced the fundamental equations governing a large number of physical phenomena. It is formulated here in terms of the dynamics of a system of particles, and is readily extended by analogy to other considerations.

We consider first a single particle of mass m, moving subject to a force field. If the vector from a fixed origin to the particle at time t is denoted by $\mathbf{r}$, then, according to Newton's laws of motion, the actual path followed is governed by the vector equation

$$m\frac{d^2\mathbf{r}}{dt^2} - \mathbf{f} = \mathbf{0}, \tag{108}$$

where the vector $\mathbf{f}$ is the force acting on the particle. Now consider any *other* path $\mathbf{r} + \delta\mathbf{r}$. We require only that the *true* path and the *varied* path coincide at two distinct instants $t = t_1$ and $t = t_2$, that is, that the *variation* $\delta\mathbf{r}$ vanish at those two instants:

$$\delta\mathbf{r}\,\big|_{t_1} = \delta\mathbf{r}\,\big|_{t_2} = \mathbf{0}. \tag{109}$$

At any intermediate time t we then have to consider the true path $\mathbf{r}$ and the varied path $\mathbf{r} + \delta\mathbf{r}$.

* Sir William Rowan Hamilton (1805–1865), an Irish mathematician, is also known for his invention of *quaternions*.

The first step in the derivation consists of taking the scalar (dot) product of the variation $\delta\mathbf{r}$ into (108), and of integrating the result with respect to time over (t_1, t_2), to obtain the relation

$$\int_{t_1}^{t_2}\left(m\,\frac{d^2\mathbf{r}}{dt^2}\cdot\delta\mathbf{r}-\mathbf{f}\cdot\delta\mathbf{r}\right)dt=0. \tag{110}$$

If the first term is integrated by parts, it takes the form

$$m\int_{t_1}^{t_2}\frac{d^2\mathbf{r}}{dt^2}\cdot\delta\mathbf{r}\,dt=m\left\{\left[\frac{d\mathbf{r}}{dt}\cdot\delta\mathbf{r}\right]_{t_1}^{t_2}-\int_{t_1}^{t_2}\frac{d\mathbf{r}}{dt}\cdot\delta\frac{d\mathbf{r}}{dt}\,dt\right\}.$$

Since the variation $\delta\mathbf{r}$ vanishes at the ends of the interval, the integrated terms vanish. Also, we have the relation

$$\frac{d\mathbf{r}}{dt}\cdot\delta\frac{d\mathbf{r}}{dt}=\frac{1}{2}\,\delta\left(\frac{d\mathbf{r}}{dt}\right)^2.$$

Hence the first term in (110) is equivalent to

$$-\delta\left[\frac{1}{2}\,m\left(\frac{d\mathbf{r}}{dt}\right)^2\right]=-\delta T, \tag{111}$$

where T is the *kinetic energy* $(\frac{1}{2}mv^2)$ of the particle, and (110) becomes

$$\int_{t_1}^{t_2}(\delta T+\mathbf{f}\cdot\delta\mathbf{r})\,dt=0. \tag{112}$$

This is *Hamilton's principle* in its most general form, as applied to the motion of a single particle. However, if the force field is *conservative* equation (112) can be put in a more concise form.

To fix ideas, suppose that $\mathbf{f}$ is specified by its components X, Y, Z in the directions of the rectangular xyz coordinates. We recall that a force field $\mathbf{f}$ is conservative if and only if

$$\mathbf{f}\cdot d\mathbf{r}\equiv X\,dx+Y\,dy+Z\,dz$$

is the differential $d\Phi$ of a single-valued function Φ. The force $\mathbf{f}$ is then the gradient of Φ. The function Φ is called the *force potential* and its *negative*, say V, is called the *potential energy*. Clearly, Φ and V each involve an irrelevant arbitrary additive constant.

It follows that $\mathbf{f}$ is conservative if there exists a single-valued function Φ such that

$$\mathbf{f}\cdot\delta\mathbf{r}=\delta\Phi. \tag{113}$$

In terms of the xyz components of $\mathbf{f}$, this means that

$$X\,\delta x+Y\,\delta y+Z\,\delta z=\delta\Phi \tag{114}$$

where

$$X = \frac{\partial\Phi}{\partial x}, \qquad Y = \frac{\partial\Phi}{\partial y}, \qquad Z = \frac{\partial\Phi}{\partial z} \tag{115}$$

Thus if Φ is the potential function, equation (112) becomes

$$\delta\int_{t_1}^{t_2}(T+\Phi)\,dt = 0. \tag{116}$$

In place of the potential function Φ, it is customary to use the *potential energy* function V,

$$V = -\Phi, \tag{117}$$

so that Hamilton's principle takes the form

$$\delta\int_{t_1}^{t_2}(T-V)\,dt = 0 \tag{118}$$

when a potential function exists, that is, when the forces acting are conservative.

For such a problem Hamilton's principle states that the motion is such that the integral of the difference between the kinetic and potential energies is stationary for the true path. It can be shown further that actually this integral is a *minimum* when compared with that corresponding to any neighboring path having the same terminal configurations, at least over a sufficiently short time interval. Thus we may say that "nature tends to equalize the kinetic and potential energies over the motion."

The energy difference

$$L = T - V$$

is sometimes called the *kinetic potential* or the *Lagrangian function*. In terms of this function, (118) becomes merely

$$\delta\int_{t_1}^{t_2} L\,dt = 0. \tag{119}$$

If nonconservative forces are present, the potential energy function generally does not exist, and recourse must be had to (112). We may notice, however, that in any case $\mathbf{f}\cdot\delta\mathbf{r}$ is the element of *work* done *by* the force $\mathbf{f}$ in a small *displacement* $\delta\mathbf{r}$. In particular, when the force *is* conservative this element of work is equivalent to $\delta\Phi = -\delta V$.

The above derivation is extended to a *system* of N particles by summation, and to a *continuous* system by integration. Thus if the kth particle is of mass m_k, is specified by the vector $\mathbf{r}_k$, and is subject to a force $\mathbf{f}_k$, the total kinetic energy is given by

$$T = \sum_{k=1}^{N}\frac{1}{2}m_k\left(\frac{d\mathbf{r}_k}{dt}\right)^2 = \sum_{k=1}^{N}\frac{1}{2}m_k{v_k}^2 \tag{120a}$$

while the total work done by the forces acting is given by

$$\sum_{k=1}^{N} \mathbf{f}_k \cdot \delta\mathbf{r}_k. \tag{120b}$$

Finally, the principle applies equally well to a general dynamical system consisting of particles and rigid bodies subject to interconnections and constraints. We notice that the derivation is independent of the coordinates specifying the system.

2.11. Lagrange's equations. In a dynamical system with n degrees of freedom it is usually possible to choose n independent geometrical quantities which uniquely specify the position of all components of the system. These quantities are known as *generalized coordinates.* For example, in the case of

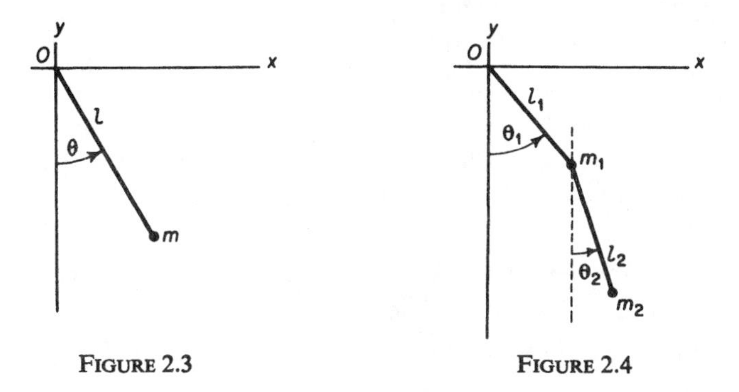

FIGURE 2.3 FIGURE 2.4

a pendulum consisting of a point mass m suspended by an inextensible string of length l, the position of the mass is completely determined by the angle θ between the deflected and equilibrium positions of the string (Figure 2.3). If xy coordinates were used, x and y would not be independent since the constraint equation $x^2 + y^2 = l^2$ would have to be imposed. Similarly, the compound pendulum of Figure 2.4 has two degrees of freedom and the indicated angles θ_1 and θ_2 are suitable generalized coordinates. In rectangular coordinates, if the quantities x_1, y_1 and x_2, y_2 representing the positions of m_1 and m_2 were used, two equations of constraint would be needed.

In the general case, the total *kinetic energy* T may depend upon the generalized coordinates, say $q_1, q_2, \cdots, q_n$, as well as upon their time rates of change or so-called *generalized velocities* $\dot{q}_1, \dot{q}_2, \cdots, \dot{q}_n$.* For a *conservative* system the total *potential energy* V is a function only of position and hence does *not* depend upon the generalized *velocities.*

* Here and elsewhere, a dot indicates *time* differentiation.

Also, the work done *by* the force system involved when the q's are given small displacements is

$$-\delta V = +\delta\Phi = Q_1\,\delta q_1 + Q_2\,\delta q_2 + \cdots + Q_n\,\delta q_n, \tag{121}$$

where

$$Q_1 = -\frac{\partial V}{\partial q_1} = \frac{\partial \Phi}{\partial q_1}, \quad \cdots, \quad Q_n = -\frac{\partial V}{\partial q_n} = \frac{\partial \Phi}{\partial q_n}. \tag{122}$$

The quantity $Q_i\,\delta q_i$ is the work done *by* the forces in a displacement δq_i. Since the Q's may or may not have the dimension of true force they are called *generalized forces*. Thus if q_i is a linear displacement, then Q_i is truly a force, while if q_i is an *angular* displacement, then Q_i is a *torque*. In other applications the q's may represent electric charges, currents, areas, volumes, and so forth, and the nature of the Q's is determined accordingly, in such a way that $Q\,\delta q$ has the dimensions of *work*.

In applications of Hamilton's principle,

$$\delta \int_{t_1}^{t_2} (T - V)\, dt = 0, \tag{123}$$

to *conservative* systems we may thus suppose that the total kinetic energy T is expressed in terms of the n q's and the n $\dot{q}$'s, while the total potential energy V is expressed in terms of the q's only. The associated Euler equations then become

$$\frac{d}{dt}\left[\frac{\partial(T - V)}{\partial \dot{q}_i}\right] - \frac{\partial(T - V)}{\partial q_i} = 0 \qquad (i = 1, 2, \cdots, n). \tag{124a}$$

Since $\partial V/\partial \dot{q}_i \equiv 0$, we may write each equation alternatively in the form

$$\frac{d}{dt}\frac{\partial T}{\partial \dot{q}_i} - \frac{\partial T}{\partial q_i} + \frac{\partial V}{\partial q_i} = 0 \tag{124b}$$

or, using (122), in the form

$$\frac{d}{dt}\frac{\partial T}{\partial \dot{q}_i} - \frac{\partial T}{\partial q_i} = Q_i. \tag{124c}$$

The three forms (124a,b,c) are completely equivalent for a conservative system, and are usually called *Lagrange's equations*. One equation is obtained for each independent q.

In illustration, for the simple pendulum of Figure 2.3 the kinetic energy is given by

$$T = \tfrac{1}{2}m(l\dot{\theta})^2. \tag{125a}$$

If damping is neglected, the work done *by* gravity in lifting the mass from its equilibrium position to the position θ is *negative* and is given by $-mgl(1 - \cos\theta)$, so that the potential energy is of the form

$$V = +mgl(1 - \cos\theta) + \text{constant}. \tag{125b}$$

Thus, with $q_1 = \theta$, equation (124b) becomes

$$\frac{d}{dt}(ml^2\dot{\theta}) - 0 + mgl \sin \theta = 0$$

or

$$\ddot{\theta} + \frac{g}{l} \sin \theta = 0. \tag{126}$$

This is the well-known equation of motion for such a pendulum.

For the compound pendulum of Figure 2.4 we may proceed as follows. If the rectangular coordinates of m_1 and m_2 are taken as (x_1, y_1) and (x_2, y_2) there follows

$$x_1 = l_1 \sin \theta_1, \qquad y_1 = -l_1 \cos \theta_1;$$
$$x_2 = l_1 \sin \theta_1 + l_2 \sin \theta_2, \qquad y_2 = -l_1 \cos \theta_1 - l_2 \cos \theta_2.$$

The total kinetic energy T is then

$$T = \tfrac{1}{2}m_1(\dot{x}_1^2 + \dot{y}_1^2) + \tfrac{1}{2}m_2(\dot{x}_2^2 + \dot{y}_2^2).$$

Hence there follows, by substitution and simplification,

$$T = \tfrac{1}{2}(m_1 + m_2)l_1^2\dot{\theta}_1^2 + m_2 l_1 l_2 \dot{\theta}_1 \dot{\theta}_2 \cos(\theta_1 - \theta_2) + \tfrac{1}{2}m_2 l_2^2 \dot{\theta}_2^2. \tag{127a}$$

The total potential energy is given by

$$V = m_1 g y_1 + m_2 g y_2 + \text{constant}$$

or

$$V = -(m_1 + m_2)g l_1 \cos \theta_1 - m_2 g l_2 \cos \theta_2 + \text{constant}. \tag{127b}$$

Use of equation (124b) then leads to the two equations of motion

$$(m_1 + m_2)l_1\ddot{\theta}_1 + m_2 l_2(\ddot{\theta}_2 \cos \alpha + \dot{\theta}_2^2 \sin \alpha) + (m_1 + m_2)g \sin \theta_1 = 0 \tag{128a}$$

and

$$l_1\ddot{\theta}_1 \cos \alpha + l_2\ddot{\theta}_2 - l_1\dot{\theta}_1^2 \sin \alpha + g \sin \theta_2 = 0, \tag{128b}$$

where

$$\alpha = \theta_1 - \theta_2. \tag{129}$$

We notice that it is not necessary to evaluate the constraints exerted by tensions in the strings supporting the masses, since they do no *work*.

For a *nonconservative* force field, Lagrange's equations must be based on the form (112), rather than (118). In this case it is still possible to express the work done by the force system in small displacements $\delta q_1, \cdots, \delta q_n$, in the form

$$\sum_{k=1}^{N} \mathbf{f}_k \cdot \delta \mathbf{r}_k = Q_1\, \delta q_1 + Q_2\, \delta q_2 + \cdots + Q_n\, \delta q_n, \tag{130}$$

as in (121). However, *the generalized forces Q_i are now in general not derivable from a potential function* as in (122). (In certain nonconservative systems such a function *may* exist, depending upon *time* as well as position.) To determine the Q's by *physical* considerations, we need only notice that, as before, $Q_i\,\delta q_i$ is the work done *by* the force system when q_i is changed to $q_i + \delta q_i$ *and the other* q's *are held fixed.*

For an *analytical* determination, in the case of a system of N particles, we may suppose that the components X_k, Y_k, and Z_k of the force $\mathbf{f}_k$ acting on the kth particle of the system are known in the directions of the x, y, and z axes. Then (130) gives the relation

$$Q_1\,\delta q_1 + Q_2\,\delta q_2 + \cdots + Q_n\,\delta q_n = \sum_{k=1}^{N}(X_k\,\delta x_k + Y_k\,\delta y_k + Z_k\,\delta z_k). \quad (131)$$

Since x_k, y_k, and z_k are functions of the coordinates $q_1, q_2, \cdots, q_n$, there follows also

$$\left.\begin{aligned} \delta x_k &= \frac{\partial x_k}{\partial q_1}\,\delta q_1 + \cdots + \frac{\partial x_k}{\partial q_n}\,\delta q_n, \\ \delta y_k &= \frac{\partial y_k}{\partial q_1}\,\delta q_1 + \cdots + \frac{\partial y_k}{\partial q_n}\,\delta q_n, \\ \delta z_k &= \frac{\partial z_k}{\partial q_1}\,\delta q_1 + \cdots + \frac{\partial z_k}{\partial q_n}\,\delta q_n \end{aligned}\right\}. \quad (132)$$

Equation (131) must hold for arbitrary choices of the δq's. In particular, if we require that *all* the δq's *except* δq_i vanish, (132) then gives

$$\delta x_k = \frac{\partial x_k}{\partial q_i}\,\delta q_i, \qquad \delta y_k = \frac{\partial y_k}{\partial q_i}\,\delta q_i, \qquad \delta z_k = \frac{\partial z_k}{\partial q_i}\,\delta q_i \quad (133a)$$

and (131) becomes

$$Q_i\,\delta q_i = \sum_{k=1}^{N}(X_k\,\delta x_k + Y_k\,\delta y_k + Z_k\,\delta z_k), \quad (133b)$$

in this case. By introducing (133a) into (133b), we then obtain the desired relation

$$Q_i = \sum_{k=1}^{N}\left(X_k\frac{\partial x_k}{\partial q_i} + Y_k\frac{\partial y_k}{\partial q_i} + Z_k\frac{\partial z_k}{\partial q_i}\right). \quad (134)$$

This result is clearly valid whether or not the system is conservative.

Hamilton's principle (112) then states that

$$\delta\int_{t_1}^{t_2} T\,dt + \int_{t_1}^{t_2}(Q_1\,\delta q_1 + Q_2\,\delta q_2 + \cdots + Q_n\,\delta q_n)\,dt = 0.$$

By calculating the variation of the first integral in the usual way, we obtain the condition

$$\int_{t_1}^{t_2}\left\{\left[\frac{\partial T}{\partial q_1} - \frac{d}{dt}\left(\frac{\partial T}{\partial \dot{q}_1}\right) + Q_1\right]\delta q_1 + \cdots + \left[\frac{\partial T}{\partial q_n} - \frac{d}{dt}\left(\frac{\partial T}{\partial \dot{q}_n}\right) + Q_n\right]\delta q_n\right\}dt = 0. \quad (135)$$

The vanishing of the coefficients of the independent variations leads again to the equations (124c).

Thus *the equations*

$$\frac{d}{dt}\left(\frac{\partial T}{\partial \dot{q}_i}\right) - \frac{\partial T}{\partial q_i} = Q_i \qquad (i = 1, 2, \cdots, n) \tag{136}$$

are valid whenever the variations of the n q's are independent. For a *conservative* system the Q's are derivable from a potential function and (124a) or (124b) can be used alternatively.

2.12. Generalized dynamical entities. Before considering the definition of additional generalized dynamical entities, it is desirable to emphasize the fact that the so-called *generalized velocity* $\dot{q}_i$, associated with a generalized coordinate q_i, is merely the *time rate of change of that coordinate*. Thus, for example, in polar coordinates (r, θ) the generalized velocities associated with r and θ are merely $\dot{r}$ and $\dot{\theta}$, respectively. We notice that $\dot{\theta}$ is *not* the component of the velocity vector in the circumferential direction $(r\dot{\theta})$. Similarly, the so-called *generalized accelerations* $\ddot{r}$ and $\ddot{\theta}$, associated with r and θ, are *not* the respective components of the acceleration vector in the radial and circumferential directions. It will be recalled that these latter quantities are of the forms $\ddot{r} - r\dot{\theta}^2$ and $r\ddot{\theta} + 2\dot{r}\dot{\theta}$.

In the remainder of this section we deal *always* with *generalized* forces, velocities, accelerations, and momenta. For brevity, the adjective "generalized" frequently will be omitted.

In rectangular coordinates (x, y, z) the quantities

$$p_x = m\dot{x}, \qquad p_y = m\dot{y}, \qquad p_z = m\dot{z}$$

are called the components of *momentum*. Since we have the relation

$$T = \tfrac{1}{2}m(\dot{x}^2 + \dot{y}^2 + \dot{z}^2),$$

it follows that

$$\frac{\partial T}{\partial \dot{x}} = p_x, \qquad \frac{\partial T}{\partial \dot{y}} = p_y, \qquad \frac{\partial T}{\partial \dot{z}} = p_z.$$

In generalized coordinates, we call the quantity $\partial T/\partial \dot{q}_i$ the *generalized momentum* associated with q_i, and write

$$p_i = \frac{\partial T}{\partial \dot{q}_i}. \tag{137}$$

The associated equation of motion (136) then becomes

$$\frac{dp_i}{dt} = Q_i + \frac{\partial T}{\partial q_i}. \tag{138}$$

Hence, *the rate of change of the* ith *generalized momentum is equal to the sum of the* ith *generalized force* Q_i *and the quantity* $\partial T/\partial q_i$. In *rectangular* coordinates the "corrective terms" $\partial T/\partial q_i$ are absent.

In motion specified by plane polar coordinates (r, θ) we have

$$T \equiv \frac{1}{2} m \left(\frac{ds}{dt}\right)^2 = \frac{1}{2} m(\dot{r}^2 + r^2\dot{\theta}^2)$$

and hence

$$p_r = m\dot{r}, \qquad p_\theta = mr^2\dot{\theta}; \qquad \frac{\partial T}{\partial r} = mr\dot{\theta}^2, \qquad \frac{\partial T}{\partial \theta} = 0. \tag{139}$$

The equations of motion (138) are then of the form

$$\frac{dp_r}{dt} = Q_r + mr\dot{\theta}^2, \tag{140a}$$

$$\frac{dp_\theta}{dt} = Q_\theta. \tag{140b}$$

Here Q_r is the impressed radial *force*, while the generalized θ force Q_θ is a *torque*. If a particle moves in such a way that the (generalized) momentum associated with r is constant and hence, from (139), dr/dt is constant, (140a) shows that a net force $Q_r = -mr\dot{\theta}^2$ then must be exerted externally (e.g., by a spring) in the r direction. More generally, in so far as change in the r momentum is involved, the mass behaves as though a force $+mr\dot{\theta}^2 \equiv \partial T/\partial r$ were acting in the r direction in addition to the *actual* external force Q_r. The fictitious force is recognized as the so-called *centrifugal force*.

Since such quantities are not true physical forces, they are often called *inertia forces*. Their presence or absence depends, not upon the particular problem at hand, but *upon the coordinate system chosen*.

In general, we see that if T involves the coordinate q_i explicitly, the quantity $\partial T/\partial q_i$ can be considered as an associated inertia force. Thus, if we denote this quantity by P_i,

$$P_i = \frac{\partial T}{\partial q_i}, \tag{141}$$

the ith equation of motion (138) becomes

$$\frac{dp_i}{dt} = Q_i + P_i. \tag{142}$$

We shall refer to the quantity P_i as a *momental* inertia force.*

We may notice next that while the quantity dp_i/dt will in general contain terms involving the *generalized acceleration* $\ddot{q}_i$, its expansion may also involve

* Various terminologies, some of which are at variance with this one, are in use.

nonaccelerational terms. Thus, if *accelerations* associated with generalized coordinates are to be of prime interest (as is usually the case), these latter terms may be conveniently transferred to the right in (142) and considered as additional (generalized) inertia forces. Such inertia forces are often said to be of the *Coriolis* type.

Thus a Coriolis inertia force is equivalent to an impressed force associated with q_i which tends to change the generalized velocity $\dot{q}_i$, but which does *not* tend to change the generalized momentum, when *actual* external forces are omitted. On the other hand, an inertia force of the "momental" type (e.g., a centrifugal force) is equivalent to an impressed force which tends to change both p_i and $\dot{q}_i$ in the absence of true external forces.

Since, from (139), we have $dp_r/dt \equiv m\ddot{r}$, no such terms are present in (140a) and we have

$$m\ddot{r} = Q_r + mr\dot{\theta}^2. \tag{143a}$$

However, since $dp_\theta/dt \equiv mr^2\ddot{\theta} + 2mr\dot{r}\dot{\theta}$, the second term can be conveniently transferred to the right in (140b) to give

$$mr^2\ddot{\theta} = Q_\theta - 2mr\dot{r}\dot{\theta}. \tag{143b}$$

The generalized " Coriolis force" in (143b) is clearly a *torque*.

We notice that the momental (centrifugal) term $mr\dot{\theta}^2$ is an inertia force with regard to change in both r momentum and r velocity, while the Coriolis term $-2mr\dot{r}\dot{\theta}$ is an inertia "force" (torque) only with regard to change in θ velocity. That is, one may say that "a θ velocity tends to change the r velocity and the r momentum, whereas simultaneous r and θ velocities tend to change the θ velocity, in the absence of actual impressed forces."

As a further example, we consider motion specified by spherical coordinates $(q_1, q_2, q_3) \equiv (r, \theta, \phi)$ in space, where r is distance from the origin, θ is polar angle, and ϕ is "cone angle" (Figure 2.5), so that

$$x = r\cos\theta\sin\phi, \qquad y = r\sin\theta\sin\phi, \qquad z = r\cos\phi.$$

Since the element of arc length is $ds^2 = dr^2 + r^2\sin^2\phi\, d\theta^2 + r^2\, d\phi^2$, there follows from $T = \frac{1}{2}m\left(\frac{ds}{dt}\right)^2$ the result

$$T = \tfrac{1}{2}m(\dot{r}^2 + r^2\dot{\theta}^2\sin^2\phi + r^2\dot{\phi}^2).$$

Thus, with the notation of (137) and (141), we have

$$p_r = m\dot{r}, \qquad p_\theta = mr^2\dot{\theta}\sin^2\phi, \qquad p_\phi = mr^2\dot{\phi} \tag{144a}$$

and

$$P_r = mr\dot{\theta}^2\sin^2\phi + mr\dot{\phi}^2,$$
$$P_\theta = 0, \quad P_\phi = mr^2\dot{\theta}^2\sin\phi\cos\phi. \tag{144b}$$

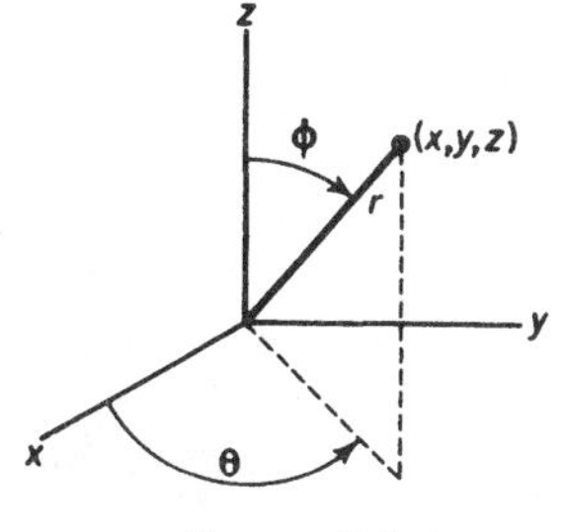

FIGURE 2.5

The equations of motion in the form (142) then become

$$\left.\begin{aligned} \frac{d}{dt}(m\dot{r}) &= Q_r + mr\dot{\theta}^2 \sin^2 \phi + mr\dot{\phi}^2, \\ \frac{d}{dt}(mr^2\dot{\theta} \sin^2 \phi) &= Q_\theta, \\ \frac{d}{dt}(mr^2\dot{\phi}) &= Q_\phi + mr^2\dot{\theta}^2 \sin \phi \cos \phi \end{aligned}\right\} . \tag{145a,b,c}$$

In particular, if a particle is constrained to move *on the surface of a sphere of radius a*, there follows $r = a$, $\dot{r} = \ddot{r} = 0$, and hence, from (145a), the necessary *physical constraint* normal to the sphere surface is given by

$$Q_r = -ma\dot{\theta}^2 \sin^2 \phi - ma\dot{\phi}^2. \tag{146}$$

That is, the total centrifugal inertia force $(P_r)_{r=a}$ must be balanced by a physical constraint equal to the negative of that quantity. Equations (145b,c) then give the equations of motion

$$ma^2 \frac{d}{dt}(\dot{\theta} \sin^2 \phi) = Q_\theta$$

or

$$ma^2 \sin^2 \phi\, \ddot{\theta} = Q_\theta - 2ma^2\dot{\theta}\dot{\phi} \sin \phi \cos \phi \tag{147a}$$

and

$$ma^2\ddot{\phi} = Q_\phi + ma^2\dot{\theta}^2 \sin \phi \cos \phi. \tag{147b}$$

The inertia force (torque) associated with θ is of Coriolis type, while that associated with ϕ is of momental type.

Similarly, if the particle is constrained to move *on the surface of the cone* $\phi = \alpha$ the constraint is given by (145c) in the form

$$Q_\phi = -mr^2\dot{\theta}^2 \sin \alpha \cos \alpha, \tag{148}$$

and the equations of motion are

$$m\ddot{r} = Q_r + mr\dot{\theta}^2 \sin^2 \alpha \tag{149a}$$

and

$$m \sin^2 \alpha \frac{d}{dt}(r^2\dot{\theta}) = Q_\theta,$$

or

$$mr^2 \sin^2 \alpha\, \ddot{\theta} = Q_\theta - 2mr\dot{r}\dot{\theta} \sin^2 \alpha. \tag{149b}$$

The inertia r force is momental, whereas the inertia θ torque is of Coriolis type.

In illustration, suppose that a bead of mass m is sliding without friction under gravity along a wire, inclined at an angle α to the vertical and rotating with constant angular velocity ω (Figure 2.6). Then m must move on the surface of the cone $\phi = \alpha$, in such a way that

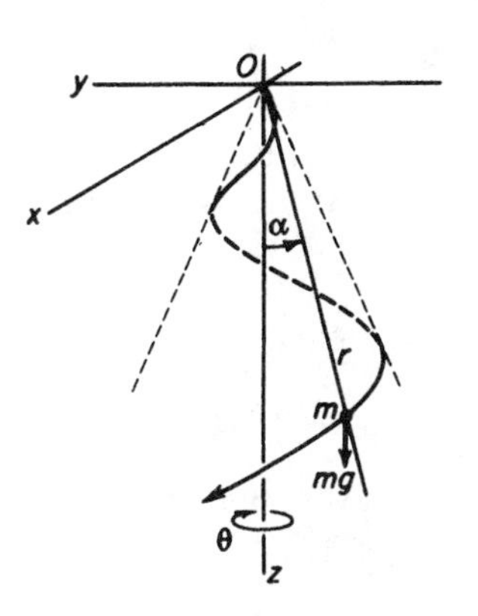

FIGURE 2.6

$$\theta = \omega t, \tag{150}$$

if we measure t from a time when $\theta = 0$. Also we have

$$Q_r = mg \cos \alpha.$$

Hence, with $\dot{\theta} = \omega$, equation (149a) is the equation of motion determining the coordinate r which, together with (150) and the relation $\phi = \alpha$, specifies the position of m. This equation takes the form

$$m\ddot{r} - mr\omega^2 \sin^2 \alpha = mg \cos \alpha, \tag{151}$$

from which there follows

$$r = c_1 \sinh (\omega t \sin \alpha) + c_2 \cosh (\omega t \sin \alpha) - \frac{g \cos \alpha}{\omega^2 \sin^2 \alpha} .$$

If the bead is released from rest at the origin at the instant $t = 0$, the evaluation of the constants gives

$$r = \frac{g \cos \alpha}{\omega^2 \sin^2 \alpha} [\cosh (\omega t \sin \alpha) - 1]. \tag{152}$$

With this expression for r, the generalized forces (torques) associated with ϕ and θ are then found from (148) and (149b), in the form

$$Q_\phi = -mr^2\omega^2 \sin\alpha \cos\alpha, \qquad Q_\theta = 2m\omega r\dot{r} \sin^2 \alpha. \tag{153a,b}$$

These results may be interpreted as follows: The forces acting on m in the positive ϕ direction are the gravity component $-mg \sin \alpha$ and the reaction component R_ϕ exerted by the wire. In a linear displacement $\delta s_\phi = r\, \delta\phi$, in which r and θ are considered to be held fixed, the work done by these forces would be $(R_\phi - mg \sin \alpha)r\, \delta\phi$. By equating this work to $Q_\phi\, \delta\phi$, we obtain the reaction R_ϕ in the form

$$R_\phi = mg \sin \alpha - mr\omega^2 \sin\alpha \cos\alpha.$$

Further, in a linear displacement $\delta s_\theta = r \sin \alpha\, \delta\theta$, in which r and ϕ are considered to be held fixed, the work done by the reaction component R_θ

would be $R_\theta\, r \sin \alpha\, \delta\theta$. Since the force of gravity has no component in the θ direction, there follows $R_\theta\, r \sin \alpha\, \delta\theta = Q_\theta\, \delta\theta$. Hence, the circumferential force exerted by the wire is given by

$$R_\theta = 2m\omega\dot{r} \sin \alpha.$$

In this example ϕ is fixed and, if we think of θ as prescribed, the system actually has only one degree of freedom. *The reactions R_ϕ and R_θ were obtained by considering the system as a degenerate case of a system with three degrees of freedom.*

If the reactions were *not* required, we would write directly

$$x = r \cos \omega t \sin \alpha, \qquad y = r \sin \omega t \sin \alpha, \qquad z = r \cos \alpha,$$

and so obtain T immediately in terms of the one independent coordinate r, in the form

$$T = \tfrac{1}{2}m(\dot{r}^2 + r^2\omega^2 \sin^2 \alpha).$$

Lagrange's equation for that coordinate would then be obtained as

$$\frac{d}{dt}(m\dot{r}) - mr\omega^2 \sin^2 \alpha = Q_r \equiv mg \cos \alpha,$$

in accordance with (151).

2.13. Constraints in dynamical systems. In some cases it is inconvenient or impossible to choose d independent generalized coordinates to specify the configuration of a system having d degrees of freedom. Instead, if n coordinates are used, we may have k independent auxiliary equations relating these n coordinates. The system is then usually said to have $n - k$ degrees of freedom, and the k restrictive equations are known as the *equations of constraint*. If these equations are of the form

$$\left.\begin{aligned} \phi_1(q_1, q_2, \cdots, q_n) &= 0, \\ \phi_2(q_1, q_2, \cdots, q_n) &= 0, \\ \cdots\cdots&\cdots\cdots \\ \phi_k(q_1, q_2, \cdots, q_n) &= 0 \end{aligned}\right\}. \tag{154}$$

then two possible procedures are available.

Clearly, if the k equations *can* be conveniently resolved to express k of the q's in terms of the $n - k$ remaining q's, then these latter q's are independent and the governing differential equations are determined by the methods of Section 2.11.

If this procedure is not convenient, the method of *Lagrange multipliers*, as described in Section 2.7, is useful. The method consists of first forming the variational conditions which follow from (154):

$$\left.\begin{array}{l}\dfrac{\partial \phi_1}{\partial q_1}\,\delta q_1 + \cdots + \dfrac{\partial \phi_1}{\partial q_n}\,\delta q_n = 0, \\ \cdots\cdots\cdots\cdots\cdots\cdots\cdots\cdots \\ \dfrac{\partial \phi_k}{\partial q_1}\,\delta q_1 + \cdots + \dfrac{\partial \phi_k}{\partial q_n}\,\delta q_n = 0\end{array}\right\}. \tag{155}$$

These equations are then multiplied respectively by functions $\lambda_1, \cdots, \lambda_k$, integrated over (t_1, t_2), and added to the equation of Hamilton's principle. Then, as was indicated in Section 2.7, we obtain n equations, each of the form

$$\frac{d}{dt}\left(\frac{\partial T}{\partial \dot{q}_i}\right) - \frac{\partial T}{\partial q_i} = Q_i + \lambda_1 \frac{\partial \phi_1}{\partial q_i} + \cdots + \lambda_k \frac{\partial \phi_k}{\partial q_i} \quad (i = 1, 2, \cdots, n). \tag{156}$$

These n equations, together with the k equations (154), then comprise $n + k$ equations in the $n + k$ unknown quantities $q_1, \cdots, q_n$ and $\lambda_1, \cdots, \lambda_k$. If the λ's are eliminated, the resultant n equations serve to determine the n q's.

In equation (156), Q_i is, as before, determined by the fact that $Q_i\,\delta q_i$ is the work done by the external forces when q_i is varied by δq_i and the remaining q's are held fixed. However, here it is important to notice that *such a variation may violate the physical constraint conditions*, which may require that a displacement δq_i actually be accompanied by changes in certain of the other q's. It is useful to notice that, for a *conservative* system, (156) is obtained by replacing V by $V - \Sigma\,\lambda_r \phi_r$ in (124b). This last quantity is sometimes called the *reduced potential energy*.

From (156) it is apparent that a term $\lambda_k\,\partial\phi_k/\partial q_i$ is of the nature of a generalized force, due to the kth constraint and associated with the ith coordinate. Each constraint thus may contribute an additional generalized force to each of the equations of motion.

However, we notice that the work done in *any* set of displacements by the force due to the kth constraint is given by

$$\lambda_k \frac{\partial \phi_k}{\partial q_1}\,\delta q_1 + \lambda_k \frac{\partial \phi_k}{\partial q_2}\,\delta q_2 + \cdots + \lambda_k \frac{\partial \phi_k}{\partial q_n}\,\delta q_n.$$

Hence, by virtue of equations (155), *the work done by the (fixed) constraint vanishes if the displacements satisfy the constraint conditions.* Displacements which are compatible with the constraint conditions are often called *virtual displacements.*

In certain cases, a constraint condition may *not* be expressible in the form (154), but may be of the form

$$C_1\,\delta q_1 + \cdots + C_n\,\delta q_n = 0, \tag{157}$$

where the left-hand member is not proportional to the variation of *any* function. In such a case the constraint is said to be *nonholonomic*. If k nonholonomic constraints are involved, it is not *possible* to eliminate certain of the q's by solving equations similar to (154), so that n coordinates are still needed to specify the configuration. Nevertheless, the system is usually said to possess only $n - k$ degrees of freedom.

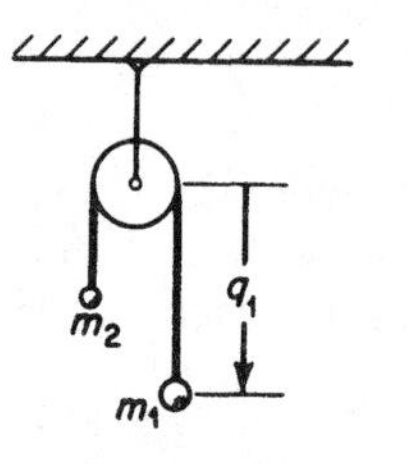

FIGURE 2.7

In any case, it is clear that the method of Lagrange multipliers (which involves only the *variations* of the coordinates) is again directly applicable in that the functions $\partial\phi_1/\partial q_1, \cdots, \partial\phi_1/\partial q_n$ are merely replaced by the functions $C_1, \cdots, C_n$. The general problem of the rolling of a disk on a plane is found to be one involving nonholonomic constraints, and is solvable by these methods.*

The basic ideas of this section may be illustrated by two elementary examples. The simple pulley of Figure 2.7 possesses one degree of freedom, and q_1 is a suitable coordinate. The kinetic energy of the system, neglecting the weight of the cord, is

$$T = \tfrac{1}{2}(m_1 + m_2)\dot{q}_1^2.$$

If q_1 is increased by δq_1, the work done by gravity is given by

$$m_1 g\, \delta q_1 - m_2 g\, \delta q_1$$

and hence

$$Q_1 = (m_1 - m_2)g.$$

Thus the equation of motion is

$$(m_1 + m_2)\ddot{q}_1 = (m_1 - m_2)g, \tag{158}$$

as is also obvious from other considerations.

Suppose, however, *to illustrate the preceding developments in a simple case*, that the *two* coordinates q_1 and q_2 of Figure 2.8 are used. These two coordinates clearly are not independent since, if the total length of the cord (assumed to be inextensible) is l, the constraint

$$q_1 + q_2 = l \tag{159}$$

must be imposed. In terms of q_1 and q_2, the kinetic energy of the system is

$$T = \tfrac{1}{2}m_1\dot{q}_1^2 + \tfrac{1}{2}m_2\dot{q}_2^2. \tag{160}$$

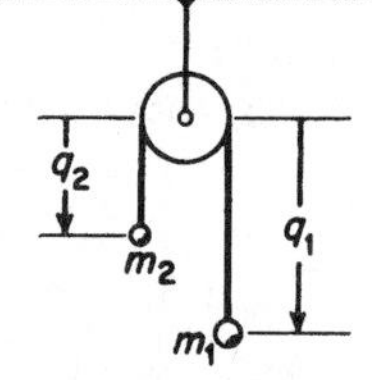

FIGURE 2.8

If q_1 is increased by δq_1 *and q_2 is held fixed* (violating the constraint condition), the work done by gravity in the displacement δq_1 is $m_1 g\, \delta q_1$. Thus, we must have

$$Q_1 = m_1 g$$

* See, for example, Reference 10.

and, similarly,

$$Q_2 = m_2 g.$$

From (159) we have also

$$\delta q_1 + \delta q_2 = 0. \tag{161}$$

This condition clearly requires that the displacements satisfy the constraint condition, that is, that they be *virtual* displacements, so that the work done by the constraining tension vanishes.

With the introduction of a Lagrange multiplier, the equations corresponding to (156) become

$$\left.\begin{aligned} m_1\ddot{q}_1 &= m_1 g + \lambda, \\ m_2\ddot{q}_2 &= m_2 g + \lambda \end{aligned}\right\}. \tag{162a,b}$$

Equations (162a,b) and (159) are the desired three equations in q_1, q_2, and λ. The elimination of λ between (162a,b) gives

$$m_1\ddot{q}_1 - m_2\ddot{q}_2 = (m_1 - m_2)g, \tag{163}$$

and the elimination of q_2 between (163) and (159) then leads to (158).

From (162) we see that λ is the force exerted by the tension in the cord on each of the masses. By eliminating q_1 between (158) and (162a), we find that

$$\lambda = -2\left(\frac{m_1 m_2}{m_1 + m_2}\right)g. \tag{164}$$

The negative sign corresponds to the fact that the tensile force acts in the negative direction relative to q_1 and q_2.

As a second example, involving two constraints, we consider the rolling of a right circular cylinder of mass m on another such cylinder, assuming the axes of the cylinders to be parallel. We choose the angles θ_1 and θ_2 of Figure 2.9 and the distance r between the centers as coordinates, noticing in advance that *so long as the cylinders are in contact* θ_1 and θ_2 are not independent, and r is actually constrained to remain constant. We notice that the rolling cylinder rotates through an angle $\theta_2 - \theta_1$ as the angle θ_1 is generated by the line of centers, and also that the kinetic energy of the rolling cylinder is composed of two parts: one of the form $\frac{1}{2}m(\dot{r}^2 + r^2\dot{\theta}_1^2)$ due to translation of the center of gravity, and one of the form $\frac{1}{2}(\frac{1}{2}mR_2^2)\dot{\theta}_2^2$ due to rotation *about* the center of gravity. Hence we have

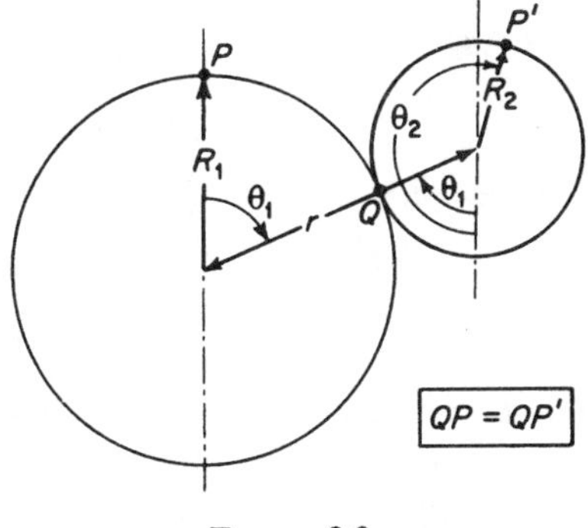

FIGURE 2.9

$$T = \tfrac{1}{2}m(\dot{r}^2 + r^2\dot{\theta}_1^2 + \tfrac{1}{2}R_2^2\dot{\theta}_2^2). \tag{165}$$

The potential energy clearly can be taken as

$$V = mgr\cos\theta_1 + \text{constant}. \tag{166}$$

The requirement of contact leads to the constraint equation

$$r - (R_1 + R_2) = 0, \tag{167}$$

while, if *pure rolling* is present, the condition $R_1\theta_1 = R_2(\theta_2 - \theta_1)$ leads to the frictional constraint equation

$$(R_1 + R_2)\theta_1 - R_2\theta_2 = 0. \tag{168}$$

The variational forms of (167) and (168) are then

$$\left.\begin{aligned} 1\cdot\delta r + 0\cdot\delta\theta_1 + 0\cdot\delta\theta_2 &= 0,\\ 0\cdot\delta r + (R_1 + R_2)\,\delta\theta_1 - R_2\,\delta\theta_2 &= 0 \end{aligned}\right\}. \tag{169a,b}$$

With the introduction of two Lagrange multipliers λ_1 and λ_2, the Lagrange equations (156) take the form

$$\left.\begin{aligned} m\ddot{r} &= -mg\cos\theta_1 + mr\dot{\theta}_1^2 + \lambda_1,\\ \frac{d}{dt}(mr^2\dot{\theta}_1) &= mgr\sin\theta_1 + \lambda_2(R_1 + R_2),\\ \tfrac{1}{2}mR_2^2\ddot{\theta}_2 &= -\lambda_2 R_2 \end{aligned}\right\}, \tag{170a,b,c}$$

the coefficients $\{1, 0, 0\}$ of λ_1 and $\{0, R_1 + R_2, -R_2\}$ of λ_2, in successive equations, being read from (169a,b).

If r and θ_2 are eliminated by use of (167) and (168), equations (170a,c) give

$$\lambda_1 = mg\cos\theta_1 - m(R_1 + R_2)\dot{\theta}_1^2, \tag{171a}$$

$$\lambda_2 = -\tfrac{1}{2}mR_2\ddot{\theta}_2 = -\tfrac{1}{2}m(R_1 + R_2)\ddot{\theta}_1, \tag{171b}$$

and the combination of (170b) and (171b) gives

$$m(R_1 + R_2)^2\ddot{\theta}_1 = mg(R_1 + R_2)\sin\theta_1 - \tfrac{1}{2}m(R_1 + R_2)^2\ddot{\theta}_1$$

or

$$\tfrac{3}{2}(R_1 + R_2)\ddot{\theta}_1 - g\sin\theta_1 = 0. \tag{172}$$

This is the required equation of motion, valid so long as contact persists.

From (170a) we see that λ_1 is the normal force (in the r direction) exerted by the stationary cylinder on the rolling cylinder. Equation (171a) shows that this force is *positive* only when

$$g\cos\theta_1 > (R_1 + R_2)\dot{\theta}_1^2,$$

after which contact ceases and the constraints are removed. Equations (170a,b,c) then hold with $\lambda_1 = \lambda_2 = 0$.

From (170c) it follows that $-\lambda_2$ is the frictional force exerted on the rolling cylinder ($-\lambda_2 R_2$ is the corresponding torque about its center). By combining (171b) and (172), we obtain

$$-\lambda_2 = \tfrac{1}{3} mg \sin \theta_1. \tag{173}$$

If the constraints were of no interest, we would more economically introduce (167) and (168) directly into (165) and (166), to obtain

$$T = \tfrac{3}{4} m(R_1 + R_2)^2 \dot{\theta}_1^{\,2}$$

and

$$V = mg(R_1 + R_2) \cos \theta_1 + \text{constant}.$$

Since θ_1 is now an independent coordinate, equation (172) follows immediately as the relevant Lagrange equation. It should be noticed that no information as to the range of validity of (172) or as to the nature of the subsequent motion would then be obtained.

While the examples just considered, in which the constraint equations were of sufficiently simple form to permit direct elimination of superfluous coordinates, do not illustrate the *efficiency* of the method of Lagrange multipliers in more involved problems, they may serve to illustrate the *technique* involved. Further, they indicate the fact that such multipliers are very often capable of useful physical interpretation, and that their use in connection with superfluous coordinates may lead to the determination of unknown constraints when they are of interest.

At the same time, it should be noticed that the simplicity involved in the use of Lagrange's equations stems from the fact that unknown constraints generally can be omitted from consideration when they are *not* of interest.

2.14. Small vibrations about equilibrium. Normal coordinates. In many problems in dynamics we deal with a system for which there exists a stable *equilibrium* configuration in which the system can remain permanently at rest, and such that motions with small displacements and velocities can persist near the equilibrium state. If the system is specified by n generalized coordinates $q_1, q_2, \cdots, q_n$, we can choose these coordinates in such a way that they are all zero at an equilibrium position. For simplicity, we consider here only the case when $n = 2$, so that the system possesses two degrees of freedom and is completely specified by the coordinates q_1 and q_2. The results to be obtained are readily generalized.

In the case of a *conservative* system with two degrees of freedom, there exists a potential energy function V which depends only upon q_1 and q_2, and which generally can be expanded in a power series, near $q_1 = q_2 = 0$, of the form

$$V(q_1, q_2) = V_0 + \left(\frac{\partial V}{\partial q_1}\right)_0 q_1 + \left(\frac{\partial V}{\partial q_2}\right)_0 q_2$$
$$+ \frac{1}{2}\left[\left(\frac{\partial^2 V}{\partial q_1^{\,2}}\right)_0 q_1^{\,2} + 2\left(\frac{\partial^2 V}{\partial q_1\,\partial q_2}\right)_0 q_1 q_2 + \left(\frac{\partial^2 V}{\partial q_2^{\,2}}\right)_0 q_2^{\,2}\right] + \cdots, \tag{174}$$

where a zero subscript indicates evaluation at the equilibrium position $q_1 = q_2 = 0$. But, since V must be stationary at equilibrium (see Problem 73), the linear terms must vanish. The constant V_0 is irrelevant, and it can be taken to be zero. Hence, if the terms of order greater than two in the expansion of V are neglected, we can write

$$V = \tfrac{1}{2}(a_{11}q_1^2 + 2a_{12}q_1q_2 + a_{22}q_2^2), \tag{175}$$

where the a's are constants defined by comparison with (174), so that (to a first approximation) V is a homogeneous quadratic function of q_1 and q_2, with constant coefficients.

The kinetic energy T is of the form

$$T = \tfrac{1}{2}(b_{11}\dot{q}_1^2 + 2b_{12}\dot{q}_1\dot{q}_2 + b_{22}\dot{q}_2^2), \tag{176}$$

where the b's may depend upon q_1 and q_2. For small departures from equilibrium, and small velocities, these coefficients may be replaced by their values when $q_1 = q_2 = 0$. Hence, in such cases, (176) expresses T approximately as a homogeneous quadratic function of $\dot{q}_1$ and $\dot{q}_2$, with constant coefficients.

With these approximations, Lagrange's equations in the form (124b) become merely

$$\frac{d}{dt}\left(\frac{\partial T}{\partial \dot{q}_i}\right) + \frac{\partial V}{\partial q_i} = 0 \qquad (i = 1, 2) \tag{177}$$

or, with the notation of (175) and (176),

$$\left.\begin{aligned} b_{11}\ddot{q}_1 + b_{12}\ddot{q}_2 + a_{11}q_1 + a_{12}q_2 &= 0, \\ b_{12}\ddot{q}_1 + b_{22}\ddot{q}_2 + a_{12}q_1 + a_{22}q_2 &= 0 \end{aligned}\right\}. \tag{178a,b}$$

It is important to notice that these differential equations are *linear, with constant coefficients.*

If the equilibrium state is to be *stable*, V must possess a relative *minimum* at $q_1 = q_2 = 0$, so that the form (175) must be *positive* unless $q_1 = q_2 = 0$. Also, the kinetic energy T cannot be negative, and cannot vanish unless all the *velocities* vanish. Quadratic forms having this property are said to be *positive definite* forms (see Section 1.19).

If we consider the coordinates q_1 and q_2 as the components of a *vector* $\mathbf{q}$, equations (178a,b) can be combined into the matrix equation

$$\mathbf{B}\,\ddot{\mathbf{q}} + \mathbf{A}\,\mathbf{q} = \mathbf{0}. \tag{179}$$

Following the usual procedure for solving such sets of equations, we seek solutions of the form

$$\mathbf{q} = \mathbf{x}\cos(\omega t + \gamma), \tag{180}$$

where the elements of $\mathbf{x}$ are the amplitudes of the required solutions, and are to be independent of t. Equation (179) then takes the form

$$\mathbf{A}\,\mathbf{x} = \omega^2\mathbf{B}\,\mathbf{x}. \tag{181}$$

In this way we are led to a characteristic-value problem of the type considered in Section 1.25, with ω^2 identified with the parameter λ of that section. The characteristic values of ω^2 thus are the roots of the characteristic equation

$$|\mathbf{A} - \omega^2\mathbf{B}| \equiv \begin{vmatrix} a_{11} - \omega^2 b_{11} & a_{12} - \omega^2 b_{12} \\ a_{12} - \omega^2 b_{12} & a_{22} - \omega^2 b_{22} \end{vmatrix} = 0. \tag{182}$$

Since $\mathbf{A}$ and $\mathbf{B}$ are symmetric and positive definite, the results of Section 1.25 show that the roots of this quadratic equation in ω^2 are *real and positive*. However, they need not be *distinct*. Corresponding to each distinct root, the problem possesses a nontrivial vector solution. Furthermore, to a double root there correspond two linearly independent solutions which can be specified in infinitely many ways. Two solution vectors $\mathbf{v}_1$ and $\mathbf{v}_2$ which correspond to distinct characteristic values of ω^2 are orthogonal with respect to both $\mathbf{A}$ and $\mathbf{B}$; that is, we have the relations

$$\mathbf{v}_1^T\,\mathbf{A}\,\mathbf{v}_2 = 0, \qquad \mathbf{v}_1^T\,\mathbf{B}\,\mathbf{v}_2 = 0 \tag{183}$$

in this case. Furthermore, the two linearly independent solution vectors corresponding to a repeated root of (182) can be so orthogonalized by the generalized Gram-Schmidt procedure, if this is desirable.

Thus if $\mathbf{v}_1 = \{v_{11}, v_{12}\}$ and $\mathbf{v}_2 = \{v_{21}, v_{22}\}$ are linearly independent characteristic vectors corresponding to ω_1^2 and ω_2^2, respectively, where ω_1 and ω_2 need not be distinct, the solutions corresponding to the assumption (180) are then given by the expressions

$$\mathbf{q}_1 = \mathbf{v}_1 \cos(\omega_1 t + \gamma_1), \qquad \mathbf{q}_2 = \mathbf{v}_2 \cos(\omega_2 t + \gamma_2) \tag{184}$$

where γ_1 and γ_2 are arbitrary constants. The most general solution is then obtained by superposition, in the vector form

$$\mathbf{q} = c_1\mathbf{v}_1 \cos(\omega_1 t + \gamma_1) + c_2\mathbf{v}_2 \cos(\omega_2 t + \gamma_2), \tag{185}$$

or in the expanded form

$$\left.\begin{aligned} q_1 &= c_1 v_{11} \cos(\omega_1 t + \gamma_1) + c_2 v_{21} \cos(\omega_2 t + \gamma_2), \\ q_2 &= c_1 v_{12} \cos(\omega_1 t + \gamma_1) + c_2 v_{22} \cos(\omega_2 t + \gamma_2) \end{aligned}\right\}, \tag{186a,b}$$

where c_1 and c_2 are also arbitrary constants.

It follows that *the most general motion of the specified system is a superposition of two simple-harmonic motions*, the *natural frequencies* of which are $\omega_i/2\pi$. The fact that this statement is true even in the case when (182) possesses repeated roots is of particular importance.

The general solution (185) or (186a,b) can be written in the matrix form

$$\begin{Bmatrix} q_1 \\ q_2 \end{Bmatrix} = \begin{bmatrix} v_{11} & v_{21} \\ v_{12} & v_{22} \end{bmatrix} \begin{Bmatrix} c_1 \cos(\omega_1 t + \gamma_1) \\ c_2 \cos(\omega_2 t + \gamma_2) \end{Bmatrix} \tag{187a}$$

or

$$\mathbf{q} = \mathbf{M}\,\mathbf{h}, \tag{187b}$$

where $\mathbf{M}$ is a *modal matrix* having the components of successive characteristic vectors as the elements of its successive *columns*, and $\mathbf{h}$ denotes the vector multiplied by $\mathbf{M}$ in (187a). If the equal members of (187b) are premultiplied by $\mathbf{M}^{-1}$, there follows simply

$$\boldsymbol{\alpha} = \mathbf{h}, \tag{188}$$

where the vector $\boldsymbol{\alpha}$ is defined by the relation

$$\mathbf{q} = \mathbf{M}\,\boldsymbol{\alpha}, \qquad \boldsymbol{\alpha} = \mathbf{M}^{-1}\,\mathbf{q}. \tag{189a,b}$$

Thus the new coordinates α_1 and α_2 so defined are such that the general solution of (178a,b) or (179) can be expressed in the simple form

$$\alpha_1 = c_1 \cos(\omega_1 t + \gamma_1), \qquad \alpha_2 = c_2 \cos(\omega_2 t + \gamma_2), \tag{190}$$

the two modes of vibration then being *uncoupled*. These new coordinates are often called *normal coordinates* of the problem.

Since the characteristic vectors $\mathbf{v}_1$ and $\mathbf{v}_2$ are each determined only within a multiplicative arbitrary constant, the modal matrix $\mathbf{M}$ and the normal coordinates α_1, α_2 are not uniquely defined. For some purposes it is convenient to *normalize* the vectors relative to either $\mathbf{A}$ or $\mathbf{B}$.* In particular, we may determine multiples of $\mathbf{v}_1$ and $\mathbf{v}_2$, say $\mathbf{e}_1$ and $\mathbf{e}_2$, which are normalized relative to $\mathbf{B}$, so that

$$\mathbf{e}_i^T\,\mathbf{B}\,\mathbf{e}_j = \delta_{ij} \tag{191}$$

where δ_{ij} is the Kronecker delta. It then follows that the *orthonormal modal matrix* $\mathbf{M}$ made up of these *normalized* characteristic vectors has the property that it satisfies the equation

$$\mathbf{M}^T\,\mathbf{B}\,\mathbf{M} = \mathbf{I}, \tag{192}$$

where $\mathbf{I}$ is the *unit matrix*. By postmultiplying both members of (192) by $\mathbf{M}^{-1}$, there then follows $\mathbf{M}^{-1} = \mathbf{M}^T\,\mathbf{B}$, so that (189) takes the more convenient form

$$\mathbf{q} = \mathbf{M}\,\boldsymbol{\alpha}, \qquad \boldsymbol{\alpha} = \mathbf{M}^T\,\mathbf{B}\,\mathbf{q} \tag{193a,b}$$

in this case. Furthermore, it then follows (as in Section 1.25) that also

$$\mathbf{M}^T\,\mathbf{A}\,\mathbf{M} = \begin{bmatrix} \omega_1^2 & 0 \\ 0 & \omega_2^2 \end{bmatrix}. \tag{194}$$

* In some references, the coordinates α_1 and α_2 are called *natural* coordinates, and are said to be *normal* coordinates only when they are *normalized* in a certain way, as the terminology suggests.

Consequently, with the substitution (193), the potential and kinetic energies then become

$$\left.\begin{aligned} V &= \tfrac{1}{2}\mathbf{q}^T\,\mathbf{A}\,\mathbf{q} = \tfrac{1}{2}\boldsymbol{\alpha}^T(\mathbf{M}^T\,\mathbf{A}\,\mathbf{M})\boldsymbol{\alpha} = \tfrac{1}{2}(\omega_1{}^2\alpha_1{}^2 + \omega_2{}^2\alpha_2{}^2), \\ T &= \tfrac{1}{2}\dot{\mathbf{q}}^T\,\mathbf{B}\,\dot{\mathbf{q}} = \tfrac{1}{2}\dot{\boldsymbol{\alpha}}^T(\mathbf{M}^T\,\mathbf{B}\,\mathbf{M})\dot{\boldsymbol{\alpha}} = \tfrac{1}{2}(\dot{\alpha}_1{}^2 + \dot{\alpha}_2{}^2) \end{aligned}\right\}. \quad (195\text{a,b})$$

The corresponding Lagrangian equations are then simply

$$\ddot{\alpha}_1 + \omega_1{}^2\alpha_1 = 0, \qquad \ddot{\alpha}_2 + \omega_2{}^2\alpha_2 = 0, \tag{196}$$

and equations (190) follow immediately.

More generally, if the columns of $\mathbf{M}$ are *not* necessarily normalized relative to $\mathbf{B}$, it is easily shown that the substitution $\mathbf{q} = \mathbf{M}\,\boldsymbol{\alpha}$ reduces V and T to the forms

$$V = \tfrac{1}{2}(\omega_1{}^2 f_1{}^2\alpha_1{}^2 + \omega_1{}^2 f_2{}^2\alpha_2{}^2), \qquad T = \tfrac{1}{2}(f_1{}^2\dot{\alpha}_1{}^2 + f_2{}^2\dot{\alpha}_2{}^2) \quad (197\text{a,b})$$

where

$$f_i{}^2 = \mathbf{v}_i{}^T\,\mathbf{B}\,\mathbf{v}_i. \tag{198}$$

The corresponding Lagrangian equations,

$$f_1{}^2(\ddot{\alpha}_1 + \omega_1{}^2\alpha_1) = 0, \qquad f_2{}^2(\ddot{\alpha}_2 + \omega_2{}^2\alpha_2) = 0$$

are seen to be of the same form as (196), in accordance with the previously established validity of (190). However, unless (192) is true the relation (193b) is not a consequence of (193a), and (189b) must be used instead, for the purpose of expressing α_1 and α_2 explicitly in terms of q_1 and q_2. Unless it is desirable to actually reduce the expressions for the potential and kinetic energies to the standard forms (195a,b), it is clear that only this advantage is gained by normalizing the characteristic vectors in the developments under consideration.

In the more general case, the impressed force system may consist of a conservative part, derivable from a potential energy V, and also of a *dissipative* (nonconservative) part. In particular, there may exist resistive forces R_1 and R_2, associated with q_1 and q_2, which are proportional to the velocities, and hence are expressible in the forms

$$R_1 = -(r_{11}\dot{q}_1 + r_{12}\dot{q}_2), \qquad R_2 = -(r_{21}\dot{q}_1 + r_{22}\dot{q}_2). \tag{199}$$

These terms would then be added to the right-hand members of (178a,b), thus introducing velocity terms into the linearized equations of motion. In the special case when $r_{21} = r_{12}$ (in particular, when these two coupling coefficients are zero), if we define the function

$$F = \tfrac{1}{2}(r_{11}\dot{q}_1{}^2 + 2r_{12}\dot{q}_1\dot{q}_2 + r_{22}\dot{q}_2{}^2), \tag{200}$$

we see that

$$R_i = -\frac{\partial F}{\partial \dot{q}_i} \qquad (i = 1, 2). \tag{201}$$

The function F is then known as *Rayleigh's dissipation function.*

If we denote any external forces associated with q_1 and q_2 which are dissipative but *not* derivable from a dissipation function by Q_1' and Q_2', respectively, equation (177) must be modified to read

$$\frac{d}{dt}\left(\frac{\partial T}{\partial \dot{q}_i}\right) + \frac{\partial V}{\partial q_i} + \frac{\partial F}{\partial \dot{q}_i} = Q_i' \qquad (i = 1, 2), \tag{202}$$

to include both conservative and dissipative forces. With the notation of (175), (176), and (200), this set of equations can be combined in the matrix form

$$\mathbf{B}\,\ddot{\mathbf{q}} + \mathbf{R}\,\dot{\mathbf{q}} + \mathbf{A}\,\mathbf{q} = \mathbf{q}' \tag{203}$$

where **B**, **R**, and **A** are symmetric square matrices and $\mathbf{q}'$ is the column vector $\{Q_1', Q_2'\}$.

The coefficients b_{ij} in (176) are often known as the *inertia coefficients* associated with q_1 and q_2, and the coefficients a_{ij} in (175) as the *stiffness coefficients*. The coefficients r_{ij} in (200), which govern deviations from simple-harmonic motions when suitable dissipative forces are present, are often called the associated *resistance coefficients*.

2.15. Numerical example. To illustrate the results of the preceding section, we consider the determination of natural modes of small vibration of the compound pendulum of Section 2.11. We obtain the form (175),

$$V = \tfrac{1}{2}(m_1 + m_2)gl_1\theta_1^2 + \tfrac{1}{2}m_2gl_2\theta_2^2, \tag{204}$$

by retaining leading terms in the expansion of (127b), and the form (176),

$$T = \tfrac{1}{2}(m_1 + m_2)l_1^2\dot{\theta}_1^2 + m_2l_1l_2\dot{\theta}_1\dot{\theta}_2 + \tfrac{1}{2}m_2l_2^2\dot{\theta}_2^2, \tag{205}$$

by setting $\theta_1 = \theta_2 = 0$ in (127a). That $\theta_1 = \theta_2 = 0$ actually specifies an equilibrium state (as is clear from physical considerations) follows mathematically from the fact that $\partial V/\partial\theta_1 = \partial V/\partial\theta_2 = 0$ when $\theta_1 = \theta_2 = 0$.

The resultant equations of small oscillations,

$$(m_1 + m_2)l_1^2\ddot{\theta}_1 + m_2l_1l_2\ddot{\theta}_2 + (m_1 + m_2)gl_1\theta_1 = 0 \tag{206a}$$

and

$$m_2l_1l_2\ddot{\theta}_1 + m_2l_2^2\ddot{\theta}_2 + m_2gl_2\theta_2 = 0, \tag{206b}$$

are equivalent to the results of linearizing (128a,b) in the displacements and velocities.

We now consider explicitly the special case in which

$$m_1 = m_2 \equiv m, \qquad l_1 = l_2 \equiv l, \tag{207}$$

so that, after removing a factor ml^2 from the equal members of the governing equations, there follows

$$\left.\begin{aligned} 2\ddot{\theta}_1 + \ddot{\theta}_2 + 2\frac{g}{l}\theta_1 = 0, \\ \ddot{\theta}_1 + \ddot{\theta}_2 + \frac{g}{l}\theta_2 = 0 \end{aligned}\right\}. \tag{208a,b}$$

Corresponding to the assumption

$$\boldsymbol{\theta} = \mathbf{x} \cos(\omega t + \gamma), \tag{209}$$

the equation corresponding to (181) is obtained in the form

$$\mathbf{A}\,\mathbf{x} = \bar{\omega}^2 \mathbf{B}\,\mathbf{x}, \tag{210}$$

where

$$\mathbf{A} = \begin{bmatrix} 2 & 0 \\ 0 & 1 \end{bmatrix}, \qquad \mathbf{B} = \begin{bmatrix} 2 & 1 \\ 1 & 1 \end{bmatrix} \tag{211a,b}$$

and where $\bar{\omega}$ is a *dimensionless* parameter defined by the relation

$$\bar{\omega}^2 = \frac{l}{g}\,\omega^2. \tag{212}$$

The matrices **A** and **B** so defined differ from those defined in the preceding section only in that their elements have been made dimensionless.

We have then to deal with the matrix

$$\mathbf{A} - \bar{\omega}^2\mathbf{B} = \begin{bmatrix} 2(1 - \bar{\omega}^2) & -\bar{\omega}^2 \\ -\bar{\omega}^2 & 1 - \bar{\omega}^2 \end{bmatrix}, \tag{213}$$

the vanishing of the determinant of which leads to the characteristic equation

$$\bar{\omega}^4 - 4\bar{\omega}^2 + 2 = 0. \tag{214}$$

Corresponding to the smaller root,

$$\bar{\omega}_1{}^2 = 0.586, \tag{215}$$

the matrix (213) takes the form

$$\mathbf{A} - \bar{\omega}_1{}^2\mathbf{B} = \begin{bmatrix} 0.828 & -0.586 \\ -0.586 & 0.414 \end{bmatrix}$$

and the elements of the modal column are proportional to the *cofactors* of the elements in either *row* of this matrix. By choosing the first row, we may take $\mathbf{v}_1$ in the form

$$\mathbf{v}_1 = \begin{Bmatrix} 0.414 \\ 0.586 \end{Bmatrix}. \tag{216}$$

Similarly, in correspondence with the other root of (214),

$$\bar{\omega}_2{}^2 = 3.414, \tag{217}$$

the modal column $\mathbf{v}_2$ may be taken in the form

$$\mathbf{v}_2 = \begin{Bmatrix} -2.414 \\ 3.414 \end{Bmatrix}. \tag{218}$$

The general solution of (208a,b) then can be expressed by the relations

$$\left.\begin{aligned} q_1 &= 0.414c_1 \cos(\omega_1 t + \gamma_1) - 2.414c_2 \cos(\omega_2 t + \gamma_2), \\ q_2 &= 0.586c_1 \cos(\omega_1 t + \gamma_1) + 3.414c_2 \cos(\omega_2 t + \gamma_2) \end{aligned}\right\}, \tag{219}$$

where

$$\omega_1 = 0.765\sqrt{\frac{g}{l}}, \qquad \omega_2 = 1.848\sqrt{\frac{g}{l}}, \tag{220}$$

and where c_1, c_2, γ_1, and γ_2 are arbitrary constants.

The modal matrix made up of the elements of (216) and (218) is then of the form

$$\mathbf{M} = \begin{bmatrix} 0.414 & -2.414 \\ 0.586 & 3.414 \end{bmatrix}, \tag{221}$$

and the corresponding normal coordinates are defined by the matrix equation

$$\boldsymbol{\alpha} = \mathbf{M}^{-1}\,\mathbf{q} = \begin{bmatrix} 1.207 & 0.854 \\ -0.207 & 0.146 \end{bmatrix} \mathbf{q}. \tag{222}$$

The modal columns (216) and (218) happen to be normalized in such a way that the elements in each column add to unity. By multiplying each column by a suitable constant, these columns can be normalized in other ways, and multiples of the present α_1 and α_2 (in terms of which the natural modes are *also* uncoupled) are then obtained.

2.16. Variational problems for deformable bodies. General variational principles have been established in connection with the theory of elasticity,* as well as in many other fields. In this section no attempts are made to establish such general theories. Instead, it is shown in what way a variational problem can be derived *from a differential equation and the associated boundary conditions,* in certain illustrative cases. In later sections it is shown that such formulations are readily adapted to *approximate* analysis.

We start with the problem of determining *small deflections of a rotating string* of length l. The governing differential equation is then of the form

$$\frac{d}{dx}\left(F\frac{dy}{dx}\right) + \rho\omega^2 y + p = 0, \tag{223}$$

where $y(x)$ is the displacement of a point from the axis of rotation, $F(x)$ is the tension, $\rho(x)$ the linear mass density, ω the angular velocity of rotation, and $p(x)$ is the intensity of a distributed radial load. Suitable end conditions are also to be prescribed.

In order to formulate a corresponding variational problem, we first multiply both sides of (223) by a variation δy and integrate the result over

* See Reference 7.

$(0, l)$ to obtain

$$\int_0^l \frac{d}{dx}\left(F\frac{dy}{dx}\right)\delta y\,dx + \int_0^l \rho\omega^2 y\,\delta y\,dx + \int_0^l p\,\delta y\,dx = 0. \tag{224}$$

The second and third integrands are the variations of $\frac{1}{2}\rho\omega^2 y^2$ and py, respectively. If the first integral is transformed by integration by parts, it takes the form

$$\left[F\frac{dy}{dx}\,\delta y\right]_0^l - \int_0^l F\frac{dy}{dx}\,\delta\frac{dy}{dx}\,dx,$$

and the integrand in this form is the variation of $\frac{1}{2}F(dy/dx)^2$. Thus the left-hand member of (224) can be transformed to the left-hand member of the equation

$$\delta\int_0^l\left[\frac{1}{2}\rho\omega^2y^2 + py - \frac{1}{2}F\left(\frac{dy}{dx}\right)^2\right]dx + \left[F\frac{dy}{dx}\,\delta y\right]_0^l = 0. \tag{225}$$

If we impose at each of the two ends one of the conditions

$$y = y_0 \quad or \quad F\frac{dy}{dx} = 0 \qquad (when\ x = 0, l) \tag{226}$$

where y_0 is a prescribed constant, the integrated terms in (225) vanish and the condition becomes

$$\delta\int_0^l\left[\frac{1}{2}\rho\omega^2y^2 + py - \frac{1}{2}F\left(\frac{dy}{dx}\right)^2\right]dx = 0. \tag{227}$$

Conversely, (223) is the Euler equation [(17a)] corresponding to (227). That is, if y renders the integral in (227) stationary it must satisfy (223), while if y satisfies (223) and end conditions of the type required in (226), then y renders the integral in (227) stationary.

The end conditions (226) or, equivalently,

$$\left[F\frac{dy}{dx}\,\delta y\right]_0^l = 0, \tag{228}$$

are the so-called *natural boundary conditions* of the variational problem (227). If we recall that (in the linearized theory) the product $F\,dy/dx$ is the component of the tensile force normal to the axis of rotation, we see that (228) *requires that the end tensions do no work*. This situation exists if no end motion is permitted, so that $\delta y = 0$, or if no end restraint (normal to the axis of rotation) is present, so that $F\,dy/dx = 0$.

Thus, *if the end tensions do no work*, we conclude that *of all functions* $y(x)$ *which satisfy the relevant end conditions, that one which also satisfies the relevant differential equation* (223) *renders the integral in* (227) *stationary.*

It is clear that the term $\frac{1}{2}\rho(\omega y)^2$ in (227) represents the *kinetic energy* of the string per unit length, since the speed of an element of the string is given by ωy. Also, since $p\,\delta y\,dx$ is the element of work done by p on an element dx in a displacement δy, the term $-py$ is *potential energy* per unit length due to the radial force distribution $p(x)$. To identify the remaining term, we notice that an element of original length dx stretches into an element of length

$$ds = \left[1 + \left(\frac{dy}{dx}\right)^2\right]^{1/2} dx.$$

The work per unit length done *against* the tensile force is then

$$F\frac{ds - dx}{dx} = F\left\{\left[1 + \left(\frac{dy}{dx}\right)^2\right]^{1/2} - 1\right\}$$

$$= F\left[1 + \frac{1}{2}\left(\frac{dy}{dx}\right)^2 + \cdots - 1\right)$$

$$\approx \frac{1}{2}F\left(\frac{dy}{dx}\right)^2,$$

if higher powers of the slope dy/dx (assumed to be small) are neglected. Thus this term represents *potential energy* per unit length due to the tension in the string, to a first approximation.

Finally, (227) requires that the difference between the total kinetic and total potential energies be stationary, in analogy with Hamilton's principle.*

In the case of a *yielding support* at the end $x = 0$, the end condition at that point would be of the form

$$\left(F\frac{dy}{dx}\right)_{x=0} = k(y)_{x=0} \tag{229}$$

where k is the modulus of the support. There would then follow

$$\left(F\frac{dy}{dx}\,\delta y\right)_{x=0} = (ky\,\delta y)_{x=0} = \delta\left(\frac{1}{2}ky^2\right)_{x=0}.$$

Since this term would not vanish, (227) would be replaced by

$$\delta\left\{\int_0^l \left[\frac{1}{2}\rho\omega^2 y^2 + py - \frac{1}{2}F\left(\frac{dy}{dx}\right)^2\right] dx - \left(\frac{1}{2}ky^2\right)_{x=0}\right\} = 0, \tag{230}$$

the additional term corresponding to the potential energy stored in the support.

* For a complete analogy, we should require that the *time integral* of this difference over (t_1, t_2) be stationary. In the present case, however, the energy difference is *independent* of time.

If the *slope* of the string at the end $x = 0$ were prescribed as

$$y'(0) = \alpha, \tag{231}$$

where α is small, the deflection $y(0)$ then being unknown, there would follow

$$\left(F \frac{dy}{dx}\,\delta y\right)_{x=0} = (F\alpha\,\delta y)_{x=0},$$

and (227) would be replaced by

$$\delta\left\{\int_0^l \left[\frac{1}{2}\rho\omega^2 y^2 + py - \frac{1}{2}F\left(\frac{dy}{dx}\right)^2\right] dx - \alpha F(0)y(0)\right\} = 0, \tag{232}$$

the additional term corresponding to work done by the component of the tension normal to the x axis ($F \sin\alpha \approx F\alpha$) in the end displacement $y(0)$.

In both (230) and (232), admissible functions must satisfy the single end condition $y(l) = 0$.

As a second example, we consider the case of *small deflections of a rotating shaft* of length l, subject to an axial end load P and to distributed transverse loading of intensity $p(x)$. The deflection $y(x)$ is then governed by the differential equation

$$\frac{d^2}{dx^2}\left(EI\frac{d^2y}{dx^2}\right) + P\frac{d^2y}{dx^2} - \rho\omega^2 y - p = 0, \tag{233}$$

where EI is the bending stiffness of the shaft. We first form the equation

$$\int_0^l (EIy'')''\,\delta y\,dx + P\int_0^l y''\,\delta y\,dx - \int_0^l (\rho\omega^2 y + p)\,\delta y\,dx = 0. \tag{234}$$

If the first term is integrated twice by parts, it becomes

$$\Big[(EIy'')'\,\delta y - EIy''\,\delta y'\Big]_0^l + \int_0^l EIy''\,\delta y''\,dx,$$

and the new integrand is recognized as the variation of $\frac{1}{2}EI(y'')^2$. After one integration by parts, the second term in (234) becomes

$$\Big[Py'\,\delta y\Big]_0^l - P\int_0^l y'\,\delta y'\,dx = \Big[Py'\,\delta y\Big]_0^l - \delta\int_0^l \frac{1}{2}P(y')^2\,dx.$$

Thus (234) implies the equation

$$\delta\int_0^l \left[\frac{1}{2}EI(y'')^2 - \frac{1}{2}P(y')^2 - \frac{1}{2}\rho\omega^2 y^2 - py\right] dx$$
$$+ \Big[\{(EIy'')' + Py'\}\,\delta y - EIy''\,\delta y'\Big]_0^l = 0. \tag{235}$$

The value of $[(EIy'')' + Py']\,\delta y$ at the end $x = l$ can be interpreted as the work done on the total transverse shear at that point in a *displacement* δy, while the value of $EIy''\,\delta y'$ is the work done by the end bending moment EIy'' in a *rotation* (change of *slope*) $\delta y'$. The total work done by end forces and moments will vanish in case of satisfaction of the *natural boundary conditions*

$$\Big[\{(EIy'')' + Py'\}\,\delta y - EIy''\,\delta y'\Big]_0^l = 0$$

or, explicitly,

$$\left.\begin{aligned} &y = y_0 \quad or \quad (EIy'')' + Py' = 0 \\ and \qquad &y' = y_0' \quad or \quad EIy'' = 0 \end{aligned}\right\} \quad (when\, x = 0, l). \tag{236}$$

In any case when such conditions are to be satisfied, the variational problem (235) reduces to the form

$$\delta \int_0^l \left[\frac{1}{2}\,EI(y'')^2 - \frac{1}{2}\,P(y')^2 - \frac{1}{2}\,\rho\omega^2 y^2 - py\right] dx = 0. \tag{237}$$

Here $\frac{1}{2}\rho\omega^2 y^2$ is the kinetic energy per unit length, and the remaining terms $\frac{1}{2}EI(y'')^2$, $-\frac{1}{2}P(y')^2$, and $-py$ can be identified with potential energies per unit length due to bending and to the end thrust and lateral loading, respectively.

If end supports do work in bending the shaft, additional terms must be added to the integral in (237) in analogy with (230) and (232).

As a third example, we consider *small steady-state forced vibration of a membrane*. The basic equation is of the form

$$\frac{\partial}{\partial x}\left(F\,\frac{\partial u}{\partial x}\right) + \frac{\partial}{\partial y}\left(F\,\frac{\partial u}{\partial y}\right) + P = \rho\,\frac{\partial^2 u}{\partial t^2}$$

where u is displacement, F tension, ρ is surface mass density, and $P(x, y, t)$ is the impressed periodic normal force. If P is of the form $P = p(x, y) \sin(\omega t + \alpha)$, we may write the steady-state displacement u in the form

$$u = w(x, y) \sin(\omega t + \alpha),$$

where w is the amplitude of the oscillation. The amplitude w then must satisfy the equation

$$(Fw_x)_x + (Fw_y)_y + \rho\omega^2 w + p = 0. \tag{238}$$

After multiplying by the variation $\delta w(x, y)$ and integrating over the region $\mathscr{R}$ of the xy plane covered by the undeformed membrane, there follows

$$\iint_{\mathscr{R}} (Fw_x)_x\,\delta w\,dx\,dy + \iint_{\mathscr{R}} (Fw_y)_y\,\delta w\,dx\,dy + \delta \iint_{\mathscr{R}} \left(\frac{1}{2}\,\rho\omega^2 w^2 + pw\right) dx\,dy = 0. \tag{239}$$

The first integrand can be written as

$$(Fw_x\,\delta w)_x - Fw_x\,\delta w_x$$

after which, by use of (53a), the first integral can be transformed to the form

$$\oint_{\mathscr{C}} Fw_x \cos \nu \, ds - \delta \iint_{\mathscr{R}} \frac{1}{2} F w_x^{\,2} \, dx\, dy,$$

where $\mathscr{C}$ is the boundary of $\mathscr{R}$. When the second integral is similarly transformed, and use is made of the general relation

$$\frac{\partial \phi}{\partial x} \cos \nu + \frac{\partial \phi}{\partial y} \sin \nu = \frac{\partial \phi}{\partial n}, \tag{240}$$

where $\partial\phi/\partial n$ is the derivative of ϕ in the direction of the outward normal to $\mathscr{C}$, the result can be written in the form

$$\oint_{\mathscr{C}} F \frac{\partial w}{\partial n} \delta w \, ds - \delta \iint_{\mathscr{R}} \left[\frac{1}{2} F(\nabla w)^2 - \frac{1}{2} \rho\omega^2 w^2 - pw \right] dA = 0, \tag{241}$$

which is independent of the coordinate system.

The single integral in (241) represents the total work of the transverse component of the edge tension, and vanishes if the membrane is fixed along $\mathscr{C}$ or, more generally, if the restraining transverse force $F\,\partial w/\partial n$ vanishes along all parts of the boundary which are not fixed. In all such cases the relevant variational problem becomes

$$\delta \iint_{\mathscr{R}} \left[\frac{1}{2} F(\nabla w)^2 - \frac{1}{2} \rho\omega^2 w^2 - pw \right] dA = 0. \tag{242}$$

The term $\frac{1}{2}\rho\omega^2 w^2$ is the kinetic energy (per unit area) corresponding to positions of maximum displacement, while the term $\frac{1}{2}F(w_x^{\,2} + w_y^{\,2})$ represents the potential energy stored in the membrane at such instants as a result of the stretching, and $-pw$ is the corresponding potential energy due to the loading.

If, instead of prescribing w along part of the boundary $\mathscr{C}$, and requiring that $F\,\partial w/\partial n$ *vanish* along the remainder of the boundary, we require that $\partial w/\partial n = \psi(s)$ along the portion $\mathscr{C}'$ where w is not prescribed, equation (241) shows that the term $-\delta \int_{\mathscr{C}'} F\psi w \, ds$ must be added to the left-hand member of (242).

When $F = 1$ and $\omega = 0$, equation (238) is *Poisson's equation*. When also $p = 0$, this equation becomes *Laplace's equation*. The variational form of the *Dirichlet problem*, where $w = \phi(s)$ along $\mathscr{C}$, then takes the form

$$\delta \iint_{\mathscr{R}} \frac{1}{2} (\nabla w)^2 \, dA = 0,$$

where the varied functions are to take on the prescribed values along $\mathscr{C}$. The variational form of the *Neumann problem*, where $\partial w/\partial n = \psi(s)$ along $\mathscr{C}$, becomes

$$\delta \left\{ \iint_{\mathscr{R}} \frac{1}{2} (\nabla w)^2 \, dA - \oint_{\mathscr{C}} \psi w \, ds \right\} = 0,$$

where the varied functions are *unrestricted* along $\mathscr{C}$.

2.17. Useful transformations. Certain formulas of frequent use in transformations of the type considered in the preceding section are collected together in this section, for convenient reference.

The formula

$$\int_{x_1}^{x_2} (pf_x)_x \, \delta f \, dx = -\delta \int_{x_1}^{x_2} \left(\frac{1}{2} pf_x^2 \right) dx + \Big[pf_x \, \delta f \Big]_{x_1}^{x_2},$$

established by integration by parts, implies the relation

$$(pf_x)_x \, \delta f = -\delta(\tfrac{1}{2} pf_x^2) + (pf_x \, \delta f)_x. \tag{243}$$

Here p is an explicit function of x, which is not to be varied. In a similar way, the following relations can be established:

$$(sf_{xx})_{xx} \, \delta f = \delta(\tfrac{1}{2} sf_{xx}^2) + [(sf_{xx})_x \, \delta f - sf_{xx} \, \delta f_x]_x. \tag{244}$$

$$[(pf_y)_x + (pf_x)_y] \, \delta f = -\delta(pf_x f_y) + (pf_y \, \delta f)_x + (pf_x \, \delta f)_y. \tag{245}$$

$$2f_{xy} \, \delta f = -\delta(f_x f_y) + (f_y \, \delta f)_x + (f_x \, \delta f)_y. \tag{246}$$

$$2f_{xxyy} \, \delta f = \begin{cases} \delta(f_{xy}^2) + 2(f_{xyy} \, \delta f)_x + 2(f_{xxy} \, \delta f)_y - 2(f_{xy} \, \delta f)_{xy}, & (247a) \\ \delta(f_{xx} f_{yy}) + (f_{xyy} \, \delta f - f_{yy} \, \delta f_x)_x + (f_{xxy} \, \delta f - f_{xx} \, \delta f_y)_y. & (247b) \end{cases}$$

The differentiations in (243) and (244) may be total or partial. In each case, the truth of the relation can be verified directly by expanding both sides of the equation.

In each of the preceding formulas, the left-hand member is expressed as the sum of an *exact variation* and one or more *derivatives*. It is of interest to notice that $2f_{xxyy} \, \delta f$ can be expressed thus in two different ways, according to (247a,b). The two alternatives can be combined by expressing the left-hand member as $(1 - \alpha)$ times (247a) plus α times (247b), where α is a completely arbitrary constant. Thus we may write

$$\begin{aligned} 2f_{xxyy} \, \delta f = \delta[(1 - \alpha)f_{xy}^2 + \alpha f_{xx} f_{yy}] + [(2 - \alpha) f_{xyy} \, \delta f - \alpha f_{yy} \, \delta f_x]_x \\ + [(2 - \alpha) f_{xxy} \, \delta f - \alpha f_{xx} \, \delta f_y]_y - 2(1 - \alpha)[f_{xy} \, \delta f]_{xy}, \end{aligned} \tag{248}$$

where α is an arbitrary constant. This form reduces to (247a) when $\alpha = 0$, and to (247b) when $\alpha = 1$.

If we take $p = 1$ in (243), and add to this expression the result of replacing x by y, we obtain the further useful result

$$\nabla^2 f \, \delta f = -\delta[\tfrac{1}{2}(\nabla f)^2] + (f_x \, \delta f)_x + (f_y \, \delta f)_y. \tag{249}$$

As a further example of the use of these formulas, we consider the product

$$\nabla^4 f\,\delta f \equiv (f_{xxxx} + 2f_{xxyy} + f_{yyyy})\,\delta f.$$

If use is made of (244) and (248), there follows

$$\begin{aligned}\nabla^4 f\,\delta f = {} & \delta(\tfrac{1}{2}f_{xx}^2) + [f_{xxx}\,\delta f - f_{xx}\,\delta f_x]_x \\ & + \delta[(1-\alpha)f_{xy}^2 + \alpha f_{xx}f_{yy}] + [(2-\alpha)f_{xyy}\,\delta f - \alpha f_{yy}\,\delta f_x]_x \\ & + [(2-\alpha)f_{xxy}\,\delta f - \alpha f_{xx}\,\delta f_y]_y - 2(1-\alpha)[f_{xy}\,\delta f]_{xy} \\ & + \delta(\tfrac{1}{2}f_{yy}^2) + [f_{yyy}\,\delta f - f_{yy}\,\delta f_y]_y,\end{aligned}$$

or, after collecting terms,

$$\begin{aligned}\nabla^4 f\,\delta f = {} & \delta\left[\frac{1}{2}(f_{xx}^2 + f_{yy}^2) + (1-\alpha)f_{xy}^2 + \alpha f_{xx}f_{yy}\right] \\ & + \frac{\partial}{\partial x}\left\{\left[\frac{\partial \nabla^2 f}{\partial x} + (1-\alpha)f_{xyy}\right]\delta f - (f_{xx} + \alpha f_{yy})\,\delta f_x\right\} \\ & + \frac{\partial}{\partial y}\left\{\left[\frac{\partial \nabla^2 f}{\partial y} + (1-\alpha)f_{xxy}\right]\delta f - (f_{yy} + \alpha f_{xx})\,\delta f_y\right\} \\ & - \frac{\partial^2}{\partial x\,\partial y}[2(1-\alpha)f_{xy}\,\delta f]. \qquad (250)\end{aligned}$$

2.18. The variational problem for the elastic plate. As a final illustration of the preceding methods, we consider the problem of determining the amplitude w of small deflections of a thin, initially flat, elastic plate of constant thickness. If the amplitude of the periodic impressed force is denoted by $p(x, y)$ and the circular frequency by ω, it is known* that under certain simplifying assumptions the governing differential equation is of the form

$$D\,\nabla^4 w - \rho\omega^2 w - p = 0. \qquad (251)$$

Here D is a constant known as the *bending stiffness* of the plate.

We consider here only the special case of a *rectangular* plate ($0 \leqq x \leqq a$, $0 \leqq y \leqq b$). Then, by multiplying both sides of (251) by a variation δw and integrating the result over the area of the plate, there follows

$$\int_0^a\int_0^b D\,\nabla^4 w\,\delta w\,dx\,dy - \delta\int_0^a\int_0^b\left[\frac{1}{2}\rho\omega^2 w^2 + pw\right]dx\,dy = 0. \qquad (252)$$

* See Reference 9.

If use is made of equation (250), this condition is transformed immediately into the requirement that

$$\delta\int_0^a\int_0^b\left\{\frac{1}{2}D[w_{xx}{}^2+w_{yy}{}^2+2\alpha w_{xx}w_{yy}+2(1-\alpha)w_{xy}{}^2]-\frac{1}{2}\rho\omega^2w^2-pw\right\}dx\,dy$$
$$+\int_0^b\left\{\left[D\frac{\partial\nabla^2 w}{\partial x}+(1-\alpha)Dw_{xyy}\right]\delta w-D(w_{xx}+\alpha w_{yy})\,\delta w_x\right\}_{x=0}^{x=a}dy$$
$$+\int_0^a\left\{\left[D\frac{\partial\nabla^2 w}{\partial y}+(1-\alpha)Dw_{xxy}\right]\delta w-D(w_{yy}+\alpha w_{xx})\,\delta w_y\right\}_{y=0}^{y=b}dx$$
$$-\left[\left[2D(1-\alpha)w_{xy}\,\delta w\right]_{x=0}^{x=a}\right]_{y=0}^{y=b}=0. \qquad (252')$$

It is to be noticed that (252′) and (252) are equivalent for *any* constant α. In the physical problem under consideration, α is identifiable with the physical constant known as *Poisson's ratio*. Its value is between zero and one-half, and is dependent upon the plate material.

If the conditions of edge support are such that no deflection or rotation of the edges is permitted, the plate is said to be *clamped*. In this case w is prescribed as zero along the complete boundary, while $\partial w/\partial x$ must be zero on each boundary $x =$ constant and $\partial w/\partial y$ must vanish on the boundaries $y = 0$ and $y = b$. In view of the fact that the corresponding variations are to vanish when these quantities are prescribed, it follows that the partially integrated terms in (252′) vanish, and the variational problem takes the form

$$\delta\int_0^a\int_0^b\left\{\frac{1}{2}D[w_{xx}{}^2+w_{yy}{}^2+2\alpha w_{xx}w_{yy}+2(1-\alpha)w_{xy}{}^2]\right.$$
$$\left.-pw-\frac{1}{2}\rho\omega^2w^2\right\}dx\,dy=0. \qquad (253)$$

That part of the integrand which involves D is known as the *strain energy* per unit area. The term $-pw$ again represents additional potential energy per unit area due to the transverse loading, and the term $\frac{1}{2}\rho\omega^2w^2$ represents the kinetic energy per unit area, each of these quantities being evaluated at a position of maximum deflection.

The *natural boundary conditions* of the problem are obtained by equating to zero the integrands of the boundary integrals. Thus, at the boundaries $x = 0$ and $x = a$, one must have either

$$w \textit{ prescribed} \quad \textit{or} \quad D\frac{\partial\nabla^2 w}{\partial x}+(1-\alpha)Dw_{xyy}=0 \qquad (254a)$$

and

$$w_x \textit{ prescribed} \quad \textit{or} \quad D(w_{xx}+\alpha w_{yy})=0. \qquad (254b)$$

The natural boundary conditions at the boundaries $y = 0$ and $y = b$ are obtained by interchanging x and y in (254a,b).

In the theory of elasticity it is shown that the quantity $D\,\partial\nabla^2 w/\partial x$ is to be interpreted as a transverse *shearing force* (Q_x) at a boundary $x =$ constant, the quantity $-(1-\alpha)Dw_{xy}$ as corresponding *twisting moment* (M_{xy}), and the quantity $-D(w_{xx} + \alpha w_{yy})$ as corresponding *bending moment* (M_{xx}). From (252′) it follows that the effective transverse edge force associated with a deflection δw along an edge $x =$ constant must be of the form $R_x = Q_x - \partial M_{xy}/\partial y$. The discovery of this fact, by *physical* reasoning, constituted a significant advance in the theory of small deflections of elastic plates. The presence of the last expression in (252′), which involves values of M_{xy} at the four *corners* of the plate, corresponds to the possible presence of *concentrated* reactions at the corners.

The variations of appropriate boundary integrals, obtained by reference to (252′), must be added to the left-hand member of (253) when edge deflections and/or rotations are not prescribed, but edge forces and/or moments are given.

It should be noticed that the case of *static* loading is contained in the above discussion when $\omega = 0$.

The present section is intended to illustrate two important facts. First, it has been shown (in a fairly complicated physical problem) that mere knowledge of the governing *differential equation* can lead to information as to which mathematical quantities should be *prescribed at the boundary*, and hence which *mathematical* quantities *must* be of principal *physical* interest.

Second, it is seen that, once the differential equation *and* the relevant boundary conditions are known, the corresponding variational problem (if one exists) can be obtained *without specialized knowledge of the physical details of the problem involved.* On the other hand, a sufficiently general knowledge of the *physical* background of the problem would permit one to *write down* the variational problem and, if it were desirable, *derive the relevant differential equation from it.*

In the remaining sections of this chapter, it is indicated that the variational formulation of a problem is often particularly well adapted to numerical procedures for obtaining an approximate solution.

2.19. The Rayleigh-Ritz method. The so-called *Rayleigh-Ritz method* is a general procedure for obtaining approximate solutions of problems expressed in variational form. In the case when a function $y(x)$ is to be determined, the procedure usually consists essentially of assuming that the desired stationary function of a given problem can be approximated by a linear combination of suitably chosen functions, of the form

$$y(x) \approx \phi_0(x) + c_1\phi_1(x) + c_2\phi_2(x) + \cdots + c_n\phi_n(x), \tag{255}$$

where the n c's are constants to be determined. Usually the functions $\phi_k(x)$ are to be so chosen that this expression satisfies the specified boundary conditions for any choice of the c's.

Thus, if y is to take on *prescribed values* at the ends of the interval under consideration, we require that the function $\phi_0(x)$ take on the prescribed values at the end points and that *each* of the remaining functions $\phi_1(x)$, $\phi_2(x), \cdots, \phi_n(x)$ *vanish* at both end points of the interval.

Otherwise, the choice of the functions ϕ_k is to a large extent arbitrary. In physically motivated problems, the general nature of the desired solution is often known, and a set of ϕ's is chosen in such a way that *some* linear combination of them may be expected to satisfactorily approximate the solution.

The relevant quantity I is then expressed as a function of the c's, and the c's are so determined that the resultant expression is stationary. Thus, in place of using the calculus of variations in attempting to determine that function which renders I stationary with reference to *all* admissible slightly varied functions, we consider only the family of all functions of type (255), and use ordinary *differential* calculus to seek the member of that family for which I is stationary with reference to slightly modified functions belonging to the same family. It is clear that the efficiency of the procedure depends upon the choice of appropriate approximating functions ϕ_k.

A more elaborate procedure consists of obtaining a *sequence* of approximations, in which the first assumption is merely $\phi_0 + c_1\phi_1$, the second $\phi_0 + c_1\phi_1 + c_2\phi_2$, and so forth, the nth assumption being of the form (255). The relevant c's are redetermined *at each stage* of the process by the method outlined above. By comparing successive approximations, an estimate of the degree of accuracy attained at any stage of the calculation can be obtained. In order that this process converge as $n \to \infty$, the functions $\phi_0, \phi_1, \cdots, \phi_n, \cdots$ should comprise an infinite set of functions such that the unknown function $y(x)$ can with certainty be approximated to any specified degree of accuracy by *some* linear combination of the ϕ's. Frequently it is useful to choose *polynomials* of successively increasing degree, satisfying the specified end conditions. In certain cases the use of special functions such as sine or cosine harmonics, Bessel functions, and so forth, may afford computational advantages.

To illustrate such procedures in a simple case, we first consider the problem of determining small static deflections of a string fixed at the points (0, 0) and (l, h), and subject to a transverse loading (in the y direction) whose intensity varies linearly, according to the law $p(x) = -qx/l$. With $\omega = 0$, the corresponding variational problem (225) then becomes

$$\delta \int_0^l \left(\frac{1}{2} F y'^2 + q \frac{x}{l} y \right) dx = 0, \tag{256}$$

the integrated terms vanishing by virtue of the end conditions

$$y(0) = 0, \qquad y(l) = h, \tag{257a,b}$$

which require that δy vanish at the end points.

The function ϕ_0 in (255) then is to satisfy (257a,b), whereas the other coordinate functions $\phi_1, \phi_2, \cdots$ are to *vanish* when $x = 0$ and when $x = l$. If *polynomials* are to be used, for convenience, among the simplest choices are the functions

$$\phi_0(x) = \frac{h}{l}x,\ \phi_1(x) = x(x-l),\ \phi_2(x) = x^2(x-l),\ \cdots,\ \phi_n(x) = x^n(x-l),$$

which correspond to an approximation of the form

$$y(x) \approx \frac{h}{l}x + x(x-l)(c_1 + c_2x + \cdots + c_nx^{n-1}). \tag{258}$$

For simplicity, we consider here only the one-parameter approximation for which $n = 1$,

$$y(x) \approx \frac{h}{l}x + c_1x(x-l), \tag{259}$$

in correspondence with which the result of replacing $y(x)$ by its approximation in (256) becomes

$$\delta\int_0^l \left\{\frac{1}{2}F\left[\frac{h}{l} + c_1(2x-l)\right]^2 + q\frac{x}{l}\left[\frac{h}{l}x + c_1(x^2 - lx)\right]\right\} dx = 0. \tag{260}$$

When the integration is carried out, this condition takes the form

$$\delta\left[\frac{1}{2}F\left(\frac{h^2}{l} + \frac{1}{3}l^3c_1^2\right) + q\left(\frac{1}{3}hl - \frac{1}{12}c_1l^3\right)\right] = 0$$

and hence, since here only c_1 is varied,

$$\frac{1}{3}Fl^3c_1\,\delta c_1 - \frac{1}{12}ql^3\,\delta c_1 = \frac{1}{3}Fl^3\left(c_1 - \frac{q}{4F}\right)\delta c_1 = 0. \tag{261}$$

Thus, since δc_1 is arbitrary, there must follow

$$c_1 = \frac{q}{4F}, \tag{262}$$

so that the desired approximation (259) is obtained in the form

$$y(x) \approx \frac{h}{l}x + \frac{q}{4F}x(x-l). \tag{263}$$

The *exact* solution of this particular problem is readily found by elementary methods,

$$y(x) = \frac{h}{l}x + \frac{q}{6Fl}x(x^2 - l^2), \tag{264}$$

and it happens to agree exactly with the approximation (263) at the midpoint $x = l/2$, as well as at the two ends of the interval $(0, l)$. Further, it happens that in this case a *two*-parameter approximation of form (258) would have yielded the *exact* solution, with $c_1 = lc_2 = q/6F$.

It is useful to notice not only that the Euler equation of (256) is the governing differential equation

$$Fy'' - \frac{qx}{l} = 0, \tag{265}$$

from which (264) was deduced, but also that if y satisfies conditions of type (226) at the ends of the interval then (256) is, by virtue of the equivalence of (224) and (225), equivalent to the equation

$$\int_0^l \left(Fy'' - \frac{qx}{l}\right) \delta y \, dx = 0. \tag{266}$$

With the approximation of (259), this condition becomes simply

$$\int_0^l \left(2c_1 F - \frac{qx}{l}\right) x(x - l) \, \delta c_1 \, dx = 0 \tag{267}$$

or

$$\left(-\frac{1}{3} c_1 F l^3 + \frac{1}{12} q l^3\right) \delta c_1 = 0,$$

and hence yields (262) somewhat more directly. This last procedure amounts to calculating the variation of (259) *before* carrying out the integration, and it frequently involves a reduced amount of calculation.

While the technique of forming (266) directly from (265) is a particularly convenient one, and certainly is appropriate in this case, its use in other situations may be less well motivated unless it is verified first that the differential equation involved *is* indeed the Euler equation of *some* variational problem, $\delta I = 0$, whose *natural boundary conditions* include those which govern the problem at hand.

For example, the equation

$$(x^2 y')' + xy = x \qquad (0 \leqq x \leqq l)$$

is readily transformed, by the methods of the preceding sections, to the variational problem

$$\delta \int_0^l \left[-\frac{1}{2} x^2 (y')^2 + \frac{1}{2} xy^2 - xy\right] dx + \left[x^2 y' \, \delta y\right]_0^l = 0.$$

If the specified boundary conditions are such that

$$\left[x^2 y' \, \delta y\right]_0^l = 0,$$

the variational problem thus can be taken as

$$\frac{1}{2}\,\delta\int_0^l [x^2(y')^2 - xy^2 + 2xy]\,dx = 0$$

or, after calculating the variation,

$$\int_0^l [(x^2y')' + xy - x]\,\delta y\,dx = 0.$$

The basic equation, when expanded, becomes

$$x^2y'' + 2xy' + xy = x.$$

While this equation is equivalent to the equation

$$xy'' + 2y' + y = 1,$$

this last form cannot be transformed to a proper variational problem, $\delta I = 0$, merely by multiplication by δy and subsequent integration by parts. The correct multiplicative factor is seen to be $x\,\delta y$. That is, the condition

$$\int_0^l (xy'' + 2y' + y - 1)\,\delta y\,dx = 0$$

is *not* the consequence of a proper variational problem, whereas the result of introducing the *weighting function* x into the integrand *does* have this property.*

In the case of a differential equation of order greater than two, it may happen that no such weighting function *exists*. However, it is readily verified that the abbreviated procedure is valid (*when appropriate boundary conditions are prescribed*) if the governing equation is of the so-called *self-adjoint* form

$$\mathscr{L}y \equiv (py')' + qy = f \tag{268a}$$

or

$$\mathscr{L}y \equiv (sy'')'' + (py')' + qy = f, \tag{268b}$$

where p, q, and s are functions of x or constants. That is, over an interval (x_1, x_2) such an equation *is* the Euler equation of a proper variational problem $\delta I = 0$, which is equivalent to the condition

$$\int_{x_1}^{x_2} (\mathscr{L}y - f)\,\delta y\,dx = 0 \tag{269}$$

* While it cannot be asserted that the omission of such a weighting function from the appropriate integrand *necessarily* will lead to a less efficient approximation technique, the motivations, interpretations, and existent literature associated with the application of the Rayleigh-Ritz method to "proper" variational problems tend to suggest that such formulations be preferred.

when $y(x)$ satisfies the appropriate natural boundary conditions. Any linear second-order equation can be rewritten in the form (268a), by suitably defining $p(x)$ and $q(x)$. While *not all* linear equations of fourth order can be reduced to (268b), the reduction is possible in most cases which arise in practice.

It may be noticed that, if the "unnatural" condition $y'(0) = \alpha$ and the natural condition $y(l) = h$ were imposed on the solution of (265), the variational problem (225) would take the form

$$\delta\left[\int_0^1 \left(\frac{1}{2} F y'^2 + q \frac{x}{l} y\right) dx + F\alpha y(0)\right] = 0,$$

in place of (256). By integrating by parts, we may transform this condition into the problem

$$\int_0^l \left(F y'' - q \frac{x}{l}\right) \delta y \, dx + F[y'(0) - \alpha]\, \delta y(0) = 0,$$

in place of (266). Here the approximating series (255) should satisfy the condition $y(l) = h$ for all values of the c's, but it need not satisfy the condition $y'(0) = \alpha$ identically. However, if it does not do so, care should be taken that the latter condition *can* be satisfied for *some* choice of the c's.

As a second illustration of the Rayleigh-Ritz method, we consider the solution of the boundary-value problem for which $y(x)$ must satisfy the differential equation

$$\frac{d^2y}{dx^2} + xy = -x \tag{270a}$$

and the homogeneous conditions

$$y(0) = 0, \qquad y(1) = 0. \tag{270b}$$

Since (270a) is in the form of (268a), and the end conditions correspond to the requirement that δy vanish at the end points, the variational problem corresponding to (270a,b) can be expressed immediately in the reduced form

$$\int_0^1 (y'' + xy + x)\, \delta y \, dx = 0. \tag{271}$$

An appropriate assumption corresponding to (255) and satisfying (270b) is of the form

$$y = x(1 - x)(c_1 + c_2 x + c_3 x^2 + \cdots), \tag{272}$$

according to which (271) takes the form

$$\int_0^1 [(-2 + x^2 - x^3)c_1 + (2 - 6x + x^3 - x^4)c_2 + \cdots + x] \cdot [\delta c_1(x - x^2) + \delta c_2(x^2 - x^3) + \cdots]\, dx = 0.$$

The result of carrying out the indicated integration is then of the form

$$(-\tfrac{19}{60}c_1 - \tfrac{11}{70}c_2 + \cdots + \tfrac{1}{12})\,\delta c_1 + (-\tfrac{11}{70}c_1 - \tfrac{107}{840}c_2 + \cdots + \tfrac{1}{20})\,\delta c_2 + \cdots = 0. \tag{273}$$

If only a one-term approximation is assumed,

$$y^{(1)} = c_1 x(1 - x), \tag{274}$$

we have $c_2 = c_3 = \cdots = 0$, and also $\delta c_2 = \delta c_3 = \cdots = 0$, and (273) reduces to the condition

$$(-\tfrac{19}{60}c_1 + \tfrac{1}{12})\,\delta c_1 = 0.$$

By virtue of the arbitrariness of δc_1, we then must have

$$c_1 = \tfrac{5}{19}$$

and hence the "best" solution of form (274) is given by

$$y^{(1)} = 0.263x(1 - x). \tag{275}$$

Similarly, for a two-term approximation of the form

$$y^{(2)} = c_1 x(1 - x) + c_2 x^2(1 - x), \tag{276}$$

the vanishing of the coefficients of the arbitrary variations δc_1 and δc_2 in (273) leads to the simultaneous equations

$$\left.\begin{aligned} 0.317c_1 + 0.157c_2 &= 0.0833, \\ 0.157c_1 + 0.127c_2 &= 0.0500 \end{aligned}\right\}, \tag{277}$$

if only three significant figures are retained in the calculations. From these equations we obtain the numerical results

$$c_1 = 0.177, \qquad c_2 = 0.173,$$

so that the "best" solution of form (276) is given by

$$y^{(2)} = (0.177x + 0.173x^2)(1 - x). \tag{278}$$

An estimate of the accuracy of this solution is afforded by the fact that the deviation between the two successive approximations (275) and (278) is found to be smaller in magnitude than 0.009 over the interval (0, 1).

In dealing similarly with the characteristic-value problem consisting of the equation

$$\frac{d^2y}{dx^2} + \lambda xy = 0, \tag{279}$$

in place of (270a), and the boundary conditions of (270b), the assumption of (272) is found to lead to the equation

$$\left[\left(-\frac{1}{3}+\frac{\lambda}{60}\right)c_1+\left(-\frac{1}{6}+\frac{\lambda}{105}\right)c_2+\cdots\right]\delta c_1$$
$$+\left[\left(-\frac{1}{6}+\frac{\lambda}{105}\right)c_1+\left(-\frac{2}{15}+\frac{\lambda}{168}\right)c_2+\cdots\right]\delta c_2+\cdots=0.$$

Corresponding to a one-term approximation (274), we obtain the condition

$$\left(-\frac{1}{3}+\frac{\lambda}{60}\right)c_1=0,$$

since δc_1 is arbitrary, and hence obtain a nontrivial approximate solution only if

$$\lambda=\lambda_1^{(1)}=20. \tag{280}$$

The coefficient c_1 is then arbitrary.

Corresponding to a two-term approximation (276), we obtain the two conditions

$$\left.\begin{aligned}(0.333-0.0167\lambda)c_1\ +(0.167-0.00955\lambda)c_2&=0,\\(0.167-0.00955\lambda)c_1+(0.133-0.00595\lambda)c_2&=0\end{aligned}\right\}. \tag{281}$$

Hence a nontrivial approximate solution can be obtained only if the determinant of the coefficients of the c's vanishes. The expansion of this determinantal equation takes the form

$$3\lambda^2-364\lambda+5880=0, \tag{282}$$

with the two roots

$$\lambda_1^{(2)}\doteq 19.2,\qquad \lambda_2^{(1)}\doteq 102. \tag{283}$$

Thus we obtain a *second* approximation to the smallest characteristic number λ_1, and a *first* approximation to the second characteristic number λ_2. For each such value of λ, the two equations of (281) become equivalent and either can be used to express c_2 as a multiple of c_1 (with c_1 arbitrary), thus determining approximations to the corresponding characteristic functions. In more involved cases, the iterative matrix methods of Sections 1.23 to 1.25 are useful.

The true characteristic functions of the problem are arbitrary multiples of the functions

$$f_n(x)=x^{1/2}J_{1/3}(\tfrac{2}{3}\lambda_n^{1/2}x^{3/2})$$

where λ_n is the nth solution of the equation

$$J_{1/3}(\tfrac{2}{3}\lambda^{1/2})=0,$$

from which it is found that $\lambda_1\doteq 18.9$ and $\lambda_2\doteq 81.8$. As is indicated by this example, the accurate calculation of the *higher* characteristic numbers by iterative methods may involve considerable labor.

As a third example, we consider the calculation of the smallest critical frequency of a vibrating membrane in the form of an isosceles right triangle. We choose dimensionless rectangular coordinates in such a way that the vertices are at the origin and at the points (1, 0) and (0, 1) (Figure 2.10). If it is assumed that the tension in the membrane is (approximately) uniform, (238) gives the differential equation satisfied by the amplitude function w in the form

$$\nabla^2 w + \lambda w = 0, \tag{284}$$

where, if the legs of the triangle are of length a,

$$\lambda = \frac{\rho\omega^2 a^2}{F}. \tag{285}$$

If we require that the membrane be fixed along its boundary, the boundary condition is

$$w = 0 \textit{ on the boundary}. \tag{286}$$

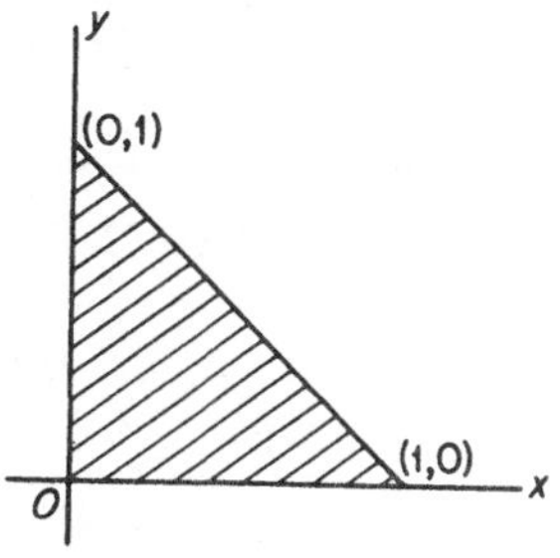

FIGURE 2.10

From the results of Section 2.16, the associated variational problem can be expressed in the form

$$\delta \iint_A \frac{1}{2}[(\nabla w)^2 - \lambda w^2]\, dx\, dy = 0 \tag{287a}$$

or, equivalently, in the reduced form

$$\iint_A (\nabla^2 w + \lambda w)\, \delta w\, dx\, dy = 0. \tag{287b}$$

Since the equation of the boundary can be written in the form

$$xy(x + y - 1) = 0,$$

appropriate approximating functions satisfying (286) are of the form

$$w = xy(x + y - 1)(c_1 + c_2 x + c_3 y + c_4 x^2 + \cdots). \tag{288}$$

For simplicity we here consider only a one-term approximation:

$$w^{(1)} = c_1 xy(x + y - 1). \tag{289}$$

When w is replaced by $w^{(1)}$ in (287b), there follows

$$c_1\, \delta c_1 \int_0^1 \int_0^{1-y} [2(x + y) + \lambda(x^2 y + xy^2 - xy)] \cdot (x^2 y + xy^2 - xy)\, dx\, dy = 0.$$

Since δc_1 is arbitrary, and $c_1 = 0$ leads to a trivial solution, the value of the double integral must vanish. The integrations are readily carried out, if use is made of the formula

$$\int_0^1 y^m(1-y)^n\,dy = \frac{m!\,n!}{(m+n+1)!}, \tag{290}$$

and there follows

$$-\frac{8}{6!}+\frac{\lambda}{7!}=0 \qquad \text{or} \qquad \lambda = 56. \tag{291}$$

Thus a first approximation to the smallest characteristic value of λ is obtained. From (285), it follows that the smallest critical frequency is $\omega_1/2\pi$, where

$$\omega_1 \approx 7.48\left(\frac{F}{\rho a^2}\right)^{1/2}. \tag{292}$$

The true value of the numerical factor in (292) is known to be $\pi\sqrt{5} \doteq 7.03$.

2.20. A semidirect method. The procedures described in the preceding section are often known as the *direct methods* in the calculus of variations. The approximating functions are completely specified at the start, and only the *constants of combination* are determined by variational methods. In the present section a modified procedure of frequent usefulness is outlined.

To fix ideas, suppose that the function w to be determined in a variational problem depends upon two independent variables x and y, and that the region of determination is the rectangle $(-a \leqq x \leqq a,\ -b \leqq y \leqq b)$. In the Rayleigh-Ritz method we would assume an approximation in the form

$$w \approx \phi_0(x, y) + c_1\phi_1(x, y) + \cdots + c_n\phi_n(x, y) \tag{293}$$

where the ϕ's are completely specified functions, satisfying appropriate boundary conditions, and the c's are to be determined by variational methods. However, in many physical problems, while the general nature of the behavior of w in, say, the x direction may be known, it may happen that the behavior in the y direction is less predictable. In such cases, it is convenient to only *partially* specify the approximating functions by writing instead

$$w \approx \phi_0(x, y) + \psi_1(x)f_1(y) + \cdots + \psi_n(x)f_n(y), \tag{294}$$

where the ψ's are suitably chosen functions of x alone, satisfying appropriate homogeneous conditions on the boundaries $x =$ constant, and the f's are unspecified functions of y alone, to be determined by a variational method. A procedure of this type, which may be called a *semidirect* method, then leads to a set of *ordinary differential equations* involving the unknown f's, together with a proper number of corresponding end conditions.

To illustrate the procedure, we consider small deflections of a square membrane ($|x| \leqq a$, $|y| \leqq a$), fixed along the edges $x = \pm a$ and along the edge $y = -a$, but unrestrained along the edge $y = a$, and subject to a distribution of static loading given by

$$p = -q(a^2 - x^2) \tag{295}$$

where q is a constant. If the tension F in the membrane is again assumed to be constant, the governing differential equation (238) is of the form

$$F\,\nabla^2 w = q(a^2 - x^2), \tag{296}$$

and the associated variational problem (241) becomes

$$-\delta\int_{-a}^{a}\int_{-a}^{a}\left[\frac{1}{2}F(\nabla w)^2 + q(a^2 - x^2)w\right] dx\,dy + \int_{-a}^{a}\left[Fw_x\,\delta w\right]_{x=-a}^{x=a} dy + \int_{-a}^{a}\left[Fw_y\,\delta w\right]_{y=-a}^{y=a} dx = 0. \tag{297}$$

The vanishing of the first boundary integral is assured if w satisfies the prescribed conditions

$$w(-a, y) = 0, \qquad w(a, y) = 0. \tag{298}$$

In order that the second boundary integral vanish, we must have

$$w(x, -a) = 0, \qquad w_y(x, a) = 0, \tag{299}$$

the first condition being the prescribed one along the edge $y = -a$, and the second representing that approximation to the requirement that no work be done by constraints along the edge $y = a$ which is in accordance with the approximations upon which the present formulation is based. Having thus formulated the boundary conditions appropriate to the variational problem, we are at liberty to use either (297) or the more convenient reduced form

$$\int_{-a}^{a}\int_{-a}^{a}[F\,\nabla^2 w - q(a^2 - x^2)]\,\delta w\,dx\,dy = 0. \tag{300}$$

In view of the nature of the loading (295), it may be suspected that the deflection $w(x, y)$ will be approximately parabolic in the x direction, with the maximum deflection (along the y axis) varying as a function of y. Thus a simple approximation of the form

$$w = (a^2 - x^2)f(y) \tag{301}$$

may be assumed. This approximation satisfies (298), regardless of the form of $f(y)$; in order that it satisfy (299) for all values of x, $f(y)$ must satisfy the end conditions

$$f(-a) = 0, \qquad f'(a) = 0. \tag{302}$$

If w is replaced by its approximation, there follows

$$\nabla^2 w = (a^2 - x^2)f''(y) - 2f(y)$$

and

$$\delta w = (a^2 - x^2)\,\delta f(y),$$

and hence (300) takes the form

$$\int_{-a}^{a}\left\{\int_{-a}^{a}[F(a^2-x^2)f''(y)-2Ff(y)-q(a^2-x^2)](a^2-x^2)\,dx\right\}\delta f(y)\,dy=0. \tag{303}$$

The integration *with respect to* x now can be carried out explicitly, to express (303) in the form

$$\int_{-a}^{a}\left\{F\left[\frac{16}{15}a^5f''(y)-\frac{8}{3}a^3f(y)\right]-\frac{16}{15}a^5q\right\}\delta f(y)\,dy=0. \tag{304}$$

From the arbitrariness of $\delta f(y)$ inside the interval $(-a, a)$, it follows that the quantity in braces must vanish, so that $f(y)$ must satisfy the differential equation

$$f''(y)-\frac{5}{2a^2}f(y)=\frac{q}{F}. \tag{305}$$

The solution of (305), satisfying (302), is found to be

$$f(y)=-\frac{2a^2q}{5F}\left[1-\frac{\cosh\frac{1}{2}\sqrt{10}(1-y/a)}{\cosh\sqrt{10}}\right] \tag{306}$$

and the introduction of (306) into (301) leads to the determination of the desired one-term approximation to the deflection.

REFERENCES

1. Bliss, G. A.: *Calculus of Variations*, Mathematical Association of America, The Open Court Publishing Company, La Salle, Ill., 1925.
2. ——————: *Lectures on the Calculus of Variations*, University of Chicago Press, Chicago, 1946.
3. Bolza, O.: *Lectures on the Calculus of Variations*, Stechert-Hafner, Inc., New York, 1946.
4. Byerly, W. E.: *Generalized Coordinates*, Ginn & Company, Boston, 1916.
5. Courant, R., and D. Hilbert: *Methods of Mathematical Physics*, Interscience Publishers, Inc., New York, 1953.
6. Fox, C.: *An Introduction to the Calculus of Variations*, Oxford University Press, Inc., New York, 1950.
7. Sokolnikoff, I. S.: *Mathematical Theory of Elasticity*, McGraw-Hill Book Company, New York, 1946 (Chapter 5).

8. Temple, G., and W. G. Bickley: *Rayleigh's Principle*, Dover Publications, Inc., New York, 1956.

9. Timoshenko, S.: *Theory of Plates and Shells*, McGraw-Hill Book Company, New York, 1940.

10. Webster, A. G.: *Dynamics*, B. G. Teubner, Leipzig, 1925.

11. Weinstock, R.: *Calculus of Variations with Applications to Physics and Engineering*, McGraw-Hill Book Company, New York, 1952.

PROBLEMS

Section 2.1.

1. Suppose that $f(x, y)$ is stationary when $x = a$ and $y = b$ [so that $\partial f/\partial x = \partial f/\partial y = 0$ at (a, b)], and that $f(x, y)$ can be expanded in power series near (a, b).

(a) Show that the relevant power series then takes the form

$$f(x, y) - f(a, b) = \tfrac{1}{2}[(x - a)^2 f_{xx}(a, b) + 2(x - a)(y - b)f_{xy}(a, b) + (y - b)^2 f_{yy}(a, b)] + \cdots,$$

where omitted terms are of degree greater than two in $(x - a)$ and $(y - b)$.

(b) Deduce that if $\partial f/\partial x = \partial f/\partial y = 0$ at (a, b), then $f(x, y)$ possesses a relative *minimum* at (a, b) if the matrix

$$\mathbf{M} \equiv \begin{bmatrix} f_{xx} & f_{xy} \\ f_{xy} & f_{yy} \end{bmatrix}$$

is positive definite when $x = a$ and $y = b$, and that it possesses a relative *maximum* if **M** is negative definite at that point. [See Section 1.19.]

(c) Use the results of Section 1.20 to show that the stationary value is a maximum if $f_{xx} < 0$ and $f_{xx}f_{yy} - f_{xy}^2 > 0$ at (a, b), and is a minimum if $f_{xx} > 0$ and $f_{xx}f_{yy} - f_{xy}^2 > 0$ at (a, b).

(d) Generalize the criterion of part (b) in the case of a function of n independent variables $x_1, x_2, \cdots, x_n$.

2. Of all rectangular parallelepipeds which have sides parallel to the coordinate planes, and which are inscribed in the ellipsoid

$$\frac{x^2}{a^2} + \frac{y^2}{b^2} + \frac{z^2}{c^2} = 1,$$

determine the dimensions of that one which has the largest possible volume.

3. Determine the lengths of the principal semiaxes of the ellipse

$$Ax^2 + 2Bxy + Cy^2 = 1,$$

where $AC > B^2$, and deduce also that the area of the ellipse is given by $\pi/\sqrt{AC - B^2}$.

4. Of all parabolas which pass through the points (0, 0) and (1, 1), determine that one which, when rotated about the x axis, generates a solid of revolution with least possible volume between $x = 0$ and $x = 1$. [Notice that the equation may be taken in the form $y = x + cx(1 - x)$, where c is to be determined.]

5. (a) If $\mathbf{x} = \{x_1, x_2, \cdots, x_n\}$ is a real vector, and $\mathbf{A}$ is a real symmetric square matrix of order n, show that the requirement that

$$F \equiv \mathbf{x}^T \mathbf{A}\, \mathbf{x} - \lambda\, \mathbf{x}^T \mathbf{x}$$

be stationary, for a prescribed $\mathbf{A}$, takes the form

$$\mathbf{A}\, \mathbf{x} = \lambda\, \mathbf{x}.$$

Deduce that the requirement that the quadratic form $A \equiv \mathbf{x}^T \mathbf{A}\, \mathbf{x}$ be stationary, subject to the constraint $B \equiv \mathbf{x}^T \mathbf{x} =$ constant, leads to the requirement $\mathbf{A}\, \mathbf{x} = \lambda\, \mathbf{x}$, where λ is a constant to be determined. [Notice that the same is true of the requirement that B be stationary, subject to the constraint $A =$ constant, with a suitable redefinition of λ.]

(b) Show that, if we write

$$\lambda = \frac{\mathbf{x}^T \mathbf{A}\, \mathbf{x}}{\mathbf{x}^T \mathbf{x}} \equiv \frac{A}{B},$$

the requirement that λ be stationary leads again to the matrix equation $\mathbf{A}\, \mathbf{x} = \lambda\, \mathbf{x}$. [Notice that the requirement $d\lambda = 0$ can be written as $(B\, dA - A\, dB)/B^2 = 0$ or $(dA - \lambda\, dB)/B = 0$.] Deduce that stationary values of the ratio $(\mathbf{x}^T \mathbf{A}\, \mathbf{x})/(\mathbf{x}^T \mathbf{x})$ are characteristic numbers of the symmetric matrix $\mathbf{A}$. (See also Problem 117 of Chapter 1.)

Section 2.2.

6. Establish the equivalence of equations (17b) and (17c).

7. Derive equation (17b) by rewriting (11) in the form

$$I = \int_{y_1}^{y_2} F\left(x, y, \frac{1}{x'}\right) x'\, dy$$

and considering x as a function of the independent variable y. [Write

$$G(x, y, x') = x'F(x, y, 1/x'),$$

notice that equation (17a) then takes the form

$$\frac{d}{dy}\left(\frac{\partial G}{\partial x'}\right) - \frac{\partial G}{\partial x} = 0,$$

and express the left-hand side of this equation in terms of F.]

8. It is required to determine the continuously differentiable function $y(x)$ which minimizes the integral $I = \int_0^1 (1 + y'^2)\, dx$, and satisfies the end conditions $y(0) = 0$, $y(1) = 1$.

(a) Obtain the relevant Euler equation, and show that the stationary function is $y = x$.

(b) With $y(x) = x$ and the special choice $\eta(x) = x(1 - x)$, and with the notation of equation (12), calculate $I(\epsilon)$ and verify directly that $dI(\epsilon)/d\epsilon = 0$ when $\epsilon = 0$.

(c) By writing $y(x) = x + u(x)$, show that the problem becomes

$$I = 2 + \int_0^1 u'^2\, dx = \text{minimum},$$

where $u(0) = u(1) = 0$, and deduce that $y(x) = x$ is indeed the required minimizing function.

9. Obtain the Euler equation relevant to the determination of extremals of the integral $\int_0^1 F(x, y, y')\, dx$ in the following cases:

(a) $F = y'^2 + yy' + y^2$.
(b) $F = xy'^2 - yy' + y$.
(c) $F = y'^2 + k^2 \cos y$.
(d) $F = a(x)y'^2 - b(x)y^2$.

10. By interchanging the order of integration, show that the extremals of the integral

$$I = \int_a^b \int_a^t K(x, t) f(x, y, y')\, dx\, dt$$

satisfy the equation

$$\frac{d}{dx}\left(\frac{\partial F}{\partial y'}\right) - \frac{\partial F}{\partial y} = 0,$$

where $F(x, y, y') = \phi(x) f(x, y, y')$ and $\phi(x) = \int_x^b K(x, t)\, dt$.

Section 2.3.

A *geodesic* on a given surface is a curve, lying on that surface, along which distance between two points is as small as possible. On a plane, a geodesic is a straight line. Determine equations of geodesics on the following surfaces:

11. Right circular cylinder. [Take $ds^2 = a^2\, d\theta^2 + dz^2$ and minimize

$$\int \sqrt{a^2 + (dz/d\theta)^2}\, d\theta \quad \text{or} \quad \int \sqrt{a^2(d\theta/dz)^2 + 1}\, dz.]$$

12. Right circular cone. [Use spherical coordinates (Figure 2.5) with

$$ds^2 = dr^2 + r^2 \sin^2 \alpha\, d\theta^2.]$$

13. Sphere. [Use spherical coordinates (Figure 2.5) with

$$ds^2 = a^2 \sin^2 \phi \, d\theta^2 + a^2 \, d\phi^2.]$$

14. Surface of revolution. [Write $x = r \cos \theta$, $y = r \sin \theta$, $z = f(r)$. Express the desired relation between r and θ in terms of an integral.]

15. The *brachistochrone* ("shortest time") *problem*, posed by Bernoulli in 1696, states that a particle of mass m starts at a point $P_1(x_1, y_1)$ with speed V and moves under gravity (acting in the negative y direction) along a curve $y = f(x)$ to a point $P_2(x_2, y_2)$, where $y_1 > y_2$ and $x_2 > x_1$, and requires the curve along which the elapsed time T is minimum.

(a) If v is the speed and s the distance travelled at time t, use the relations $dt = ds/v$ and $\frac{1}{2}mv^2 + mgy =$ constant to obtain the formulation

$$T = \frac{1}{\sqrt{2g}} \int_{x_1}^{x_2} \frac{\sqrt{1 + y'^2}}{\sqrt{\alpha - y}} \, dx = \text{minimum} \qquad \left(\alpha = y_1 + \frac{V^2}{2g}\right)$$

where $y(x_1) = y_1$ and $y(x_2) = y_2$.

(b) Deduce the relation

$$dx = -\frac{\sqrt{\alpha - y}}{\sqrt{c_1 - (\alpha - y)}} \, dy,$$

where c_1 is a constant.

(c) By writing $\alpha - y = c_1 \sin^2 \phi/2$, show that if a minimizing curve exists, then it can be defined by parametric equations of the form

$$x = \frac{1}{2} c_1(\phi - \sin \phi) + c_2, \qquad y = y_1 + \frac{V^2}{2g} - \frac{1}{2} c_1(1 - \cos \phi),$$

where c_1 and c_2 are constants to be determined such that P_1 and P_2 are on the curve.

[The extremals are *cycloids*. It is known that c_1 and c_2 can be uniquely determined so that P_1 and P_2 are on the same "arch" of one such cycloid and that the corresponding arc P_1P_2 then truly minimizes T.]

Section 2.4.

16. Determine the stationary functions associated with the integral

$$I = \int_0^1 (y'^2 - 2\alpha yy' - 2\beta y') \, dx,$$

where α and β are constants, in each of the following situations:

(a) The end conditions $y(0) = 0$ and $y(1) = 1$ are preassigned.
(b) Only the end condition $y(0) = 0$ is preassigned.
(c) Only the end condition $y(1) = 1$ is preassigned.
(d) No end conditions are preassigned.

Note any exceptional cases.

17. Determine the natural boundary conditions associated with the determination of extremals in each of the cases considered in Problem 9.

18. Determine the stationary function associated with the integral

$$I = \int_0^1 y'^2 f(x)\,dx$$

when $y(0) = 0$ and $y(1) = 1$, where

$$f(x) = \begin{cases} -1 & (0 \leq x < \frac{1}{4}), \\ +1 & (\frac{1}{4} < x \leq 1). \end{cases}$$

Section 2.5.

19. If $I(y) = \int_0^1 \sqrt{1 + y'^2}\,dx$, calculate $I(x)$ and $I(\cosh x)$.

20. If $F = 1 + x + y + y'^2$, calculate the following quantities for $x = 0$:

(a) dF where $y = \sin x$ and $dx = \epsilon$.
(b) δF where $y = \sin x$ and $\delta y = \epsilon(x + 1)$.

21. If $I = \int_0^1 (x^2 - y^2 + y'^2)\,dx$, calculate both ΔI and δI when $y = x$ and $\delta y = \epsilon x^2$.

22. Let $y = 1 + x^2$, where x and y are functions of an independent variable t. Calculate $\delta \dfrac{dy}{dx}$ and $\dfrac{d}{dx}\delta y$ when $x = \dfrac{1}{t}$ and $\delta x = \epsilon t^2$, and verify the validity of equation (44′) in this case.

23. Determine the stationary function $y(x)$ for the problem

$$\begin{cases} \delta\left\{\int_0^1 y'^2\,dx + [y(1)]^2\right\} = 0, \\ y(0) = 1. \end{cases}$$

24. Obtain the Euler equation and the associated natural boundary conditions for the problem

$$\delta\left[\int_a^b F(x, y, y')\,dx - \beta y(b) + \alpha y(a)\right] = 0,$$

where α and β are given constants but $y(a)$ and $y(b)$ are *not* preassigned.

Section 2.6.

25. Determine the geodesics in the xy plane by writing

$$ds = \left[\left(\frac{dx}{dt}\right)^2 + \left(\frac{dy}{dt}\right)^2\right]^{1/2} dt,$$

where t is a parameter, rather than by *assuming* that y is a single-valued function of x on a geodesic. (Notice that the lines $x =$ constant are not exceptional here.)

26. Derive the Euler equation of the problem

$$\delta \int_{x_1}^{x_2} F(x, y, y', y'')\, dx = 0$$

in the form

$$\frac{d^2}{dx^2}\left(\frac{\partial F}{\partial y''}\right) - \frac{d}{dx}\left(\frac{\partial F}{\partial y'}\right) + \frac{\partial F}{\partial y} = 0,$$

and show that the associated natural boundary conditions are

$$\left[\left(\frac{d}{dx}\frac{\partial F}{\partial y''} - \frac{\partial F}{\partial y'}\right)\delta y\right]_{x_1}^{x_2} = 0 \quad \text{and} \quad \left[\frac{\partial F}{\partial y''}\,\delta y'\right]_{x_1}^{x_2} = 0.$$

27. Specialize the results of Problem 26 in the case of the problem

$$\delta \int_{x_1}^{x_2} [a(x)y''^2 - b(x)y'^2 + c(x)y^2]\, dx = 0.$$

28. Derive the Euler equation of the problem

$$\delta \iint_{\mathscr{R}} F(x, y, u, u_x, u_y)\, dx\, dy = 0$$

in the form

$$\frac{\partial}{\partial x}\left(\frac{\partial F}{\partial u_x}\right) + \frac{\partial}{\partial y}\left(\frac{\partial F}{\partial u_y}\right) - \frac{\partial F}{\partial u} = 0,$$

subject to the requirement that $u(x, y)$ is prescribed along the closed boundary $\mathscr{C}$ of the region $\mathscr{R}$.

29. Obtain the natural boundary condition relevant to Problem 28 in the form

$$\left(\frac{\partial F}{\partial u_x}\cos\nu + \frac{\partial F}{\partial u_y}\sin\nu\right)\delta u = 0 \quad \text{on } \mathscr{C},$$

where ν is the angle from the positive x axis to the outward normal at a point of $\mathscr{C}$.

30. Specialize the results of Problems 28 and 29 in the case of the problem

$$\delta \iint_{\mathscr{R}} [a(x, y)u_x^2 + b(x, y)u_y^2 - c(x, y)u^2]\, dx\, dy = 0.$$

In particular, show that if $b(x, y) = a(x, y)$ the natural boundary condition takes the form

$$a\frac{\partial u}{\partial n}\delta u = 0 \quad \text{on } \mathscr{C}$$

where $\partial u/\partial n$ is the normal derivative of u on $\mathscr{C}$.

31. Derive the Euler equation of the problem

$$\delta \int_{x_1}^{x_2}\int_{y_1}^{y_2} F(x, y, u, u_x, u_y, u_{xx}, u_{xy}, u_{yy})\, dx\, dy = 0,$$

where x_1, x_2, y_1, and y_2 are constants, in the form

$$\frac{\partial^2}{\partial x^2}\left(\frac{\partial F}{\partial u_{xx}}\right) + \frac{\partial^2}{\partial x\,\partial y}\left(\frac{\partial F}{\partial u_{xy}}\right) + \frac{\partial^2}{\partial y^2}\left(\frac{\partial F}{\partial u_{yy}}\right) - \frac{\partial}{\partial x}\left(\frac{\partial F}{\partial u_x}\right) - \frac{\partial}{\partial y}\left(\frac{\partial F}{\partial u_y}\right) + \frac{\partial F}{\partial u} = 0,$$

and show that the associated natural boundary conditions are then

$$\left[\left(\frac{\partial}{\partial x}\frac{\partial F}{\partial u_{xx}} + \frac{\partial}{\partial y}\frac{\partial F}{\partial u_{xy}} - \frac{\partial F}{\partial u_x}\right)\delta u\right]_{x_1}^{x_2} = 0, \qquad \left[\frac{\partial F}{\partial u_{xx}}\,\delta u_x\right]_{x_1}^{x_2} = 0,$$

and

$$\left[\left(\frac{\partial}{\partial y}\frac{\partial F}{\partial u_{yy}} + \frac{\partial}{\partial x}\frac{\partial F}{\partial u_{xy}} - \frac{\partial F}{\partial u_y}\right)\delta u\right]_{y_1}^{y_2} = 0, \qquad \left[\frac{\partial F}{\partial u_{yy}}\,\delta u_y\right]_{y_1}^{y_2} = 0.$$

32. Specialize the results of Problem 31 in the case of the problem

$$\delta\int_{x_1}^{x_2}\int_{y_1}^{y_2}\left[\frac{1}{2}\,u_{xx}^{\;2} + \frac{1}{2}\,u_{yy}^{\;2} + \alpha u_{xx}u_{yy} + (1-\alpha)u_{xy}^{\;2}\right]dx\,dy = 0,$$

where α is a constant. [Show that the Euler equation is of the form $\nabla^4 u = 0$, regardless of the value of α, whereas the natural boundary conditions are dependent upon α.]

33. Consider the problem of determining a stationary function associated with the integral

$$I = \iint_{\mathscr{R}} u_{xx}^{\;2}\,dx\,dy,$$

where $\mathscr{R}$ is a simple region in the xy plane, bounded by a smooth curve $\mathscr{C}$.

(a) Show that the condition $\delta I = 0$ can be transformed to the requirement

$$\oint_{\mathscr{C}} u_{xx}\cos\nu\,\delta u_x\,ds - \oint_{\mathscr{C}} u_{xxx}\cos\nu\,\delta u\,ds + \iint_{\mathscr{R}} u_{xxxx}\,\delta u\,dx\,dy = 0.$$

(b) Obtain the relations

$$\frac{\partial u}{\partial x} = \frac{\partial u}{\partial n}\cos\nu - \frac{\partial u}{\partial s}\sin\nu, \qquad \frac{\partial u}{\partial y} = \frac{\partial u}{\partial n}\sin\nu + \frac{\partial u}{\partial s}\cos\nu,$$

where $\partial u/\partial n$ and $\partial u/\partial s$ are the derivatives of u normal to $\mathscr{C}$ and along $\mathscr{C}$, respectively, and use the first relation to transform the first boundary integral in part (a) to the form

$$\oint_{\mathscr{C}} u_{xx}\cos\nu\left(\delta\frac{\partial u}{\partial n}\cos\nu - \delta\frac{\partial u}{\partial s}\sin\nu\right)ds = \oint_{\mathscr{C}} u_{xx}\cos^2\nu\,\delta\frac{\partial u}{\partial n}\,ds$$
$$- \oint_{\mathscr{C}} \frac{\partial}{\partial s}(u_{xx}\sin\nu\cos\nu\,\delta u)\,ds + \oint_{\mathscr{C}} \frac{\partial}{\partial s}(u_{xx}\sin\nu\cos\nu)\,\delta u\,ds.$$

(c) Noticing that the second integral on the right in the preceding relation vanishes, deduce the Euler equation

$$u_{xxxx} = 0$$

and the associated natural boundary conditions

$$u_{xx} \cos^2 \nu \, \delta \frac{\partial u}{\partial n} = \left[u_{xxx} \cos \nu - \frac{\partial}{\partial s} (u_{xx} \sin \nu \cos \nu) \right] \delta u = 0 \quad \text{on } \mathscr{C}.$$

[Notice that this result agrees with the result of Problem 31 when $F = u_{xx}^2$ and $\mathscr{R}$ is a rectangle with sides parallel to the axes. Similar procedures can be used to obtain the natural boundary conditions associated with a more general integrand, involving partial derivatives of order two or more, on the boundary of a region $\mathscr{R}$.]

34. Make use of Green's theorem in the form

$$\iiint_{\mathscr{R}} \nabla f \cdot \nabla g \, d\tau = -\iiint_{\mathscr{R}} g \, \nabla^2 f \, d\tau + \oint\!\!\oint_{\mathscr{S}} g \frac{\partial f}{\partial n} \, d\sigma,$$

where $\mathscr{S}$ is the boundary of $\mathscr{R}$ and $d\tau$ and $d\sigma$ are elements of volume and surface area, to establish the result

$$\delta \iiint_{\mathscr{R}} |\nabla \phi|^2 \, d\tau = -2 \iiint_{\mathscr{R}} (\nabla^2 \phi) \, \delta\phi \, d\tau + 2 \oint\!\!\oint_{\mathscr{S}} \frac{\partial \phi}{\partial n} \, \delta\phi \, d\sigma,$$

where $|\nabla \phi|^2 \equiv \nabla \phi \cdot \nabla \phi$.

Section 2.7.

35. Suppose that the constraint (64) is replaced by the more general condition

$$\phi(x, u, v, u_x, v_x) = 0. \qquad (*)$$

Noticing that this condition implies the constraint

$$\int_a^b \lambda(x) \phi(x, u, v, u_x, v_x) \, dx = 0,$$

where λ is any function of x, deduce that when (*) is imposed on (63) the two Euler equations

$$\frac{d}{dx}\left(\frac{\partial F}{\partial u_x} + \lambda \frac{\partial \phi}{\partial u_x} \right) - \frac{\partial F}{\partial u} - \lambda \frac{\partial \phi}{\partial u} = 0, \qquad \frac{d}{dx}\left(\frac{\partial F}{\partial v_x} + \lambda \frac{\partial \phi}{\partial v_x} \right) - \frac{\partial F}{\partial v} - \lambda \frac{\partial \phi}{\partial v} = 0$$

must be satisfied by the three unknown functions u, v, and λ, in addition to the *original* constraint condition (*).

36. A particle moves on the surface $\phi(x, y, z) = 0$ from the point (x_1, y_1, z_1) to the point (x_2, y_2, z_2) in the time T. Show that if it moves in such a way that the integral of its kinetic energy over that time is a minimum, its coordinates must also satisfy the equations $\ddot{x}/\phi_x = \ddot{y}/\phi_y = \ddot{z}/\phi_z$. [Minimize $\int_0^T \frac{1}{2} (\dot{x}^2 + \dot{y}^2 + \dot{z}^2) \, dt$, subject to the constraint $\phi = 0$.]

37. Specialize Problem 36 in the case when the particle moves on the unit sphere $x^2 + y^2 + z^2 - 1 = 0$, from $(0, 0, 1)$ to $(0, 0, -1)$, in time T. [Show first that the motion must be described by the equations $r = \sqrt{x^2 + y^2} = \sin \frac{n\pi t}{T}$, $z = \cos \frac{n\pi t}{T}$, $\theta = \tan^{-1} \frac{y}{x} = \text{const.}$, where n is an odd integer, so that motion is along a great circle of the sphere. Then show that the integrated kinetic energy is least when $n = 1$, and is then given by $\pi^2/(2T)$.]

38. Determine the equation of the shortest arc in the first quadrant, which passes through the points (0, 0) and (1, 0) and encloses a prescribed area A with the x axis, where $A \leq \pi/8$. [Reduce the problem of determining the arbitrary constants to the solution of a transcendental equation.]

39. (a) Show that the extremals of the problem

$$\delta \int_{x_1}^{x_2} [p(x)y'^2 - q(x)y^2]\, dx = 0, \qquad \int_{x_1}^{x_2} r(x)y^2\, dx = 1,$$

where $y(x_1)$ and $y(x_2)$ are prescribed, are solutions of the equation

$$\frac{d}{dx}\left(p\frac{dy}{dx}\right) + (q + \lambda r)y = 0,$$

where λ is a constant.

(b) Show that the associated natural boundary conditions are of the form

$$\left[p\frac{dy}{dx}\,\delta y\right]_{x_1}^{x_2} = 0,$$

so that the same result follows if py' is required to vanish at an end point where y is not prescribed.

40. Specialize Problem 39 in the following special case:

$$\delta \int_0^\pi y'^2\, dx = 0, \qquad \int_0^\pi y^2\, dx = 1;$$

$$y(0) = 0, \qquad y(\pi) = 0.$$

[Show that the stationary functions are of the form $y = \sqrt{2/\pi}\sin nx$, where n is an integer other than zero.]

41. Show that, if the constraint $\int_0^\pi y^2\, dx = 1$ is omitted in Problem 40, the only stationary function is the trivial one $y \equiv 0$.

42. (a) Show that the extremals of the problem

$$\delta \int_{x_1}^{x_2} [s(x)y''^2 - p(x)y'^2 + q(x)y^2]\, dx = 0,$$

$$\int_{x_1}^{x_2} r(x)y^2\, dx = 1,$$

where $y(x_1)$, $y'(x_1)$, $y(x_2)$, and $y'(x_2)$ are prescribed, are solutions of the equation

$$\frac{d^2}{dx^2}\left(s\frac{d^2y}{dx^2}\right) + \frac{d}{dx}\left(p\frac{dy}{dx}\right) + (q - \lambda r)y = 0,$$

where λ is a constant.

(b) By considering the associated natural boundary conditions, show that the same result follows if $(sy'')' + py'$ is required to vanish at an end point where y is not prescribed, and sy'' is required to vanish at an end point where y' is not prescribed.

43. Specialize Problem 42 in the following special case:

$$\delta\int_0^\pi y''^2\,dx = 0, \qquad \int_0^\pi y^2\,dx = 1;$$

$$y(0) = y''(0) = 0, \qquad y(\pi) = y''(\pi) = 0.$$

[Show that the end conditions are appropriate, and that the stationary functions are of the form $y = \sqrt{2/\pi}\sin nx$, where n is an integer other than zero.]

44. Show that, if the constraint $\int_0^\pi y^2\,dx = 1$ is omitted in Problem 43, the only stationary function is the trivial one $y \equiv 0$.

Section 2.8.

45. If F depends upon x_2, show that the transversality condition (90) must be replaced by the condition

$$\left[F + (g' - y')\frac{\partial F}{\partial y'}\right]_{x=x_2} + \int_{x_1}^{x_2}\frac{\partial F}{\partial x_2}\,dx = 0.$$

46. Show that if (84) is replaced by the end condition $y(x_1) = f(x_1)$, and if F may depend upon x_1, then the transversality condition at $x = x_1$ is

$$\left[F + (f' - y')\frac{\partial F}{\partial y'}\right]_{x=x_1} - \int_{x_1}^{x_2}\frac{\partial F}{\partial x_1}\,dx = 0.$$

47. If l is not preassigned, show that the stationary functions corresponding to the problem

$$\begin{cases}\delta\displaystyle\int_0^l y'^2\,dx = 0,\\ y(0) = 2, \qquad y(l) = \sin l\end{cases}$$

are of the form $y = 2 + 2x\cos l$, where l satisfies the transcendental equation $2 + 2l\cos l - \sin l = 0$. Also verify that the smallest positive value of l is between $\pi/2$ and $3\pi/4$.

48. If l is not preassigned, show that the stationary functions corresponding to the problem

$$\begin{cases} \delta \int_0^l [y'^2 + 4(y - l)]\,dx = 0, \\ y(0) = 2, \qquad y(l) = l^2 \end{cases}$$

are of the form $y = x^2 - 2(x/l) + 2$, where l is one of the two real roots of the equation $2l^4 - 2l^3 - 1 = 0$.

49. Show that if the brachistochrone problem (Problem 15) is modified in such a way that the terminal point P_2 is required only to lie on the curve $y = g(x)$, then the end condition at P_2 is of the form $y'(x_2)g'(x_2) = -1$, and hence requires that the cycloid intersect the curve $y = g(x)$ orthogonally. [Notice that the treatment of the case when P_1 lies on a given curve is somewhat more complicated since then the integrand involves y_1.]

50. Use the calculus of variations to find the shortest distance between the line $y = x$ and the parabola $y^2 = x - 1$.

Section 2.9.

51. Verify that the Euler equation relevant to the problem $\delta\lambda = 0$, where

$$\lambda = \frac{\int_{x_1}^{x_2} (sy''^2 - py'^2 + qy^2)\,dx}{\int_{x_1}^{x_2} ry^2\,dx},$$

is of the form

$$\frac{d^2}{dx^2}\left(s\frac{d^2y}{dx^2}\right) + \frac{d}{dx}\left(p\frac{dy}{dx}\right) + (q - \lambda r)y = 0,$$

and that the relevant natural boundary conditions at $x = x_1$ and $x = x_2$ are the following:

$$(sy'')' + py' = 0 \text{ or } y \text{ prescribed} \qquad \text{and} \qquad sy'' = 0 \text{ or } y' \text{ prescribed.}$$

[Compare Problem 42.]. Deduce, in particular, that when *homogeneous* natural boundary conditions are prescribed, stationary values of the ratio λ are characteristic values of the associated boundary-value problem.

52. The deflection y of a beam executing small free vibrations of frequency ω satisfies the differential equation

$$\frac{d^2}{dx^2}\left(EI\frac{d^2y}{dx^2}\right) - \rho\omega^2 y = 0,$$

where EI is the flexural rigidity and ρ the linear mass density. Deduce from Problem

51 that the deflection modes are stationary functions of the problem

$$\delta\omega^2 \equiv \delta\left[\frac{\int_0^l EIy''^2\,dx}{\int_0^l \rho y^2\,dx}\right] = 0,$$

when appropriate homogeneous end conditions are prescribed, and where l is the length of the beam, and that stationary values of the ratio are squares of the natural frequencies. [The *bending moment* M is given (approximately) by $M = EIy''$, and the transverse *shearing force* S by $S = M' = (EIy'')'$. Notice that the natural boundary conditions are satisfied if either $S = 0$ or y is prescribed *and* either $M = 0$ or y' is prescribed at each end of the beam. It can be shown that the *smallest* stationary value of ω^2 is truly the *minimum* value of the ratio.]

53. Suppose that the tension F and linear density ρ of a freely vibrating string of length l are *nearly uniform*, and that the string is fixed at the ends $x = 0$ and $x = l$. Recalling that the natural vibration modes for a *uniform* string are multiples of the functions

$$y_n(x) = \sin\left(\frac{n\pi x}{l}\right) \qquad (n = 1, 2, \cdots),$$

motivate the approximate formula

$$\omega_n \approx \frac{n\pi}{l}\sqrt{\frac{\int_0^l F\cos^2(n\pi x/l)\,dx}{\int_0^l \rho\sin^2(n\pi x/l)\,dx}} \qquad (n = 1, 2, \cdots),$$

for the nth natural frequency, in the case under consideration. If the small deviations in F and ρ from uniformity are assumed to be *linear*, show also that the approximate values of the natural frequencies take the form

$$\omega_n \approx \frac{n\pi}{l}\sqrt{\frac{\bar{F}}{\bar{\rho}}} \qquad (n = 1, 2, \cdots),$$

where $\bar{F}$ and $\bar{\rho}$ are the *mean values* of F and ρ.

54. Obtain formulas analogous to those of Problem 53, in the case of the freely vibrating beam of Problem 52, both ends of which are *hinged* in such a way that both y and M vanish.

55. Let ω_i^2 represent the ith characteristic value of ω^2 for the problem consisting of the equation $(Fy')' + \rho\omega^2 y = 0$ and of specific end conditions which require that at each end of the interval $(0, l)$ either y or Fy' vanishes, and denote the corresponding characteristic function by $\phi_i(x)$. Suppose also that the ϕ's are *normalized* in such a way that

$$\int_0^l \rho\phi_i\phi_j\,dx = \delta_{ij}$$

(see Section 1.29), and that the ω^2's are arranged in increasing order of magnitude.

(a) Show that

$$\int_0^l F\phi_i'^2\,dx = -\int_0^l (F\phi_i')'\phi_i\,dx = \omega_i^2.$$

(b) By making use of the fact that any continuously differentiable function $y(x)$ which satisfies the prescribed end conditions can be expressed in the form

$$y(x) = \sum_{k=1}^{\infty} c_k\phi_k(x) \qquad (0 \leqq x \leqq l),$$

where the series converges uniformly, and by taking into account the orthogonality of the ϕ's relative to ρ, show that the relation

$$\omega^2 = \frac{\int_0^l Fy'^2\,dx}{\int_0^l \rho y^2\,dx}$$

takes the form

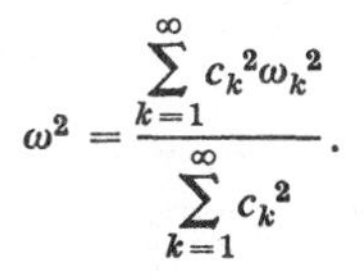

$$\omega^2 = \frac{\sum_{k=1}^{\infty} c_k^2\omega_k^2}{\sum_{k=1}^{\infty} c_k^2}.$$

(c) Show that $\omega^2 - \omega_1^2 \geqq 0$, and that $\omega^2 = \omega_1^2$ when $y(x) = \phi_1(x)$. Hence deduce that *the smallest characteristic value of* ω^2 *is the minimum value of the ratio n* (106) *for all admissible functions.*

(d) Show that, if $c_1 = c_2 = \cdots = c_{r-1} = 0$, there follows $\omega^2 - \omega_r^2 \geqq 0$, and that $\omega^2 = \omega_r^2$ when $y(x) = \phi_r(x)$. Hence deduce that the rth *characteristic value of* ω^2 *is the minimum value of the ratio in* (106) *for all admissible functions which are orthogonal to the first* $r - 1$ *characteristic functions.* (Compare Problem 117 of Chapter 1.)

Section 2.10.

56. A particle of mass m is falling vertically, under the action of gravity. If x is distance measured downward, and no resistive forces are present, show that the Lagrangian function is

$$L = T - V = m(\tfrac{1}{2}\dot{x}^2 + gx) + \text{constant},$$

and verify that the Euler equation of the problem $\delta\int_{t_1}^{t_2} L\,dt = 0$ is the proper equation of motion of the particle.

57. A particle of mass m is moving vertically, under the action of gravity and a resistive force numerically equal to k times the displacement x from an equilibrium position. Show that the equation of Hamilton's principle is of the form

$$\delta \int_{t_1}^{t_2} \left(\frac{1}{2} m\dot{x}^2 + mgx - \frac{1}{2} kx^2 \right) dt = 0,$$

and obtain the Euler equation.

58. A particle of mass m is falling vertically under the action of gravity, and its motion is resisted by a force numerically equal to a constant c times its velocity $\dot{x}$. Show that the equation of Hamilton's principle takes the form

$$\delta \int_{t_1}^{t_2} \left(\frac{1}{2} m\dot{x}^2 + mgx \right) dt - \int_{t_1}^{t_2} c\dot{x}\,\delta x\, dt = 0.$$

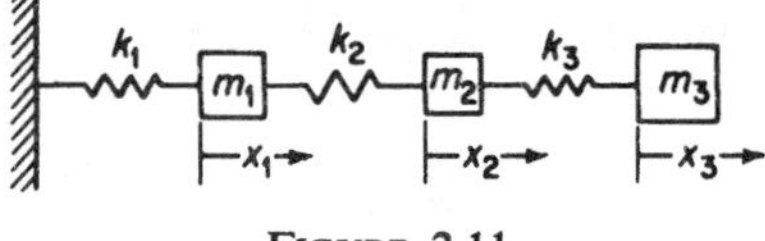

FIGURE 2.11

59. Three masses are connected in series to a fixed support, by linear springs. Assuming that only the spring forces are present, and using the notation of Figure 2.11, show that the Lagrangian function of the system is

$$L = \tfrac{1}{2}[m_1\dot{x}_1^2 + m_2\dot{x}_2^2 + m_3\dot{x}_3^2 - k_1x_1^2 - k_2(x_2 - x_1)^2 - k_3(x_3 - x_2)^2] + \text{const.},$$

where the x_i represent displacements from equilibrium. [Notice that if the x_i are given increments δx_i the total work done *by* the springs is given by

$$\begin{aligned}\delta\Phi = -\delta V &= [k_2(x_2 - x_1) - k_1x_1]\,\delta x_1 \\ &\quad + [k_3(x_3 - x_2) - k_2(x_2 - x_1)]\,\delta x_2 + [-k_3(x_3 - x_2)]\,\delta x_3.]\end{aligned}$$

Section 2.11.

60. Obtain the Lagrangian equations relevant to the mechanical system of Problem 59.

61. A mass $4m$ is attached to a string which passes over a smooth pulley. The other end of the string is attached to a smooth pulley of mass m, over which passes a second string attached to masses m and $2m$. If the system starts from rest, determine the motion of the mass $4m$, using the coordinates q_1 and q_2 indicated in Figure 2.12.

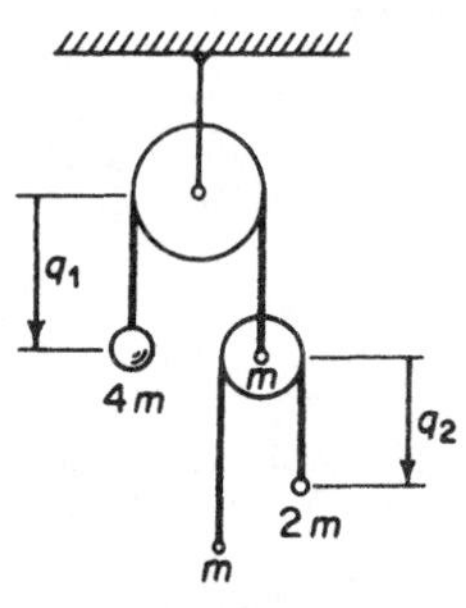

FIGURE 2.12

62. Obtain the Lagrangian equations for a triple pendulum consisting of three weights of equal mass m, connected in series to a fixed support by inextensible strings of equal length a, taking as the coordinates the angles θ_1, θ_2, and θ_3 made with the vertical by the three strings. Show also that, for small deviations from equilibrium, and small velocities, the Lagrangian function takes the approximate form

$$L = \frac{ma^2}{2}(3\dot{\theta}_1^2 + 2\dot{\theta}_2^2 + \dot{\theta}_3^2 + 4\dot{\theta}_1\dot{\theta}_2 + 2\dot{\theta}_2\dot{\theta}_3 + 2\dot{\theta}_1\dot{\theta}_3$$

$$- \frac{mga}{2}(3\theta_1^2 + 2\theta_2^2 + \theta_3^2) + \text{const.}$$

63. Two particles of equal mass m are connected by an inextensible string which passes through a hole in a smooth horizontal table, the first particle resting on the table, and the second particle being suspended vertically. Initially, the first particle is caused to describe a circular path about the hole, with an angular velocity $\omega = \sqrt{g/a}$, where a is the radius of the path, so that the suspended mass is held at equilibrium. At the instant $t = 0$, the suspended mass is pulled downward a short distance and is released, while the first mass continues to rotate.

(a) If x represents the distance of the second mass below its equilibrium position at time t, and θ represents angular position of the first particle at time t, show that the Lagrangian function is given by

$$L = m[\dot{x}^2 + \tfrac{1}{2}(a - x)^2\dot{\theta}^2 + gx] + \text{const.},$$

and obtain the equations of motion.

(b) Show that the first integral of the θ equation is of the form

$$(a - x)^2\dot{\theta} = a\sqrt{ag},$$

and that the result of eliminating $\dot{\theta}$ between this equation and the x equation becomes

$$2\ddot{x} + \left[\frac{1}{(1 - x/a)^3} - 1\right]g = 0.$$

(c) In the case when the displacement of the suspended mass from equilibrium is small, show that the suspended mass performs small vertical oscillations of period $2\pi\sqrt{2a/3g}$.

Section 2.12.

64. (a) In terms of Lagrange's function $L(q_1, \cdots, q_n; \dot{q}_1, \cdots, \dot{q}_n)$, such that $L = T - V$, show that the equations of motion become

$$\frac{d}{dt}\left(\frac{\partial L}{\partial \dot{q}_i}\right) = \frac{\partial L}{\partial q_i} \qquad (i = 1, 2, \cdots, n).$$

(b) Show that the generalized momentum p_i corresponding to the ith coordinate q_1 is given by

$$p_i = \frac{\partial T}{\partial \dot{q}_i} = \frac{\partial L}{\partial \dot{q}_i}.$$

65. By noticing that T is a homogeneous quadratic form in the n $\dot{q}$'s, establish the identity

$$\sum_{k=1}^{n} p_k \dot{q}_k = 2T.$$

[Compare Problem 56 of Chapter 1.]

66. The *Hamiltonian function* H, of a conservative system, is defined as the *sum* of the kinetic and potential energies: $H = T + V$.

(a) By making use of the result of Problem 65, show that one may write

$$H = \sum_{k=1}^{n} p_k \dot{q}_k - L.$$

(b) Let the $3n$ variables $p_1, \cdots, p_n; q_1, \cdots, q_n; \dot{q}_1 \cdots, \dot{q}_n$ be considered independent. By using the result of Problem 64(b), show that

$$\frac{\partial H}{\partial \dot{q}_i} = 0,$$

so that H *is a function only of the* p's *and* q's, and is independent of the $\dot{q}$'s.

67. (a) By noticing that L is a function only of the q's and $\dot{q}$'s, use the result of Problem 66(a) to show that

$$\frac{\partial H}{\partial p_i} = \dot{q}_i \qquad (i = 1, 2, \cdots, n).$$

(b) By combining the results of Problem 66(a) and 64(a), show that the equations of motion can be written in the form

$$\frac{\partial H}{\partial q_i} = -\dot{p}_i \qquad (i = 1, 2, \cdots, n).$$

[The two sets of equations obtained in parts (a) and (b) are known as *Hamilton's canonical equations.*]

(c) By multiplying the ith equation of part (a) by $\dot{p}_i$, the ith equation of part (b) by $\dot{q}_i$, adding, and summing the results over i, deduce the equation of *conservation of total energy*, $dH/dt = 0$.

68. (a) For the simple pendulum of Figure 2.3, show that the Hamiltonian function (expressed in terms of coordinates and momenta) is of the form

$$H = \frac{p^2}{2ml^2} - mgl \cos \theta + \text{const.},$$

where p is the generalized momentum associated with the generalized coordinate $q = \theta$.

(b) Obtain Hamilton's canonical equations in the form

$$\frac{p}{ml^2} = \dot{\theta}, \qquad mgl \sin \theta = -\dot{p},$$

and show that they imply equation (126).

69. For a harmonic oscillator with one degree of freedom, show that the Hamiltonian function is of the form

$$H = \frac{p^2}{2m} + \frac{kq^2}{2} + \text{const.},$$

where k is the stiffness constant of the system. Show also that the canonical equations take the form $\dot{p} = -kq$ and $\dot{q} = p/m$.

70. A mass m moves in the xy plane, under the action of a central force directed along the radius from the origin. If the position of the mass is specified by the polar coordinates r and θ, and the potential energy function is denoted by $V(r)$, express the Hamiltonian function in terms of r, θ, p_r, and p_θ, and obtain the four relevant canonical equations.

Section 2.13.

71. Solve the problem of the simple pendulum by taking as *two* coordinates of the mass m the distance r from the support to the mass and the angle θ of Figure 2.3, subject to the constraint $r = l$, and making use of a Lagrange multiplier. [Show that the equations corresponding to (156) become

$$m(\ddot{r} - r\dot{\theta}^2 - g \cos \theta) = \lambda,$$
$$m(r^2\ddot{\theta} + 2r\dot{r}\dot{\theta} + gr \sin \theta) = 0.$$

By introducing the relation $r = l$, obtain equation (126) and deduce further that the tension S in the string is given by $S = -\lambda = mg \cos \theta + ml\dot{\theta}^2$.]

72. Suppose that the oscillations of a simple pendulum are not restricted to a plane. By appropriately introducing the spherical coordinates of Figure 2.5, obtain the equations of motion in the form

$$\ddot{\phi} - \dot{\theta}^2 \sin \phi \cos \phi + \frac{g}{l} \sin \phi = 0, \qquad \dot{\theta} \sin^2 \phi = C,$$

and show also that the tension in the string is given by

$$S = mg \cos \phi + ml(\dot{\phi}^2 + \dot{\theta}^2 \sin^2 \phi).$$

Sections 2.14, 2.15.

73. Prove that the potential energy V must be stationary when the relevant system is at equilibrium. [Notice that the kinetic energy T is a homogeneous quadratic form in the $\dot{q}$'s, with coefficients depending upon the q's, and set each $\dot{q}$ and $\ddot{q}$ equal to zero in Lagrange's equations.]

74. *Potential energy of a linear spring.* Suppose that the force exerted by a spring is directed along the spring, and is proportional to its stretch e beyond its "natural length" l_0.

(a) Prove that the potential energy V_s stored in the spring is given by

$$V_s = \frac{k}{2} e^2 + \text{const.},$$

where k is the "spring constant" of proportionality. [Calculate the work done in stretching the spring from the length l_0 to the length $l_0 + e$.]

(b) Suppose that an unstretched spring of length l_0 coincides with the vector $a\mathbf{i} + b\mathbf{j} + c\mathbf{k}$, where $a^2 + b^2 + c^2 = l_0^2$, and that the subsequent displacement of one end relative to the other is defined by the vector $u\mathbf{i} + v\mathbf{j} + w\mathbf{k}$. Show that the potential energy is of the form

$$V_s = \frac{k}{2}[\sqrt{(a+u)^2 + (b+v)^2 + (c+w)^2} - l_0]^2 + \text{const.}$$

(c) Under the assumption of small displacements, obtain the expansion

$$V_s = \frac{k}{2}\left(\frac{au + bv + cw}{l_0}\right)^2 + \cdots + \text{const.}$$

$$= \frac{k}{2}(\hat{l}u + \hat{m}v + \hat{n}w)^2 + \cdots + \text{const.},$$

where omitted terms involve powers of u, v, and w greater than two, and $\hat{l}$, $\hat{m}$, and $\hat{n}$, are the direction cosines of the line of action of the spring in its natural position, so that *the contents of the parentheses comprise the component of the relative displacement vector in the direction of the natural position of the spring.*

75. A mass m is elastically restrained in space by a number of springs with spring constants k_i, which are attached to the points P_i. The mass is at equilibrium at the origin O, the springs then being of natural length (Figure 2.13). If the direction cosines of the radii OP_i are denoted by $(\hat{l}_i\ \hat{m}_i,\ \hat{n}_i)$, and small displacements are assumed, obtain the potential energy stored in the springs when the mass is displaced to the position (x, y, z) in the form

$$V_s = \sum_i \frac{k_i}{2}(\hat{l}_i x + \hat{m}_i y + \hat{n}_i z)^2.$$

[Use the result of Problem 74(c).] Also, obtain the equations of motion in the absence of external forces.

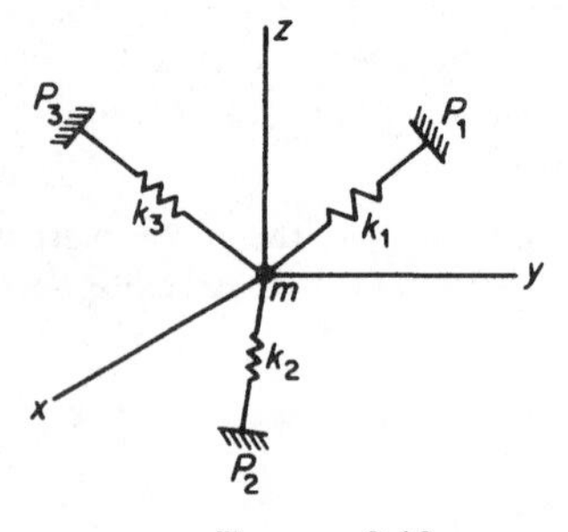

FIGURE 2.13

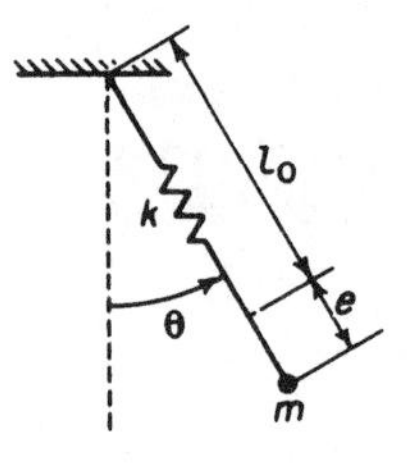

FIGURE 2.14

76. Suppose that a pendulum, vibrating in a plane, consists of a mass m attached to a fixed support by a linear spring with spring constant k (Figure 2.14).

(a) Show that the potential energy is given by

$$V = \frac{k}{2} e^2 - mg(l_0 + e) \cos \theta + \text{const.},$$

where l_0 is the natural length of the spring and e is its stretch.

(b) Show that the position of equilibrium is specified by $e = mg/k$, $\theta = 0$. With the introduction of the new coordinate $s = e - mg/k$, such that s is the stretch beyond the length $l = l_0 + mg/k$ assumed by the loaded spring in equilibrium under the action of gravity, obtain the relevant energy functions in the form

$$T = \frac{m}{2} [\dot{s}^2 + (l + s)^2 \dot{\theta}^2],$$

$$V = \frac{k}{2}\left(\frac{mg}{k} + s\right)^2 - mg(l + s) \cos \theta + \text{const.},$$

and deduce the equations of motion in the form

$$m\ddot{s} + ks - m(l + s)\dot{\theta}^2 + mg(1 - \cos \theta) = 0,$$

$$m \frac{d}{dt} [(l + s)^2 \dot{\theta}] + mg(l + s) \sin \theta = 0.$$

(c) Assuming small stretch and deflection, obtain the approximations

$$T = \frac{m}{2} (\dot{s}^2 + l^2 \dot{\theta}^2), \qquad V = \frac{k}{2} s^2 + \frac{1}{2} mgl\theta^2 + \text{const.},$$

and deduce that in the linear theory the extensional and deflectional vibration modes are uncoupled, with frequencies $\sqrt{k/m}$ and $\sqrt{g/l}$, respectively, so that s and θ are normal coordinates.

77. In Problem 76, use as coordinates the components x and y of the displacement of the mass m from equilibrium position, in the horizontal and vertical directions, respectively. Show that there follows

$$T = \frac{m}{2} (\dot{x}^2 + \dot{y}^2),$$

$$V = \frac{k}{2} [\sqrt{x^2 + (y - l)^2} - l_0]^2 + mgy + \text{const.},$$

where $l_0 = l - \dfrac{mg}{k}$; obtain the expansion $V = \dfrac{k}{2} y^2 + \dfrac{mg}{2l} x^2 + \cdots + \text{const.}$, relevant to small oscillations; and compare the corresponding linearized equations of motion with the results of Problem 76(c). [Notice that the results of Problem 74(c) are not applicable here, since x and y are not measured from a position corresponding to zero stretch.]

78. The point of suspension of a simple pendulum is completely restrained from vertical motion, and is partially restrained from horizontal motion by a spring system which exerts a restoring force equal to $-kx$ when the horizontal displacement of that point is x (Figure 2.15). Obtain the equations of motion of the suspended mass m, assuming the string to be inextensible and of length l. Show that for small displacements there follows approximately

$$x = \frac{mg}{k}\,\theta \qquad \text{and} \qquad \left(l + \frac{mg}{k}\right)\ddot{\theta} + g\theta = 0,$$

so that the system is then equivalent to a simple pendulum of length $l + \dfrac{mg}{k}$ with a fixed support.

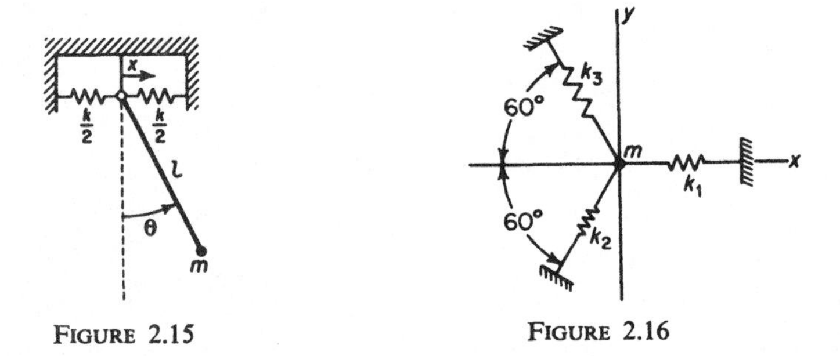

FIGURE 2.15

FIGURE 2.16

79. A mass m is attached to three symmetrically placed supports by linear springs. With the notation of Figure 2.16, the mass is at equilibrium at the origin, equidistant from the three supports (and in their plane), the springs then being unstretched.

(a) Assuming small oscillations, show that the potential energy stored in the springs, corresponding to a displacement (x, y) in the plane of the supports, is of the form

$$V_s = \tfrac{1}{8}[(4k_1 + k_2 + k_3)x^2 + 2\sqrt{3}(k_2 - k_3)xy + 3(k_2 + k_3)y^2],$$

and obtain the corresponding equations of motion of the mass.

(b) In the special case when $k_1 = 2k$, $k_2 = (2 + \sqrt{3})k$, and $k_3 = (2 - \sqrt{3})k$, determine the natural frequencies and the natural modes of small oscillations. Show also that the coordinates $\alpha_1 = (x + y)/\sqrt{2}$, $\alpha_2 = (x - y)/\sqrt{2}$ are then normal coordinates, and express the kinetic and potential energies of the system in terms of them.

80. A mass m under the action of gravity executes small oscillations near the origin on a frictionless paraboloid

$$z = \tfrac{1}{2}(Ax^2 + 2Bxy + Cy^2),$$

where $B^2 < AC$, $A > 0$, and where the z axis is directed upward. Obtain the characteristic equation determining the natural frequencies. [Use x and y as the Lagrangian coordinates.]

81. A mass m under the action of gravity executes small oscillations near the origin on a frictionless ellipsoid

$$\frac{x^2}{A^2} + \frac{y^2}{B^2} + \frac{(z - C)^2}{C^2} = 1,$$

where the z axis is directed upward. Show that the coordinates x and y are normal coordinates, and that the natural frequencies are $\frac{\sqrt{gC}}{A}$ and $\frac{\sqrt{gC}}{B}$. [Show that, near the origin, there follows $z = \frac{C}{2}\left(\frac{x^2}{A^2} + \frac{y^2}{B^2}\right)$ + terms of higher order in x and y.]

82. From the analogy between coupled mechanical systems and coupled electric networks, in which linear displacement x corresponds to charge $Q = \int_{t_0}^{t} I\,dt$, where I is current, and where mass m corresponds to inductance L, spring constant k to reciprocal capacity $1/C$, damping coefficient r to resistance R, and impressed force F to impressed voltage E, deduce that to the potential energy V there must correspond the "electromagnetic energy"

$$V = \frac{1}{2}\sum_{i=1}^{n}\frac{1}{C_i}Q_i^2 + \frac{1}{2}\sum_{i=1}^{n}\sum_{j=1}^{i-1}\frac{1}{C_{ij}}(Q_i - Q_j)^2 - \sum_{i=1}^{n}E_iQ_i,$$

where $\dot{Q}_i = I_i$ is the current flowing in the ith circuit, C_i is the capacitance of that circuit which is not in common with other circuits, $C_{ij} = C_{ji}$ is mutual capacitance in common with the ith and jth circuits, and E_i is the impressed voltage (positive in the positive direction of I_i) in the ith circuit. Show also that to the kinetic energy T there must correspond the "magnetic energy"

$$T = \frac{1}{2}\sum_{i=1}^{n}L_i\dot{Q}_i^2 + \frac{1}{2}\sum_{i=1}^{n}\sum_{j=1}^{i-1}L_{ij}(\dot{Q}_i - \dot{Q}_j)^2,$$

where L_i and $L_{ij} = L_{ji}$ are coefficients of self-inductance and mutual inductance, respectively. Finally, show that to the Rayleigh dissipation function there must correspond the "heat dissipation function"

$$F = \frac{1}{2}\sum_{i=1}^{n}R_i\dot{Q}_i^2 + \frac{1}{2}\sum_{i=1}^{n}\sum_{j=1}^{i-1}R_{ij}(\dot{Q}_i - \dot{Q}_j)^2,$$

where R_i and $R_{ij} = R_{ji}$ are the resistances, after which the circuit equations are obtained in the Lagrangian form

$$\frac{d}{dt}\left(\frac{\partial T}{\partial \dot{Q}_i}\right) + \frac{\partial V}{\partial Q_i} + \frac{\partial F}{\partial \dot{Q}_i} = 0 \qquad (i = 1, 2, \cdots, n).$$

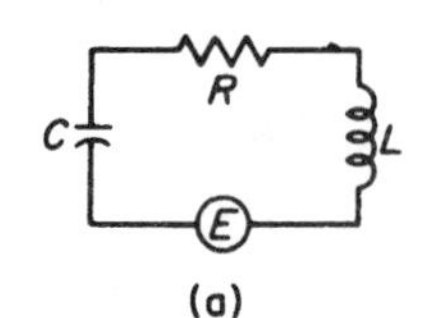

(b)

FIGURE 2.17

83. Derive the circuit equations relevant to the two networks of Figure 2.17 by the Lagrangian method of Problem 82, and verify the results by use of Kirchhoff's laws. Also, investigate the natural frequencies of small oscillating currents in each of the two networks when the resistances are neglected. [In the second case, merely express the characteristic equation in terms of the vanishing of a determinant.]

Section 2.16.

84. Suppose that $y(x)$ satisfies the differential equation

$$(sy'')'' + (py')' + qy = f$$

everywhere in the interval (x_1, x_2) except at an interior point ξ, and that one or more of the functions s, p, q, and f may be defined by different analytic expressions over the two subintervals (x_1, ξ) and (ξ, x_2).

(a) If y and y' are required to be continuous at $x = \xi$, obtain the relation

$$\delta \int_{x_1}^{x_2} \left(\frac{1}{2} sy''^2 - \frac{1}{2} py'^2 + \frac{1}{2} qy^2 - fy \right) dx + \Big[\{(sy'')' + py'\}\, \delta y - (sy'')\, \delta y' \Big]_{x_1}^{x_2}$$
$$- \Big[(sy'')' + py' \Big]_{\xi-}^{\xi+} \delta y(\xi) + \Big[sy'' \Big]_{\xi-}^{\xi+} \delta y'(\xi) = 0.$$

(b) Deduce the *natural transition conditions*

$$y(\xi+) = y(\xi-), \qquad y'(\xi+) = y'(\xi-),$$
$$[sy'']_{\xi+} = [sy'']_{\xi-}, \qquad [(sy'')' + py']_{\xi+} = [(sy'')' + py']_{\xi-}.$$

(c) Suppose that the conditions $y = y_1$ and $y' = y_1'$ are prescribed at $x = x_1$, and that the conditions $(sy'')' + py' = S_2$ and $sy'' = M_2$ are prescribed at $x = x_2$. Further, suppose that it is required that $(sy'')' + py'$ possess a *jump* of A, and sy'' a jump of B as the point $x = \xi$ is crossed in the positive direction. Show that the variational problem takes the form

$$\delta \Bigg[\int_{x_1}^{x_2} \left(\frac{1}{2} sy''^2 - \frac{1}{2} py'^2 + \frac{1}{2} qy^2 - fy \right) dx$$
$$+ S_2 y(x_2) - M_2 y'(x_2) - Ay(\xi) + By'(\xi) \Bigg] = 0,$$

where admissible functions are to satisfy the conditions $y(x_1) = y_1$ and $y'(x_1) = y_1'$, and are to be continuously differentiable in (x_1, x_2).

85. Specialize the results of Problem 84 in the case of the equation

$$\frac{d^2}{dx^2}\left(EI \frac{d^2 y}{dx^2} \right) - \rho\omega^2 y = p,$$

which governs the steady-state amplitude of small forced vibration of a beam, and interpret the conditions in physical terms. [See the note to Problem 52.]

86. (a) Modify the treatments of Problem 84 in the case of the second-order equation

$$(py')' + qy = f,$$

omitting the requirement that y' be continuous at $x = \xi$.

(b) Specialize the results of part (a) in the case of the equation

$$\frac{d}{dx}\left(F\frac{dy}{dx}\right) + \rho\omega^2 y + p = 0,$$

which governs the steady-state amplitude of small forced vibration of a string, and interpret the conditions in physical terms.

87. Two unknown functions $y_1(x)$ and $y_2(x)$ are governed, over an interval (a, b), by the simultaneous equations

$$(p_{11}y_1')' + (p_{12}y_2')' + r_{11}y_1 + r_{12}y_2 = f_1,$$
$$(p_{12}y_1')' + (p_{22}y_2')' + r_{12}y_1 + r_{22}y_2 = f_2,$$

where p_{ij}, r_{ij}, and f_i are prescribed functions of x. By multiplying the first equation by a variation δy_1, the second by δy_2, adding, integrating over (a, b), and simplifying the result, show that the corresponding variational problem is of the form

$$\delta\int_a^b \left[\frac{1}{2}\left(p_{11}y_1'^2 + 2p_{12}y_1'y_2' + p_{22}y_2'^2 - r_{11}y_1^2 - 2r_{12}y_1y_2 - r_{22}y_2^2\right) + f_1y_1 + f_2y_2\right] dx = 0,$$

if the prescribed boundary conditions are compatible with the following ones:

$$\left[(p_{11}y_1' + p_{12}y_2')\,\delta y_1\right]_a^b = 0, \qquad \left[(p_{12}y_1' + p_{22}y_2')\,\delta y_2\right]_a^b = 0.$$

88. Show that the equations

$$xy_1'' + 2y_1' + xy_1 - x^2y_2 = \phi_1,$$
$$xy_2'' + 4y_2' - y_1 + xy_2 = \phi_2$$

are reducible to the standard form of Problem 87, and obtain the relevant variational problem.

89. By starting with the known differential equation, or otherwise, deduce the following variational problems in the cases noted. In each case, u represents deflection at time t, and f represents the corresponding impressed force intensity.

(a) Transverse deformation of a string:

$$\delta\int_{t_1}^{t_2}\int_0^l \left[\frac{1}{2}\rho\left(\frac{\partial u}{\partial t}\right)^2 - \frac{1}{2}F\left(\frac{\partial u}{\partial x}\right)^2 + fu\right] dx\, dt = 0,$$

(b) Transverse deformation of a beam:

$$\delta \int_{t_1}^{t_2} \int_0^l \left[\frac{1}{2}\rho\left(\frac{\partial u}{\partial t}\right)^2 - \frac{1}{2}EI\left(\frac{\partial^2 u}{\partial x^2}\right)^2 + fu\right] dx\, dt = 0.$$

(c) Transverse deformation of a membrane:

$$\delta \int_{t_1}^{t_2} \iint_{\mathscr{R}} \left[\frac{1}{2}\rho\left(\frac{\partial u}{\partial t}\right)^2 - \frac{1}{2}F(\nabla u)^2 + fu\right] dA\, dt = 0.$$

(d) Longitudinal deformation of a rod:

$$\delta \int_{t_1}^{t_2} \int_0^L \left[\frac{1}{2}\rho\left(\frac{\partial u}{\partial t}\right)^2 - \frac{1}{2}EA\left(\frac{\partial u}{\partial x}\right)^2 + fu\right] dx\, dt = 0.$$

[E is Young's modulus, A the cross-sectional area.]

Section 2.17.

90. Derive equations (244), (245), and (246) by considering the integral of each left-hand member over an appropriate interval or region, and transforming the integral by integration by parts.

91. Verify equation (248), by expanding the right-hand member or otherwise.

92. A linear partial differential equation of second order, of the form

$$F[w] \equiv aw_{xx} + 2bw_{xy} + cw_{yy} + dw_x + ew_y + fw + g = 0,$$

where the coefficients may be functions of x and y, is derivable from a variational problem $\delta \iint_{\mathscr{R}} G\, dx\, dy = 0$ if and only if the left-hand member can be reduced to the left-hand member of a so-called "self-adjoint" form

$$S[w] \equiv (pw_x)_x + (qw_y)_x + (qw_x)_y + (rw_y)_y + sw + t = 0,$$

by multiplication by a function $A(x, y)$, which may be termed a "reducing factor."

(a) By requiring that $S[w]$ be identical with $AF[w]$, show that there must follow

$$p = Aa, \quad q = Ab, \quad r = Ac, \quad s = Af, \quad t = Ag,$$

and that the reducing factor A then must satisfy the simultaneous first-order partial differential equations

$$a\frac{\partial A}{\partial x} + b\frac{\partial A}{\partial y} = (d - a_x - b_y)A,$$

$$b\frac{\partial A}{\partial x} + c\frac{\partial A}{\partial y} = (e - b_x - c_y)A.$$

Unless these equations possess a common solution, the equation $F[w] = 0$ is not derivable from a variational problem.

(b) Suppose that a reducing factor A exists. By multiplying the equation $F[w] = 0$ by $A\,\delta w\, dx\, dy$, integrating the result over the relevant region $\mathscr{R}$, and making use of equations (243) and (245), show that the variational problem is of the form

$$\delta \iint_{\mathscr{R}} \left[\frac{1}{2}(aw_x^2 + 2bw_xw_y + cw_y^2 - fw^2) - gw \right] A\, dx\, dy = 0,$$

if appropriate boundary conditions are prescribed.

(c) Apply the preceding technique to the differential equation

$$x^2w_{xx} + 2xw_{xy} + x^3w_{yy} + 3xw_x + 2w_y + xw + g = 0.$$

[Show that the reducing factor must be a constant multiple of $A = x$.]

Section 2.18.

93. When the plate considered in Section 2.18 is also subjected to compressive forces N_1 and N_2, parallel to its surface and in the x and y directions, respectively, and to a shearing force S parallel to its surface, the approximate governing differential equation differs from equation (251) in that the zero right-hand member is replaced by $-(N_1w_{xx} + 2Sw_{xy} + N_2w_{yy})$. Show that the integrand of (253) is then to be modified by the addition of the expression

$$-\tfrac{1}{2}(N_1w_x^2 + 2Sw_xw_y + N_2w_y^2).$$

94. Suppose that a rectangular plate of uniform thickness is acted on only by a uniform compressive force N in the x direction.

(a) Show that the variational problem derived in Problem 93 takes the form

$$\frac{D}{2}\delta\int_0^a\int_0^b [w_{xx}^2 + w_{yy}^2 + 2\alpha w_{xx}w_{yy} + 2(1-\alpha)w_{xy}^2]\, dx\, dy$$
$$- \frac{N}{2}\delta\int_0^a\int_0^b w_x^2\, dx\, dy = 0.$$

(b) Deduce that the critical buckling loads (for which the problem possesses a nontrivial solution) are stationary values of the ratio

$$N = \frac{D\displaystyle\int_0^a\int_0^b [w_{xx}^2 + w_{yy}^2 + 2\alpha\omega_{xx}w_{yy} + 2(1-\alpha)w_{xy}^2]\, dx\, dy}{\displaystyle\int_0^a\int_0^b w_x^2\, dx\, dy},$$

where w satisfies the appropriate support conditions along the boundary. [Compare equation (94) and Problem 51. It can be shown that the *smallest* stationary value of N is the *minimum* value of the ratio.]

Section 2.19.

95. Use the Rayleigh-Ritz method to obtain an approximate solution of the problem

$$\frac{d}{dx}\left(x\frac{dy}{dx}\right) + y = x, \qquad y(0) = 0, \quad y(1) = 1,$$

in the form $y \approx x + x(1-x)(c_1 + c_2x)$.

96. Use the Rayleigh-Ritz method to find two successive approximations to the smallest characteristic value of λ in the problem

$$\frac{d}{dx}\left[(1 + x)\frac{dy}{dx}\right] + \lambda y = 0, \qquad y(0) = 0, \quad y(1) = 0,$$

assuming first $y(x) \approx c_1 x(1 - x)$, and second $y(x) \approx (c_1 + c_2 x)x(1 - x)$.

97. Suppose that a small mass m is attached at the point $x = a$ to a vibrating string of linear mass density ρ and length l. If the string is fixed at the ends $x = 0$ and $x = l$, show that the variational problem for small vibrations of frequency ω is of the form

$$\delta \int_0^l \left[\frac{1}{2}\rho\omega^2 y^2 - \frac{1}{2}F\left(\frac{dy}{dx}\right)^2\right] dx + \delta\left[\frac{1}{2}m\omega^2[y(a)]^2\right] = 0$$

or, equivalently,

$$\int_0^l \left[F\frac{d^2y}{dx^2} + \rho\omega^2 y\right] \delta y\, dx + m\omega^2 y(a)\, \delta y(a) = 0,$$

where F is the tension in the string. Assuming that F and ρ are constant, that $m \ll \rho l$, and that the deflection modes differ slightly from those in which m is absent, show that the nth natural frequency is approximately given by

$$\omega_n \approx \frac{n\pi}{l}\sqrt{\frac{F}{\rho}}\left(1 - \frac{m}{\rho l}\sin^2\frac{n\pi a}{l}\right).$$

[Compare the use of the procedure of Problem 53.]

98. A uniform square plate of length a is subject to a uniformly distributed compressive load N in the x direction, in the plane of the plate. The plate is clamped along its complete boundary ($x = 0, x = a, y = 0, y = a$). Show that the approximation

$$w \approx C\left(1 - \cos\frac{2\pi x}{a}\right)\left(1 - \cos\frac{2\pi y}{a}\right),$$

for the fundamental buckling mode, satisfies the relevant boundary conditions, and determine a corresponding approximation to the critical buckling load N_{cr}. [Use the result of Problem 94 or, equivalently, use the relation

$$D\int_0^a\int_0^a \nabla^4 w\, \delta w\, dx\, dy + N_{cr}\int_0^a\int_0^a w_{xx}\, \delta w\, dx\, dy = 0.$$

The required approximation is given by $N_{cr} \approx 32\pi^2 D/3a^2 \doteq 105D/a^2$, whereas the rounded true value is known to be $103.5D/a^2$.]

99. (a) Establish the relations

$$\int_{x_1}^{x_2} py'^2\, dx = -\int_{x_1}^{x_2} (py')'y\, dx + \Big[py'y\Big]_{x_1}^{x_2}$$

and

$$\int_{x_1}^{x_2} sy''^2\, dx = \int_{x_1}^{x_2} (sy'')''y\, dx + \Big[sy''y' - (sy'')'y\Big]_{x_1}^{x_2}.$$

(b) Use the results of part (a) to show that the expression

$$\delta \int_{x_1}^{x_2} \left(\frac{1}{2} sy''^2 - \frac{1}{2} py'^2 + \frac{1}{2} qy^2 - fy \right) dx$$

can be written, not only in the form

$$\int_{x_1}^{x_2} [(sy'')'' + (py')' + qy - f] \, \delta y \, dx + \Big[(sy'') \, \delta y' - \{(sy'')' + py'\} \, \delta y \Big]_{x_1}^{x_2},$$

but also in the form

$$\frac{1}{2} \delta \left\{ \int_{x_1}^{x_2} [(sy'')'' + (py')' + qy - 2f] \, y \, dx + \Big[(sy'')y' - \{(sy'')' + py'\}y \Big]_{x_1}^{x_2} \right\}.$$

100. Let $\mathscr{R}$ denote a region of the xy plane, with boundary $\mathscr{C}$ made up of one or more nonintersecting closed curves, and suppose that w is to satisfy Laplace's equation in $\mathscr{R}$, that w is prescribed as $\phi(s)$ along the portion $\mathscr{C}'$ of $\mathscr{C}$, and that $\partial w / \partial n$ is prescribed as $\psi(s)$ along the remainder of the boundary $\mathscr{C}''$, where s represents distance along $\mathscr{C}$. By calculating the variation, verify that the problem

$$\delta \left[\frac{1}{2} \iint_{\mathscr{R}} (\nabla w)^2 \, dx \, dy - \int_{\mathscr{C}'} (w - \phi) \frac{\partial w}{\partial n} \, ds - \int_{\mathscr{C}''} \psi w \, ds \right] = 0$$

is equivalent to the problem

$$\iint_{\mathscr{R}} \nabla^2 w \, \delta w \, dx \, dy + \int_{\mathscr{C}'} (w - \phi) \, \delta \frac{\partial w}{\partial n} \, ds - \int_{\mathscr{C}''} \left(\frac{\partial w}{\partial n} - \psi \right) \delta w \, ds = 0,$$

and hence deduce that the desired solution is a stationary function of either formulation of this variational problem, where the admissible functions are *unrestricted* along $\mathscr{C}$. [Either $\mathscr{C}'$ or $\mathscr{C}''$ may, of course, be identified with the whole of $\mathscr{C}$. Notice that, if use is made of the Rayleigh-Ritz method, the linear combination of approximating functions need not identically satisfy the prescribed conditions along either $\mathscr{C}'$ or $\mathscr{C}''$. (In two-dimensional problems, it is often inconvenient to choose approximating functions which have this property.) In particular, it is possible to choose, as approximating functions, special solutions of Laplace's equation, so that the *double integral* in the *second* form vanishes identically, and then to determine the constants of combination in such a way that the sum of the boundary integrals vanishes.]

101. Let the symbol Y represent y_1 when $x = x_1$ and y_2 when $x = x_2$. With this notation, verify that the problem

$$\delta \left\{ \int_{x_1}^{x_2} \left(\frac{1}{2} py'^2 - \frac{1}{2} qy^2 - fy \right) dx - \Big[p(y - Y)y' \Big]_{x_1}^{x_2} \right\} = 0$$

is equivalent to the problem

$$\int_{x_1}^{x_2} [(py')' + qy - f] \, \delta y \, dx + \Big[p(y - Y) \, \delta y' \Big]_{x_1}^{x_2} = 0.$$

Hence deduce that a stationary function of this problem, when the admissible functions are *unrestricted* at $x = x_1$ and $x = x_2$, is the solution of the equation $(py')' + qy = f$ for which $y(x_1) = y_1$ and $y(x_2) = y_2$.

Section 2.20.

102. The edges $x = 0$, $x = a$, and $y = 0$ of a vibrating square membrane are fixed. Whereas the edge $y = a$ is unrestrained, the thickness of the membrane is abruptly increased at that edge. Suppose that the additional material may be considered as concentrated along the edge, with linear mass density $A\rho/h$, where A is the effective cross-sectional area and h is the uniform membrane thickness.

(a) Show that the relevant variational problem is of the form

$$\delta \int_0^a \int_a^a \left[\frac{1}{2}\rho\omega^2 w^2 - \frac{1}{2}F(w_x{}^2 + w_y{}^2)\right] dx\, dy + \delta \int_0^a \left[\frac{1}{2}\frac{A\rho}{h}\omega^2 w^2\right]_{y=a} dx = 0,$$

and that this requirement is equivalent to the condition

$$\int_0^a \int_0^a [F\nabla^2 w + \rho\omega^2 w]\, \delta w\, dx\, dy - \int_0^a \left[\left(F\frac{\partial w}{\partial y} - \frac{A\rho}{h}\omega^2 w\right)\delta w\right]_{y=a} dx = 0,$$

where w is to vanish along the three fixed edges.

(b) Suppose that ρ, F, and h are considered to be constant, whereas A may vary moderately along the edge $y = a$. By assuming approximate deflection modes in the form

$$w = f_m(y) \sin\frac{m\pi x}{a} \qquad (m = 1, 2, \cdots),$$

where $f_m(0) = 0$, show that $f_m(y)$ must satisfy the differential equation

$$f_m''(y) + \left(\frac{\rho\omega^2}{F} - \frac{m^2\pi^2}{a^2}\right) f_m(y) = 0,$$

and the homogeneous end conditions

$$f_m(0) = 0, \qquad af_m'(a) - \alpha_m\omega^2 f_m(a) = 0,$$

where

$$\alpha_m = \frac{2\rho}{hF}\int_0^a A(x)\sin^2\frac{m\pi x}{a}\, dx.$$

(c) Deduce that corresponding critical frequencies are of the approximate form

$$\omega_{mn} = \sqrt{\frac{F}{\rho a^2}}\sqrt{m^2\pi^2 + k_{mn}{}^2},$$

where k_{mn} is the nth solution of the equation

$$\alpha_m \tan k = \frac{\rho a^2}{F}\frac{k}{k^2 + m^2\pi^2}\,.$$

[The approximation can be shown to be *exact* when A is constant.]

(d) Specialize this result in the two limiting cases in which the edge $y = a$ is unstiffened ($\alpha_m = 0$) and in which it is fixed ($\alpha_m = \infty$).

103. A uniform square plate is clamped along the edges $x = 0, x = a$, and $y = 0$, and completely unrestrained along the edge $y = a$, and is subject to a uniform loading $p = -p_0$ normal to its surface. If an approximate deflection

$$w = x^2(a - x)^2 f(y) \equiv \phi(x)f(y)$$

is assumed, where $f(y)$ satisfies the conditions $f(0) = f'(0) = 0$ along the edge $y = 0$, use equation (252') to show that the relevant natural boundary conditions along the edge $y = a$ take the form

$$k_1 f'''(a) + (2 - \alpha)k_2 f'(a) = 0,$$
$$k_1 f''(a) + \alpha k_2 f(a) = 0,$$

where

$$k_1 = \int_0^a \phi^2\, dx = \frac{a^9}{630}, \qquad k_2 = \int_0^a \phi\phi''\, dx = -\frac{2a^7}{105},$$

and where α is Poisson's ratio for the plate material. Show also that the differential equation governing $f(y)$ is obtained, from the condition

$$\int_0^a \left\{\int_0^a [D\,\nabla^4(\phi f) + p_0]\,\phi\, dx\right\} \delta f\, dy = 0$$

for arbitrary δf, in the form

$$k_1 f^{iv}(y) + 2k_2 f''(y) + k_3 f(y) = -k_4 \frac{p_0}{D},$$

where k_1 and k_2 are as defined above, and

$$k_3 = \int_0^a \phi\phi^{iv}\, dx = \frac{4a^5}{5}, \qquad k_4 = \int_0^a \phi\, dx = \frac{a^5}{30}.$$

CHAPTER THREE

Integral Equations

3.1. Introduction. An *integral equation* is an equation in which a function to be determined appears under an integral sign. We consider here principally *linear* equations, that is, equations in which no nonlinear functions of the unknown function are involved.

Linear integral equations of most frequent occurrence in practice are conventionally divided into two classifications. First, an equation of the form

$$\alpha(x)y(x) = F(x) + \lambda\int_a^b K(x, \xi)y(\xi)\,d\xi, \tag{1}$$

where α, F, and K are given functions and λ, a, and b are constant, is known as a *Fredholm equation*. The function $y(x)$ is to be determined. The given function $K(x, \xi)$, which depends upon the current variable x as well as the auxiliary variable ξ, is known as the *kernel* of the integral equation. If the upper limit of the integral is not a constant, but is identified instead with the current variable, the equation takes the form

$$\alpha(x)y(x) = F(x) + \lambda\int_a^x K(x, \xi)y(\xi)\,d\xi, \tag{2}$$

and is known as a *Volterra equation*.

It is clear that the constant λ could be incorporated into the kernel $K(x, \xi)$ in both (1) and (2). However, in many applications this constant represents a significant parameter which may take on various values in a particular discussion. Also, it will be seen that the introduction of this parameter is advantageous in theoretical treatments.

When $\alpha \not\equiv 0$, the above equations involve the unknown function y both inside and outside the integral. In the special case when $\alpha \equiv 0$, the unknown function appears only under the integral sign, and the equation is known as an *integral equation of the first kind*, while in the case when $\alpha \equiv 1$ the equation is said to be of the *second kind.*

In the more general case when α is not a constant, but is a prescribed function of x, the equation is sometimes called an integral equation of the *third kind.* However, by suitably redefining the unknown function and/or the kernel, it is always possible to rewrite such an equation in the form of an equation of the second kind. In particular, when the function $\alpha(x)$ is positive throughout the interval (a, b), equation (1) can be rewritten in an equivalent symmetric form

$$\sqrt{\alpha(x)}\, y(x) = \frac{F(x)}{\sqrt{\alpha(x)}} + \lambda \int_a^b \frac{K(x, \xi)}{\sqrt{\alpha(x)\alpha(\xi)}} \sqrt{\alpha(\xi)}\, y(\xi)\, d\xi, \tag{3}$$

and hence, in this form, can be considered an integral equation of the *second* kind in the unknown function $\sqrt{\alpha(x)}\, y(x)$, with a modified kernel. Whereas other similar rearrangements are clearly possible, it frequently happens that $K(x, \xi)$ is a symmetric function of x and ξ; the modified kernel in (3) then preserves this symmetry. As will be seen, symmetric *kernels* are of the same importance in the theory of linear *integral* equations as are symmetric *matrices* in the theory of sets of linear *algebraic* equations (Chapter 1).

Usually, in practice, the functions $\alpha(x)$, $F(x)$, and $K(x, \xi)$ are continuous in (a, b), and it is to be required also that the solution $y(x)$ be continuous in that interval.

In the preceding equations the unknown function depends only upon one independent variable. If an unknown function w depends upon two current variables x and y, the corresponding two-dimensional Fredholm equation is of the form

$$\alpha(x, y)w(x, y) = F(x, y) + \lambda \iint_{\mathscr{R}} K(x, y; \xi, \eta) w(\xi, \eta)\, d\xi\, d\eta. \tag{4}$$

In general, an integral equation comprises the complete formulation of the problem, in the sense that additional conditions need not and cannot be specified. That is, auxiliary conditions are, in a sense, already written into the equation.

Certain integral equations can be deduced from or reduced to differential equations. In order to accomplish the reduction, it is frequently necessary to make use of the known formula,

$$\frac{d}{dx} \int_{A(x)}^{B(x)} F(x, \xi)\, d\xi$$
$$= \int_A^B \frac{\partial F(x, \xi)}{\partial x}\, d\xi + F[x, B(x)] \frac{dB}{dx} - F[x, A(x)] \frac{dA}{dx}, \tag{5}$$

for differentiation of an integral involving a parameter.*

* The formula of equation (5) is valid if both F and $\partial F/\partial x$ are continuous functions of both x and ξ and if both $A'(x)$ and $B'(x)$ are continuous.

As a useful application of this formula, we consider the differentiation of the function $I_n(x)$ defined by the equation

$$I_n(x) = \int_a^x (x - \xi)^{n-1} f(\xi)\, d\xi, \tag{6}$$

where n is a positive integer and a is a constant. With

$$F(x, \xi) = (x - \xi)^{n-1} f(\xi),$$

equation (5) gives the derivative of (6) in the form

$$\frac{dI_n}{dx} = (n-1)\int_a^x (x - \xi)^{n-2} f(\xi)\, d\xi + [(x - \xi)^{n-1} f(\xi)]_{\xi=x}.$$

Hence, if $n > 1$, there follows

$$\frac{dI_n}{dx} = (n-1) I_{n-1} \qquad (n > 1), \tag{7}$$

while if $n = 1$, we have

$$\frac{dI_1}{dx} = f(x). \tag{8}$$

Repeated use of (7) leads to the general relation

$$\frac{d^k I_n}{dx^k} = (n-1)(n-2)\cdots(n-k) I_{n-k} \qquad (n > k). \tag{9}$$

In particular, we obtain the result

$$\frac{d^{n-1} I_n}{dx^{n-1}} = (n-1)!\, I_1(x), \tag{9a}$$

and hence by using (8), there follows

$$\frac{d^n I_n}{dx^n} = (n-1)!\, f(x). \tag{9b}$$

If we notice that $I_n(a) = 0$ when $n \geqq 1$, we conclude from (9) and (9a) that $I_n(x)$ and its first $n - 1$ derivatives all vanish when $x = a$.

Thus we may deduce that

$$I_1(x) = \int_a^x f(x_1)\, dx_1, \qquad I_2(x) = \int_a^x I_1(x_2)\, dx_2 = \int_a^x \int_a^{x_2} f(x_1)\, dx_1\, dx_2,$$

and, in the general case,

$$I_n(x) = (n-1)! \int_a^x \int_a^{x_n} \cdots \int_a^{x_3} \int_a^{x_2} f(x_1)\, dx_1\, dx_2 \cdots dx_{n-1}\, dx_n,$$

so that we have the result

$$\int_a^x \int_a^{x_n} \cdots \int_a^{x_3} \int_a^{x_2} f(x_1)\, dx_1\, dx_2 \cdots dx_{n-1}\, dx_n = \frac{1}{(n-1)!} \int_a^x (x-\xi)^{n-1} f(\xi)\, d\xi. \tag{10}$$

This relation will be useful in the work that follows.

The left-hand member of (10) can be interpreted as the result of first integrating f from a to x and then iterating this operation $n-1$ additional times, and is often represented symbolically in the form

$$\overbrace{\int_a^x \cdots \int_a^x}^{n \text{ times}} f(x)\, \overbrace{dx \cdots dx}^{n \text{ times}}.$$

3.2. Relations between differential and integral equations. We consider first the initial-value problem consisting of the linear second-order differential equation

$$\frac{d^2y}{dx^2} + A(x)\frac{dy}{dx} + B(x)y = f(x), \tag{11}$$

together with the prescribed initial conditions

$$y(a) = y_0, \qquad y'(a) = y_0'. \tag{12}$$

If we solve (11) for y'', replace x by x_1, and integrate the result with respect to x_1 over the interval (a, x), using (12), there follows

$$y'(x) - y_0' = -\int_a^x A(x_1)y'(x_1)\, dx_1 - \int_a^x B(x_1)y(x_1)\, dx_1 + \int_a^x f(x_1)\, dx_1$$

or, after integrating the first term on the right by parts,

$$y'(x) = -A(x)y(x) - \int_a^x [B(x_1) - A'(x_1)]y(x_1)\, dx_1 + \int_a^x f(x_1)\, dx_1 + A(a)y_0 + y_0'.$$

A second integration then gives the relation

$$y(x) - y_0 = -\int_a^x A(x_1)y(x_1)\, dx_1 - \int_a^x \int_a^{x_2} [B(x_1) - A'(x_1)]y(x_1)\, dx_1\, dx_2 + \int_a^x \int_a^{x_2} f(x_1)\, dx_1\, dx_2 + [A(a)y_0 + y_0'](x-a).$$

If use is made of equation (10), this equation can be put in the form

$$y(x) = -\int_a^x \{A(\xi) + (x-\xi)[B(\xi) - A'(\xi)]\}y(\xi)\, d\xi + \int_a^x (x-\xi)f(\xi)\, d\xi + [A(a)y_0 + y_0'](x-a) + y_0$$

or, equivalently,

$$y(x) = \int_a^x K(x, \xi) y(\xi)\, d\xi + F(x), \tag{13}$$

where we have written

$$K(x, \xi) = (\xi - x)[B(\xi) - A'(\xi)] - A(\xi) \tag{14a}$$

and

$$F(x) = \int_a^x (x - \xi) f(\xi)\, d\xi + [A(a) y_0 + y_0'](x - a) + y_0. \tag{14b}$$

This equation is seen to be a *Volterra equation of the second kind.* We may notice that the kernel K is a *linear* function of the current variable x. It must be assumed, of course, that the coefficients A and B and the function f are such that the indicated integrals exist.*

In illustration, the problem

$$\left.\begin{aligned} \frac{d^2y}{dx^2} + \lambda y = f(x), \\ y(0) = 1, \qquad y'(0) = 0 \end{aligned}\right\} \tag{15}$$

is transformed in this way to the integral equation

$$y(x) = \lambda \int_0^x (\xi - x) y(\xi)\, d\xi + 1 - \int_0^x (\xi - x) f(\xi)\, d\xi. \tag{16}$$

Conversely, the use of (5) permits the reduction of (13) to (11) by two differentiations. The initial conditions (12) are recovered by setting $x = a$ in (13) and in the result of the first differentiation. Thus, differentiation of (16) gives

$$\frac{dy}{dx} = -\lambda \int_0^x y(\xi)\, d\xi + \int_0^x f(\xi)\, d\xi, \tag{17}$$

and a second differentiation leads to the original differential equation. Since the integrals vanish when the upper and lower limits coincide, equations (16) and (17) supply the initial values $y(0) = 1$ and $y'(0) = 0$.

To illustrate the corresponding procedure in the case of *boundary-value* problems, we consider first a simple example. Starting with the problem

$$\left.\begin{aligned} \frac{d^2y}{dx^2} + \lambda y = 0, \\ y(0) = 0, \qquad y(l) = 0 \end{aligned}\right\}, \tag{18}$$

we obtain after a first integration over $(0, x)$ the relation

$$\frac{dy}{dx} = -\lambda \int_0^x y(x_1)\, dx_1 + C, \tag{19}$$

* An alternative procedure is suggested in Problem 7.

where C represents the unknown value of $y'(0)$. A second integration over $(0, x)$ then leads to the relation

$$y(x) = -\lambda \int_0^x (x - \xi)y(\xi)\, d\xi + Cx. \tag{20}$$

While the condition $y(0) = 0$ has been incorporated into this relation, it remains to determine C so that the second end condition $y(l) = 0$ is satisfied. When this condition is imposed on (20) there follows

$$\lambda \int_0^l (l - \xi)y(\xi)\, d\xi = Cl. \tag{21}$$

If the value of C so determined is introduced into (20), this relation takes the form

$$y(x) = -\lambda \int_0^x (x - \xi)y(\xi)\, d\xi + \lambda \frac{x}{l} \int_0^l (l - \xi)y(\xi)\, d\xi$$

or

$$y(x) = \lambda \int_0^x \frac{\xi}{l}(l - x)y(\xi)\, d\xi + \lambda \int_x^l \frac{x}{l}(l - \xi)y(\xi)\, d\xi. \tag{22}$$

With the abbreviation

$$K(x, \xi) = \begin{cases} \dfrac{\xi}{l}(l - x) & \textit{when} \quad \xi < x, \\[2ex] \dfrac{x}{l}(l - \xi) & \textit{when} \quad \xi > x, \end{cases} \tag{23}$$

equation (22) becomes

$$y(x) = \lambda \int_0^l K(x, \xi)y(\xi)\, d\xi. \tag{24}$$

Thus, the integral equation corresponding to the *boundary-value* problem (18) is a *Fredholm equation of the second kind*.

To recover (18) from (24), we differentiate the equal members of (22) twice, making use of (5), as follows:

$$\frac{dy}{dx} = \frac{\lambda}{l}\left[-\int_0^x \xi y(\xi)\, d\xi + x(l - x)y(x) + \int_x^l (l - \xi)y(\xi)\, d\xi - x(l - x)y(x)\right]$$
$$= \frac{\lambda}{l}\left[-\int_0^x \xi y(\xi)\, d\xi + \int_x^l (l - \xi)y(\xi)\, d\xi\right]$$

and

$$\frac{d^2y}{dx^2} = \frac{\lambda}{l}[-xy(x) - (l - x)y(x)] = -\lambda y(x),$$

in accordance with (18). The boundary conditions $y(0) = y(l) = 0$ follow directly from (22) by setting $x = 0$ and $x = l$.

We may notice that the kernel (23) has different analytic expressions in the two regions $\xi < x$ and $\xi > x$, but that the expressions are equivalent when $\xi = x$. Thus, if we think of K as *a function of* x, for a fixed value of ξ, then K is *continuous* at $x = \xi$. However, the derivative $\partial K/\partial x$ is given by $1 - \xi/l$ when $x < \xi$ and by $-\xi/l$ when $x > \xi$. Thus $\partial K/\partial x$ is discontinuous at $x = \xi$, and it has a finite jump of magnitude -1 as x *increases* through ξ. Further, we notice that in each region K is a linear function of x, that is, it satisfies the differential equation $\partial^2 K/\partial x^2 = 0$, and K vanishes at the end points $x = 0$ and $x = l$. Finally, it is seen that $K(x, \xi)$ is unchanged if x and ξ are interchanged; that is,

$$K(x, \xi) = K(\xi, x).$$

Kernels having this last property are said to be *symmetric*.

If analogous methods are used in the case of the more general second-order equation

$$\frac{d^2y}{dx^2} + A\frac{dy}{dx} + By = f,$$

with associated *end* conditions, the kernel so obtained usually is *discontinuous* at $x = \xi$ (see Problem 8). However, a kernel which is continuous can be obtained, in general, by a different procedure which is presented in the following section.

3.3. The Green's function. We consider first the problem consisting of the differential equation

$$\mathscr{L}y + \Phi(x) = 0, \tag{25}$$

where $\mathscr{L}$ is the differential operator

$$\mathscr{L} = \frac{d}{dx}\left(p\frac{d}{dx}\right) + q = p\frac{d^2}{dx^2} + \frac{dp}{dx}\frac{d}{dx} + q, \tag{26}$$

together with *homogeneous* boundary conditions, each of the form

$$\alpha y + \beta\frac{dy}{dx} = 0$$

for some constant values of α and β, which are imposed at the end points of an interval $a \leqq x \leqq b$.*

The function Φ may be a given direct function of x, or it may also depend upon x indirectly by also involving the unknown function $y(x)$, and so being expressible in the form

$$\Phi(x) = \phi(x, y(x)).$$

* If $p(x) = 0$ at an end point, the corresponding appropriate end condition may require merely that $y(x)$ remain *finite* at that point.

In order to obtain a convenient reformulation of this problem, we first attempt the determination of a *Green's function* G which, for a given number ξ, is given by $G_1(x)$ when $x < \xi$ and by $G_2(x)$ when $x > \xi$, and which has the four following properties:

1. The functions G_1 and G_2 satisfy the equation $\mathscr{L}G = 0$ in their intervals of definition; that is, $\mathscr{L}G_1 = 0$ when $x < \xi$, and $\mathscr{L}G_2 = 0$ when $x > \xi$.
2. The function G satisfies the homogeneous conditions prescribed at the end points $x = a$ and $x = b$; that is, G_1 satisfies the condition prescribed at $x = a$, and G_2 that corresponding to $x = b$.
3. The function G is continuous at $x = \xi$; that is, $G_1(\xi) = G_2(\xi)$.
4. The derivative of G has a discontinuity of magnitude $-1/p(\xi)$ at the point $x = \xi$; that is, $G_2'(\xi) - G_1'(\xi) = -1/p(\xi)$.

It is assumed that the function $p(x)$ is continuous and that it differs from zero inside the interval (a, b), so that the discontinuity in the derivative of G is of finite magnitude, and also that $p'(x)$ and $q(x)$ are continuous in (a, b).

We then show that, when the function $G(x, \xi)$ exists, the original formulation of the problem can be transformed to the relation

$$y(x) = \int_a^b G(x, \xi)\Phi(\xi)\, d\xi, \tag{27}$$

in the sense that then (27) defines the *solution* of the problem when Φ is a given direct function of x, whereas (27) constitutes an equivalent *integral-equation* formulation of the problem when Φ involves y.

For the purpose of determining G, let $y = u(x)$ be a nontrivial solution of the *associated* equation $\mathscr{L}y = 0$ which satisfies the prescribed homogeneous condition at $x = a$, and let $y = v(x)$ be a nontrivial solution of that equation which satisfies the condition prescribed at $x = b$. Then conditions 1 and 2 are satisfied if we write $G_1 = c_1u(x)$ and $G_2 = c_2v(x)$, where c_1 and c_2 are constants, so that

$$G = \begin{cases} c_1u(x) & \textit{when} \quad x < \xi, \\ c_2v(x) & \textit{when} \quad x > \xi. \end{cases} \tag{28}$$

Conditions 3 and 4 then determine c_1 and c_2 in terms of the value of ξ since condition 3 requires that

$$c_2v(\xi) - c_1u(\xi) = 0, \tag{29a}$$

while condition 4 gives the requirement

$$c_2v'(\xi) - c_1u'(\xi) = -\frac{1}{p(\xi)}\,. \tag{29b}$$

Equations (29a,b) possess a unique solution if the determinant

$$W[u(\xi), v(\xi)] \equiv \begin{vmatrix} u(\xi) & v(\xi) \\ u'(\xi) & v'(\xi) \end{vmatrix} = u(\xi)v'(\xi) - v(\xi)u'(\xi) \tag{30}$$

does not vanish. This quantity is called the *Wronskian* of the solutions u and v of the equation $\mathscr{L}y = 0$, and it cannot vanish unless the functions u and v are linearly dependent. According to *Abel's formula*,* this expression has the value $A/p(\xi)$, where A is a certain constant independent of ξ; that is, we have

$$u(\xi)v'(\xi) - v(\xi)u'(\xi) = \frac{A}{p(\xi)}. \tag{31}$$

With this relation, the solution of (29a,b) becomes

$$c_1 = -\frac{v(\xi)}{A}, \qquad c_2 = -\frac{u(\xi)}{A},$$

and hence (28) takes the form

$$G(x, \xi) = \begin{cases} -\dfrac{1}{A}\,u(x)v(\xi) & \textit{when} \quad x < \xi, \\[2ex] -\dfrac{1}{A}\,u(\xi)v(x) & \textit{when} \quad x > \xi, \end{cases} \tag{32}$$

where A is a *constant*, independent of x and ξ, which is determined by (31).

This determination fails if and only if A vanishes, so that u and v are linearly dependent, and hence are each multiples of a certain nontrivial function $U(x)$. In this case, the function $U(x)$ satisfies the equation $\mathscr{L}y = 0$ and *both* end conditions. Thus, for example, since the function $U(x) = 1$ solves the problem $d^2y/dx^2 = 0$, $y'(0) = y'(1) = 0$, the Green's function does not exist for the expression $\mathscr{L}y \equiv d^2y/dx^2$, relevant to the end conditions $y'(0) = y'(1) = 0$. A *generalized Green's function* which is appropriate to such exceptional situations is defined in Problem 19.

We now show that, with the definition of equation (32), the relation

$$y(x) = \int_a^b G(x, \xi)\Phi(\xi)\, d\xi \tag{33}$$

implies the differential equation

$$\mathscr{L}y + \Phi(x) = 0, \tag{34}$$

together with the prescribed boundary conditions. For this purpose, we write (33) in the explicit form

$$y(x) = -\frac{1}{A}\left[\int_a^x v(x)u(\xi)\Phi(\xi)\, d\xi + \int_x^b u(x)v(\xi)\Phi(\xi)\, d\xi\right]. \tag{35}$$

* Abel's formula may be derived as follows: The requirements that $u(x)$ and $v(x)$ satisfy $\mathscr{L}y = 0$ are $(pu')' + qu = 0$ and $(pv')' + qv = 0$. By multiplying the second equation by u and the first by v, and subtracting the results, there follows

$$u(pv')' - v(pu')' = [p(uv' - vu')]' = 0.$$

Hence we have $p(uv' - vu') = A$, where A is a constant, in accordance with (31).

Two differentiations, making use of (5), then lead to the relations

$$y'(x) = -\frac{1}{A}\left[\int_a^x v'(x)u(\xi)\Phi(\xi)\,d\xi + \int_x^b u'(x)v(\xi)\Phi(\xi)\,d\xi\right] \tag{36}$$

and

$$y''(x) = -\frac{1}{A}\left[\int_a^x v''(x)u(\xi)\Phi(\xi)\,d\xi + \int_x^b u''(x)v(\xi)\Phi(\xi)\,d\xi\right] - \frac{1}{A}[v'(x)u(x) - u'(x)v(x)]\Phi(x). \tag{37}$$

If we form the combination

$$\mathscr{L}y \equiv p(x)y''(x) + p'(x)y'(x) + q(x)y(x)$$

from these results, and make use of (31), there follows

$$\mathscr{L}y(x) = -\frac{1}{A}\left\{\int_a^x [\mathscr{L}v(x)]u(\xi)\Phi(\xi)\,d\xi + \int_x^b [\mathscr{L}u(x)]v(\xi)\Phi(\xi)\,d\xi\right\} - \frac{1}{A}\left[p(x)\cdot\frac{A}{p(x)}\cdot\Phi(x)\right].$$

But since $u(x)$ and $v(x)$ satisfy $\mathscr{L}y = 0$, the two integrands vanish identically, and this relation becomes merely

$$\mathscr{L}y(x) = -\Phi(x),$$

so that (33) implies (34). That is, a function y satisfying (33) satisfies the differential equation (34). Also, since (35) and (36) give

$$y(a) = -\frac{1}{A}u(a)\int_a^b v(\xi)\Phi(\xi)\,d\xi, \tag{38a}$$

$$y'(a) = -\frac{1}{A}u'(a)\int_a^b v(\xi)\Phi(\xi)\,d\xi, \tag{38b}$$

it follows that a function y defined by (33) satisfies the same homogeneous condition at $x = a$ as the function u. But this condition was specified as that which is imposed on the solution of (34) at $x = a$. A similar statement applies to the satisfaction of the condition prescribed at $x = b$.

Conversely, the differential equation (34) together with the associated end conditions can be shown to imply the relation (33) (see Problem 17), so that the two formulations are entirely equivalent.

In particular, by identifying $\Phi(x)$ with $\lambda r(x)y(x) - f(x)$ we may deduce the equivalence of the differential equation

$$\mathscr{L}y(x) + \lambda r(x)y(x) = f(x), \tag{39}$$

with associated homogeneous conditions imposed at the ends of the interval (a, b), and the Fredholm integral equation

$$y(x) = \lambda \int_a^b G(x, \xi) r(\xi) y(\xi)\, d\xi - \int_a^b G(x, \xi) f(\xi)\, d\xi, \tag{40}$$

where G is the relevant Green's function.

We may notice that the kernel $K(x, \xi)$ of (40) is actually the product $G(x, \xi)r(\xi)$. While the definition (32) shows that $G(x, \xi)$ is *symmetric*, the product $K(x, \xi)$ is *not* symmetric unless $r(x)$ is a constant. However, if we write

$$\sqrt{r(x)}\, y(x) = Y(x), \tag{41}$$

under the assumption that $r(x)$ is nonnegative over (a, b), as is usually the case in practice, equation (40) can be written in the form

$$Y(x) = \lambda \int_a^b \bar{K}(x, \xi) Y(\xi)\, d\xi - \int_a^b \bar{K}(x, \xi) \frac{f(\xi)}{\sqrt{r(\xi)}}\, d\xi, \tag{42}$$

where $\bar{K}$ is defined by the relation

$$\bar{K}(x, \xi) = \sqrt{r(x) r(\xi)}\, G(x, \xi), \tag{43}$$

and hence possesses the same symmetry as G. The importance of symmetry will be seen in later considerations.

The Green's function $G(x, \xi)$ defined by (32), or by the properties 1 to 4 (page 229), is often subject to a simple physical interpretation, as is illustrated in Section 3.5.

In the special case when the operator $\mathscr{L}$ and the associated end conditions are such that

$$\mathscr{L}y = y'', \qquad y(0) = y(l) = 0, \tag{44}$$

it is readily verified that the relevant Green's function G is identified with the kernel K defined by (23),

$$G(x, \xi) = \begin{cases} \dfrac{x}{l}(l - \xi) & (x < \xi), \\[2ex] \dfrac{\xi}{l}(l - x) & (x > \xi). \end{cases} \tag{45}$$

Thus, in particular, the solution of the problem

$$y'' = f(x), \qquad y(0) = y(l) = 0 \tag{46}$$

$$y(x) = -\int_0^l G(x, \xi) f(\xi)\, d\xi, \tag{47}$$

whereas the problem

$$y'' + \lambda r y = f(x), \qquad y(0) = y(l) = 0 \tag{48}$$

is equivalent to the integral equation

$$y(x) = \lambda \int_0^l G(x, \xi) r(\xi) y(\xi)\, d\xi - \int_0^l G(x, \xi) f(\xi)\, d\xi. \tag{49}$$

When the prescribed end conditions are not homogeneous, a modified procedure is needed. In this case, we denote by $G(x, \xi)$ the Green's function corresponding to the associated *homogeneous* end conditions, and attempt to determine a function $P(x)$ such that the relation

$$y(x) = P(x) + \int_a^b G(x, \xi)\Phi(\xi)\, d\xi \tag{50}$$

is equivalent to the differential equation

$$\mathscr{L}y(x) + \Phi(x) = 0, \tag{51}$$

together with the prescribed *nonhomogeneous* end conditions. Since

$$\mathscr{L} \int_a^b G(x, \xi)\Phi(\xi)\, d\xi = -\Phi(x),$$

the requirement that (50) imply (51) takes the form

$$\mathscr{L}P(x) = 0 \tag{52}$$

and, since the second term in (50) satisfies the associated *homogeneous* end conditions, it follows that *the function* $P(x)$ *in* (50) *must be the solution of* (52) *which satisfies the prescribed nonhomogeneous end conditions.* The existence of $P(x)$ is insured when $G(x, \xi)$ itself exists. The *equivalence* of the two formulations follows from the previously established results when the new unknown function $w(x) = y(x) - P(x)$ is introduced.

In illustration, in order to transform the problem

$$y'' + xy = 1, \qquad y(0) = 0, \quad y(l) = 1 \tag{53}$$

to an integral equation, we again use the Green's function (45), corresponding to the associated *homogeneous* conditions $y(0) = y(l) = 0$, and determine $P(x)$ such that $P''(x) = 0$, $P(0) = 0$, $P(l) = 1$, and hence $P(x) = x/l$. With $\Phi(x) = xy(x) - 1$, equation (50) then becomes

$$y(x) = \frac{x}{l} + \int_0^l G(x, \xi)[\xi y(\xi) - 1]\, d\xi$$

and reduces to the form

$$y(x) = \frac{x}{l} - \frac{x}{2}(l - x) + \int_0^l G(x, \xi)\xi y(\xi)\, d\xi. \tag{54}$$

As an explicit illustration of the use of (32), we consider the problem

$$\left.\begin{aligned} x^2 \frac{d^2y}{dx^2} + x\frac{dy}{dx} + (\lambda x^2 - 1)y = 0, \\ y(0) = 0, \qquad y(1) = 0 \end{aligned}\right\}. \tag{55}$$

The differential equation is first put into the form of (25),

$$\frac{d}{dx}\left(x\frac{dy}{dx}\right) + \left(-\frac{1}{x} + \lambda x\right)y = 0,$$

from which there follows

$$\mathscr{L}y = \frac{d}{dx}\left(x\frac{dy}{dx}\right) - \frac{y}{x}, \quad p = x, \quad q = -\frac{1}{x}, \quad r = x. \tag{56}$$

The general solution of the equation $\mathscr{L}y = 0$ is found to be

$$y = c_1x + c_2x^{-1}.$$

As a solution for which $y(0) = 0$ we may take $y = u(x)$, where

$$u(x) = x, \tag{57}$$

and as a solution for which $y(1) = 0$ we may take $y = v(x)$, where

$$v(x) = \frac{1}{x} - x. \tag{58}$$

The Wronskian of u and v is then given by

$$u(x)v'(x) - v(x)u'(x) = -\frac{2}{x} \equiv \frac{-2}{p(x)},$$

and hence, with the notation of (31), we have

$$A = -2. \tag{59}$$

Thus (32) becomes

$$G(x, \xi) = \begin{cases} \dfrac{x}{2\xi}(1 - \xi^2) & \textit{when} \quad x < \xi, \\[2ex] \dfrac{\xi}{2x}(1 - x^2) & \textit{when} \quad x > \xi. \end{cases} \tag{60}$$

It follows from (40) that the problem (55) then corresponds to the integral equation

$$y(x) = \lambda \int_0^1 G(x, \xi)\xi y(\xi)\, d\xi. \tag{61}$$

It is easily seen that the Bessel equation (55) has no solution other than the trivial solution $y \equiv 0$, satisfying the prescribed end conditions, unless λ satisfies the characteristic equation

$$J_1(\sqrt{\lambda_n}) = 0, \tag{62}$$

in which case the solution is

$$y = c\, J_1(\sqrt{\lambda_n}\, x), \tag{63}$$

where c is arbitrary. The same statement must then apply to the integral equation (61), with G given by (60).

If the condition $y(1) = 0$ were replaced by the condition $y(1) = k$ in (55), the function $P(x) = kx$ would be added to the right-hand member of (61). Because of the singular point of the differential equation at $x = 0$, a nonzero value of $y(0)$ could not be prescribed.

A completely analogous procedure can be used in transforming a boundary-value problem consisting of a differential equation of order n, and relevant homogeneous boundary conditions, to a Fredholm integral equation. The Green's function corresponding to $\mathscr{L}y$ in the interval (a, b) then is to possess the following properties:

1. G satisfies the equation $\mathscr{L}G = 0$ when $x < \xi$ and when $x > \xi$.
2. G satisfies the prescribed homogeneous boundary conditions.
3. G and its first $n - 2$ x-derivatives are continuous at $x = \xi$.
4. The $(n - 1)$th x-derivative of G has a jump of magnitude $-1/s(\xi)$ as x increases through ξ, where $s(x)$ is the coefficient of d^n/dx^n in $\mathscr{L}$.

With the function G so defined, the differential equation $\mathscr{L}y + \Phi(x) = 0$, subject to the relevant homogeneous end conditions, corresponds to the relation

$$y(x) = \int_a^b G(x, \xi)\Phi(\xi)\, d\xi$$

and, in particular, the problem consisting of the equation $\mathscr{L}y + \lambda ry = f$ and the prescribed homogeneous end conditions is equivalent to the Fredholm equation

$$y(x) = \lambda \int_a^b G(x, \xi)r(\xi)y(\xi)\, d\xi - \int_a^b G(x, \xi)f(\xi)\, d\xi.$$

In addition, for those *fourth-order* operators which are expressed in the *self-adjoint* form

$$\mathscr{L} = \frac{d^2}{dx^2}\left[s(x)\frac{d^2}{dx^2}\right] + \frac{d}{dx}\left[p(x)\frac{d}{dx}\right] + q(x), \tag{64}$$

it will be found that the Green's function $G(x, \xi)$ is *symmetric*. Most of the linear fourth-order operators occurring in practice can be expressed in this form.

3.4. Alternative definition of the Green's function. A useful interpretation of the above definition of the Green's function may be obtained as follows. We again consider the problem consisting of the differential equation

$$\mathscr{L}y + \Phi(x) = 0, \tag{65}$$

and suitably prescribed homogeneous boundary conditions at the ends of the interval (a, b). Suppose first that $\Phi(x)$ is replaced by a function $h_\epsilon(x)$ which is zero in (a, b) except over a small interval $(\xi - \epsilon, \xi + \epsilon)$ about a temporarily fixed point ξ, and is given by $1/(2\epsilon)$ over that interval, so that

$$\int_{\xi-\epsilon}^{\xi+\epsilon} h_\epsilon(x)\, dx = 1, \tag{66}$$

and denote the solution of the resultant problem by G_ϵ. If the equal members of the equation

$$\mathscr{L}G_\epsilon + h_\epsilon(x) = 0$$

are integrated over $(\xi - \epsilon, \xi + \epsilon)$, it follows that the solution of that equation must be such that

$$\int_{\xi-\epsilon}^{\xi+\epsilon} \mathscr{L}G_\epsilon\, dx = -1. \tag{67}$$

For explicitness, suppose that

$$\mathscr{L} = \frac{d}{dx}\left(p\frac{d}{dx}\right) + q, \tag{68}$$

where $p(x)$ and $q(x)$ are continuous and $p(x) \neq 0$ in the interval (a, b). In this case, (67) takes the form

$$p\frac{dG_\epsilon}{dx}\bigg|_{\xi-\epsilon}^{\xi+\epsilon} + \int_{\xi-\epsilon}^{\xi+\epsilon} qG_\epsilon\, dx = -1. \tag{69}$$

We are concerned with the limiting form of this relation as $\epsilon \to 0$ and G_ϵ tends to a function G.

If we ask that G_ϵ continue to satisfy the equation $\mathscr{L}G_\epsilon + h_\epsilon = 0$ throughout the interval (a, b) in the limit as $\epsilon \to 0$, the derivative of $p\, dG/dx$ must *exist* at all points of that interval and hence, in particular, the quantities $p\, dG_\epsilon/dx$ and qG_ϵ must remain *continuous* at the point $x = \xi$ as $\epsilon \to 0$. Thus the left-hand members of (69) then must tend to zero as $\epsilon \to 0$, so that (69) *cannot* be satisfied in the limit.

However, if we relax the requirement to the extent that, while G is to be continuous throughout (a, b), a discontinuity in dG/dx is permitted at the point $x = \xi$, it is seen that the limiting condition is satisfied if dG/dx has a jump of magnitude $-1/p(\xi)$ at $x = \xi$. Since we require further, in the limit, that the differential equation $\mathscr{L}G = 0$ be satisfied on both sides of this point, and that the prescribed end conditions be satisfied, we then have exactly the conditions which define the Green's function of the preceding section.

The same conclusion is readily obtained in the more general case of a linear operator $\mathscr{L}$ of order n, if we require that all derivatives of order less than $n - 1$ be continuous at $x = \xi$.

It is convenient (even though lacking in mathematical elegance) to say that, as $\epsilon \to 0$, the function $h_\epsilon(x)$ "tends to the unit singularity function, with singularity at $x = \xi$." This latter "function" is then considered to be zero throughout (a, b) except at the point $x = \xi$, and is imagined to be infinite at that point in such a way that the integral of the "function" across its singularity is unity. It is often known as the "unit impulse function" or as the "Dirac delta function."

The convention is extended to two- or three-dimensional space in an obvious way. Thus, in three dimensions, we start with a function which vanishes except inside a small *sphere* $\mathscr{S}_\epsilon$ of radius ϵ surrounding a certain point Q, and which is so defined inside that sphere that its integral over the volume of the sphere is unity for all values of ϵ. We then solve a problem, the formulation of which involves that function, and consider the limit of the *solution* as the sphere $\mathscr{S}_\epsilon$ enclosing the point Q shrinks to a point. It is then convenient to say that the limit of the solution (if it exists) is the "solution" corresponding to a "unit singularity function, with singularity at Q."

If we agree to the meaning of this convention, we may say that the Green's function, associated with a linear differential operator $\mathscr{L}$ and given boundary conditions, is the "solution" of the equation $\mathscr{L}y + \delta_Q = 0$, subject to the same boundary conditions, where δ_Q is the unit singularity function (or "delta function"), with singularity at a point Q. The Green's function thus involves the coordinates of Q, as well as the current variables representing position in the space considered.

When only one independent variable is involved, the Green's function can be obtained by the procedure outlined in the preceding section, and we have seen that if it is of the form $G(x, \xi)$ then the equation $\mathscr{L}y + \Phi(x) = 0$, subject to the relevant homogeneous boundary conditions, is equivalent to the relation

$$y(x) = \int_a^b G(x, \xi)\Phi(\xi)\, d\xi.$$

In order to indicate the plausibility of the truth of an analogous statement in the more general case, we consider the determination of a function $w(x, y, z)$ which satisfies a linear partial differential equation of the form

$$\mathscr{L}w + \Phi(x, y, z) = 0 \tag{70}$$

inside a three-dimensional region $\mathscr{R}$, together with appropriate homogeneous boundary conditions along the boundary of $\mathscr{R}$. Let $G_\epsilon(x, y, z; \xi, \eta, \zeta)$ be a function which, for any relevant fixed values of ξ, η, and ζ, satisfies the equation

$$\mathscr{L}G_\epsilon + h_\epsilon = 0$$

and the same boundary conditions, where h_ϵ, considered as a function of (x, y, z), vanishes outside a sphere $\mathscr{S}_\epsilon$ with center at the point $Q(\xi, \eta, \zeta)$ and radius ϵ, and has the property that

$$\iiint_{\mathscr{S}_\epsilon} h_\epsilon\, dx\, dy\, dz = 1.$$

Then also h_ϵ, considered as a function of (ξ, η, ζ), vanishes outside a sphere $\mathscr{S}'_\epsilon$ with center at the point $P(x, y, z)$ and radius ϵ, and has the property that

$$\iiint_{\mathscr{S}'_\epsilon} h_\epsilon \, d\xi \, d\eta \, d\zeta = 1.$$

If we then define the function

$$w_\epsilon(x, y, z) = \iiint_{\mathscr{R}} G_\epsilon(x, y, z; \xi, \eta, \zeta)\Phi(\xi, \eta, \zeta) \, d\xi \, d\eta \, d\zeta,$$

and calculate $\mathscr{L}w_\epsilon$ by *formally* differentiating under the integral sign, there follows

$$\begin{aligned} \mathscr{L}w_\epsilon &= -\iiint_{\mathscr{R}} \Phi(\xi, \eta, \zeta) \, h_\epsilon \, d\xi \, d\eta \, d\zeta \\ &= -\iiint_{\mathscr{S}'_\epsilon} \Phi(\xi, \eta, \zeta) \, h_\epsilon \, d\xi \, d\eta \, d\zeta, \end{aligned}$$

where again $\mathscr{S}'_\epsilon$ is a sphere of radius ϵ with center at the point $P(x, y, z)$. If the function Φ is continuous at P, its values in $\mathscr{S}'_\epsilon$ will approximate $\Phi(x, y, z)$ for small values of ϵ, so that it may be expected that the approximation

$$\mathscr{L}w_\epsilon \approx -\Phi(x, y, z) \iiint_{\mathscr{S}'_\epsilon} h_\epsilon \, d\xi \, d\eta \, d\zeta = -\Phi(x, y, z)$$

will, in general, tend to an equality as ϵ tends to zero. Hence, if we denote the limits of G_ϵ and w_ϵ by G and w, respectively, (and if $\mathscr{L}w_\epsilon$ tends to $\mathscr{L}w$) these formal arguments indicate that the relation

$$w(x, y, z) = \iiint_{\mathscr{R}} G(x, y, z; \xi, \eta, \zeta)\Phi(\xi, \eta, \zeta) \, d\xi \, d\eta \, d\zeta \tag{71}$$

implies the differential equation (70). The rigorous establishment of this fact, and of the fact that the right-hand member of (71) also satisfies the same homogeneous boundary conditions as does the Green's function G, is complicated by the fact that G generally becomes infinite when the points $P(x, y, z)$ and $Q(\xi, \eta, \zeta)$ coincide, but is possible in most practical cases.

As this statement implies, the function G will not satisfy the differential equation $\mathscr{L}\, G = 0$ *at* the point Q where the unit singularity is located. In the special case of the partial differential operator

$$\mathscr{L} = \frac{\partial}{\partial x}\left(p \frac{\partial}{\partial x}\right) + \frac{\partial}{\partial y}\left(p \frac{\partial}{\partial y}\right) + \frac{\partial}{\partial z}\left(p \frac{\partial}{\partial z}\right) + q, \tag{72}$$

associated with a three-dimensional region, it is found that G must behave near the point $Q(\xi, \eta, \zeta)$ in such a way that the integral of the *normal derivative* of G over the *surface* of the sphere $\mathscr{S}_\epsilon$ tends to $-1/p(\xi, \eta, \zeta)$ as the radius ϵ tends to zero:

$$\lim_{\epsilon \to 0} \oiint_{\mathscr{S}_\epsilon} \frac{\partial G}{\partial n}\, d\sigma = -\frac{1}{p(\xi, \eta, \zeta)}. \tag{73}$$

Here $\partial G/\partial n$ represents the derivative of G in the direction of the *outward* normal at a point of the spherical boundary, and $d\sigma$ is the element of surface area.

This result can be obtained by noticing first that the equation $\mathscr{L}G_\epsilon + h_\epsilon = 0$ can be written in terms of the vector differential operator ∇, in the form

$$\nabla \cdot (p \nabla G_\epsilon) + qG_\epsilon = -h_\epsilon.$$

If the equal members of this equation are integrated over the region $\mathscr{R}_\epsilon$ bounded by the sphere $\mathscr{S}_\epsilon$ with center at $Q(\xi, \eta, \zeta)$, and the condition

$$\iiint_{\mathscr{R}_\epsilon} h_\epsilon\, d\tau = 1$$

is imposed, where $d\tau$ is the element of volume, it follows that G_ϵ must satisfy the condition

$$\iiint_{\mathscr{R}_\epsilon} \nabla \cdot (p \nabla G_\epsilon)\, d\tau + \iiint_{\mathscr{R}_\epsilon} qG_\epsilon\, d\tau = -1.$$

The first volume integral on the left can be transformed to a surface integral, by use of the *divergence theorem*, so that this condition takes the form

$$\oiint_{\mathscr{S}_\epsilon} p \frac{\partial G_\epsilon}{\partial n}\, d\sigma + \iiint_{\mathscr{R}_\epsilon} qG_\epsilon\, d\tau = -1.$$

It is clear that this condition cannot be satisfied in the limit as $\epsilon \to 0$ if G_ϵ and its first partial derivatives are required to remain finite at the point Q in the limit. For small values of ϵ, the first term on the left is approximated by $4\pi\epsilon^2$ times the mean value of $p\, \partial G_\epsilon/\partial n$ on $\mathscr{S}_\epsilon$, while the second term is approximated by $4\pi\epsilon^3/3$ times the mean value of qG_ϵ in $\mathscr{R}_\epsilon$. If we represent radial distance from the point Q by the variable r, so that $\partial G_\epsilon/\partial n = \partial G_\epsilon/\partial r$ on $\mathscr{S}_\epsilon$, the first term thus approaches a finite nonzero limit as ϵ tends to zero if and only if $\partial G_\epsilon/\partial r$ becomes infinite like $1/r^2$ on $\mathscr{S}_\epsilon$ as $\epsilon \to 0$. In this case, G_ϵ becomes infinite like $1/r$ and the second term is thus small of order ϵ^2 when ϵ is small. Hence, as $\epsilon \to 0$, we must require that the first term tend to the value -1. For small values of ϵ, the function p may be evaluated at Q and the condition (73) follows.

In the case of the special operator

$$\mathscr{L} = \frac{\partial}{\partial x}\left(p \frac{\partial}{\partial x}\right) + \frac{\partial}{\partial y}\left(p \frac{\partial}{\partial y}\right) + q, \tag{74}$$

associated with a *two*-dimensional problem, we consider a *circle* $\mathscr{C}_\epsilon$, of radius ϵ, surrounding the point $Q(\xi, \eta)$. The condition corresponding to (73) then requires that the integral of $\partial G/\partial n$ around the perimeter of $\mathscr{C}_\epsilon$ tend to $-1/p(\xi, \eta)$ as ϵ tends to zero:

$$\lim_{\epsilon \to 0} \oint_{\mathscr{C}_\epsilon} \frac{\partial G}{\partial n} \, ds = - \frac{1}{p(\xi, \eta)} . \tag{75}$$

The condition (73) or (75), together with the requirements that G satisfy the equation $\mathscr{L}G = 0$ except at the point Q, as well as the prescribed boundary conditions, serves (in general) to determine the Green's function relevant to the operator (72) or (74).

In the two-dimensional case, it is convenient to represent by r the distance from the point $Q(\xi, \eta)$ to the point $P(x, y)$,

$$r = r(P, Q) = \sqrt{(x - \xi)^2 + (y - \eta)^2}. \tag{76}$$

On the circle $\mathscr{C}_\epsilon$ surrounding Q we may then write $ds = r\, d\theta$, where θ represents angular position and $r = \epsilon$. Equation (75) then can be written in the form

$$\lim_{r \to 0} \int_0^{2\pi} \frac{\partial G}{\partial r} \, r \, d\theta = - \frac{1}{p(\xi, \eta)} . \tag{75'}$$

This condition can be satisfied only if $r\, \partial G/\partial r$ tends to the value $-1/[2\pi p(\xi, \eta)]$ as r tends to zero. Thus the function G must behave like $-(\log r)/[2\pi p(\xi, \eta)]$ when $P(x, y)$ is in the neighborhood of the point $Q(\xi, \eta)$. It is sometimes convenient to indicate this fact symbolically in the form

$$G(P, Q) \sim - \frac{\log r(P, Q)}{2\pi p(Q)} \qquad (P \to Q).$$

In particular, when $p = 1$, the Green's function relevant to the operator (74) must be of the form

$$\begin{aligned} G(x, y; \xi, \eta) \equiv G(P, Q) &= - \frac{1}{2\pi} \log \sqrt{(x - \xi)^2 + (y - \eta)^2} + g(x, y; \xi, \eta) \\ &\equiv - \frac{1}{2\pi} \log r(P, Q) + g(P, Q), \end{aligned} \tag{77}$$

where g satisfies the equation

$$\mathscr{L} g \equiv \nabla^2 g + q g = \frac{1}{2\pi} q \log r$$

in the given region, and is so determined that the right-hand member of (77) satisfies the prescribed boundary conditions. When also $q = 0$ and Φ is independent of w, the equation $\mathscr{L}w + \Phi = 0$ becomes *Poisson's equation*,

$$\frac{\partial^2 w}{\partial x^2} + \frac{\partial^2 w}{\partial y^2} + \Phi = 0. \tag{78}$$

If the Green's function (77) is known for a region $\mathscr{R}$, with specified homogeneous boundary conditions, then the solution of (78) in $\mathscr{R}$ which satisfies those conditions is given by the integral*

$$w(x, y) = \iint_{\mathscr{R}} G(x, y; \xi, \eta)\Phi(\xi, \eta)\, d\xi\, d\eta. \tag{79}$$

Similar considerations lead to the fact that in the three-dimensional case of (72) the Green's function must behave like $r^{-1}/[4\pi p(\xi, \eta, \zeta)]$ near the point $Q\,(\xi, \eta, \zeta)$, where here r represents the distance

$$r = r(P, Q) = \sqrt{(x - \xi)^2 + (y - \eta)^2 + (z - \zeta)^2}. \tag{80}$$

In particular, when $p = 1$, the Green's function relevant to (72) must be of the form

$$\begin{aligned} G(x, y, z; \xi, \eta, \zeta) &\equiv G(P, Q) \\ &= \frac{1}{4\pi} \frac{1}{\sqrt{(x - \xi)^2 + (y - \eta)^2 + (z - \zeta)^2}} + g(x, y, z; \xi, \eta, \zeta) \\ &\equiv \frac{1}{4\pi r(P, Q)} + g(P, Q), \end{aligned} \tag{81}$$

where g satisfies the equation

$$\nabla^2 g + qg = -\frac{q}{4\pi r}$$

everywhere in the given region and $g + 1/(4\pi r)$ satisfies the prescribed boundary conditions.

We may notice that whereas the Green's function relevant to the one-dimensional operator (68) merely possesses a discontinuous derivative at Q, that function relevant to (74) becomes logarithmically infinite at Q, while that corresponding to (72) becomes infinite like $1/r$, where r represents distance from Q.

In the case of the operator (68), which is a one-dimensional specialization of (74), the circle $\mathscr{C}$ is replaced by a *line segment* extending from the point $\xi - \epsilon$ to the point $\xi + \epsilon$. The *outward* derivative of G at the point $x = \xi + \epsilon$ is given by the value of dG/dx at that point, whereas the outward derivative at $x = \xi - \epsilon$ is given by the *negative* of dG/dx at that point. The requirement that the *sum* of these outward derivatives tend to $-1/[p(\xi)]$ as ϵ tends to zero,

$$\lim_{\epsilon \to 0} \left[\frac{dG}{dx}\bigg|_{\xi+\epsilon} - \frac{dG}{dx}\bigg|_{\xi-\epsilon} \right] = -\frac{1}{p(\xi)},$$

is seen to be analogous to (73) and (75), and identical with condition 4 of page 229.

* See also Problem 27.

We have now seen that the Green's function can be defined, first, as the limit of the solution of a certain problem in which a prescribed function tends to a unit singularity function and, second, as a function satisfying a differential equation (with boundary conditions) except at a certain point, and having a certain prescribed behavior near that point. The two definitions are equivalent, the second usually being more convenient than the first in actual applications. In the following section a third alternative interpretation of the Green's function is presented.

3.5. Linear equations in cause and effect. The influence function. Linear integral equations arise most frequently in physical problems as a result of the possibility of superimposing the effects due to several causes. To indicate the general reasoning involved, we suppose that x and ξ are variables, each of which may take on all values in a certain common interval or region $\mathscr{R}$. We may, for example, think of x and ξ as each representing position (in space of one, two, or three dimensions) or time. We suppose further that a distribution of causes is active over the region $\mathscr{R}$, and that we are interested in studying the resultant distribution of effects in $\mathscr{R}$.

If the effect at x due to a *unit* cause concentrated at ξ is denoted by the function $G(x, \xi)$, then the differential effect at x due to a uniform distribution of causes of intensity $c(\xi)$ over an elementary region $(\xi, \xi + d\xi)$ is given by $c(\xi)G(x, \xi)\, d\xi$. Hence the effect $e(x)$ at x, due to a distribution of causes $c(\xi)$ over the entire region $\mathscr{R}$ is given by the integral

$$e(x) = \int_{\mathscr{R}} G(x, \xi)c(\xi)\, d\xi \tag{82}$$

if superposition is valid, that is, if the effect due to the sum of two separate causes is (exactly or approximately) the sum of the effects due to each of the causes.

The function $G(x, \xi)$, which represents *the effect at x due to a unit concentrated cause at ξ*, is often known as the *influence function* of the problem. As may be expected, this function is either identical with or proportional to the Green's function defined in the preceding sections, when that definition is applicable.

If the distribution of *causes* is prescribed, and if the influence function is known, (82) permits the determination of the effect by direct integration. However, if it is required to determine a distribution of causes which will produce a known or desired *effect* distribution, (82) represents a *Fredholm integral equation of the first kind* for the determination of c. The kernel is then identified with the influence function of the problem.

If, instead, the physical problem prescribes neither the cause nor the effect separately, but requires merely that they satisfy a certain linear relation of the form

$$c(x) = \phi(x) + \lambda e(x), \tag{83}$$

where ϕ is a given function or zero and λ is a constant, then the effect e can be eliminated between (82) and (83), to give the relation

$$c(x) = \phi(x) + \lambda \int_{\mathscr{R}} G(x, \xi)c(\xi)\, d\xi. \tag{84}$$

This relation is a Fredholm integral equation of the *second* kind, for the determination of the cause distribution. Alternatively, if the cause c is eliminated between (82) and (83), the equation

$$e(x) = \int_{\mathscr{R}} G(x, \xi)\phi(\xi)\, d\xi + \lambda \int_{\mathscr{R}} G(x, \xi)e(\xi)\, d\xi \tag{85}$$

serves to determine the effect distribution. Both cause and effect are determined by solving either (84) or (85), and using (83).

As an explicit example of such derivations, we consider the study of small deflections of a string fixed at the points $x = 0$ and $x = l$, under a loading distribution of intensity $p(x)$. We suppose that the string is initially so tightly stretched that nonuniformity of the tension, due to small deflections, can be neglected. If a *unit* concentrated load is applied in the y direction at an arbitrary point ξ (Figure 3.1), the string will then be deflected into two linear parts with a corner at the point $x = \xi$. If we denote the (approximately) uniform tension by T, the requirement of force equilibrium in the y direction leads to the condition

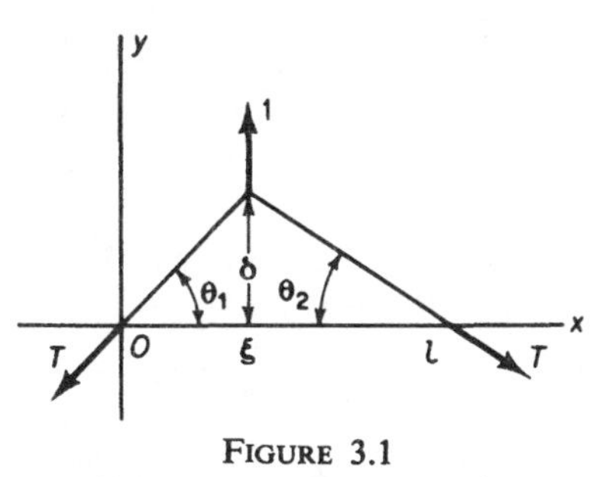

FIGURE 3.1

$$T \sin \theta_1 + T \sin \theta_2 = 1, \tag{86}$$

with the notation of Figure 3.1. For small deflections (and slopes) we have the approximations

$$\left.\begin{aligned} \sin \theta_1 &\approx \tan \theta_1 = \frac{\delta}{\xi}, \\ \sin \theta_2 &\approx \tan \theta_2 = \frac{\delta}{l - \xi} \end{aligned}\right\}, \tag{87}$$

where δ is the maximum deflection of the string, at the loaded point ξ. The introduction of these approximations into (86) leads to the relation

$$T\left(\frac{\delta}{\xi} + \frac{\delta}{l - \xi}\right) = 1,$$

and hence determines the deflection δ in the form

$$\delta = \frac{1}{Tl}\,\xi(l - \xi). \tag{88}$$

The equation of the corresponding deflection curve is then readily obtained in the form

$$y = \begin{cases} \delta\,\dfrac{x}{\xi} & \text{when} \quad x < \xi, \\[2ex] \delta\,\dfrac{l - x}{l - \xi} & \text{when} \quad x > \xi, \end{cases} \tag{89}$$

where δ is given by (88), so that the *influence function* (for small deflections) is given by the expression

$$G(x, \xi) = \begin{cases} \dfrac{x}{Tl}(l - \xi) & \text{when} \quad x < \xi, \\[2ex] \dfrac{\xi}{Tl}(l - x) & \text{when} \quad x > \xi. \end{cases} \tag{90}$$

Hence, by superposition, the deflection $y(x)$ due to a loading distribution $p(x)$ is given by

$$y(x) = \int_0^l G(x, \xi)p(\xi)\,d\xi. \tag{91}$$

If the *deflection* is prescribed, this relation constitutes an integral equation of the first kind for the determination of the necessary loading distribution.

Suppose next that the string is rotating uniformly about the x axis, with angular velocity ω, and that in addition a continuous distribution of loading $f(x)$ is imposed in the direction radially outward from the axis of revolution. If the linear mass density of the string is denoted by $\rho(x)$, the total effective load intensity can be written in the form

$$p(x) = \omega^2\rho(x)y(x) + f(x), \tag{92}$$

so that (91) takes the form

$$y(x) = \omega^2\int_0^l G(x, \xi)\rho(\xi)y(\xi)\,d\xi + \int_0^l G(x, \xi)f(\xi)\,d\xi. \tag{93}$$

It may be noticed that the influence function (90) differs from the kernel (23) of Section 3.2 only in a multiplicative factor $1/T$. By performing two differentiations, as in the reduction of (24) to (18), it is easily shown that (93) is equivalent to the more familiar formulation in terms of a differential equation with boundary conditions,

$$\left.\begin{aligned} T\frac{d^2y}{dx^2} + \rho\omega^2 y + f = 0, \\ y(0) = 0, \qquad y(l) = 0 \end{aligned}\right\}. \tag{94}$$

If we write the differential equation in the form $\mathcal{L}y + \Phi = 0$, where $\mathcal{L} = T\,d^2/dx^2$ and $\Phi = \rho\omega^2 y + f$, we may recall that the *Green's function* of the problem would then be that function which satisfies $T\,d^2y/dx^2 = 0$ except at $x = \xi$, which vanishes when $x = 0$ and $x = l$, which is continuous at $x = \xi$, and for which the radial force resultant $T\,dy/dx$ *decreases abruptly by unity* at $x = \xi$. These are precisely the conditions which determined the *influence function.*

It is of interest to notice that if, in addition, a concentrated mass m_0 were attached to the rotating string at the point $x = x_0$, the integral equation (93) would be modified by merely adding the deflection $m_0\omega^2 y(x_0)G(x, x_0)$ to the right-hand member. The corresponding modification in the differential-equation formulation would be somewhat more complicated.

In many physical problems, the Green's function is obtained empirically, and is specified only by a table of numerical values. Thus, for example, in studying small deflections of a beam of irregular cross section, subject to certain physical end restraints, a number of points $x_1, x_2, \cdots, x_n$ may be first selected along the axis of the beam. By applying loads successively at the points $\xi = x_j$ and (in each case) measuring the deflections at each of the points $x = x_i$, a table of values of the influence function $G(x, \xi)$ can be obtained, giving deflections at points $x = x_i$ due to *unit* loads applied at points $\xi = x_j$. If the beam extends from $x = 0$ to $x = l$, we thus obtain n^2 entries, specifying $G(x, \xi) = G(x_i, x_j)$ at symmetrically placed points of the square $(0 \leqq x \leqq l, 0 \leqq \xi \leqq l)$ in a fictitious $x\xi$ plane.

It is known that the deflection at a point x, due to a unit load at a point ξ, is equal to the deflection at ξ, due to a unit load at x. The truth of this *reciprocity* relation reduces the number of necessary measurements by a factor of nearly two, and shows that the relevant Green's function is *symmetric;* that is, $G(x, \xi) = G(\xi, x)$. Accordingly, the matrix of entries $G(x_i, x_j)$ is symmetrical with respect to its principal diagonal.

The determination of small deflections of the same beam when it is rotating and subject to a radial force distribution then can be based on the solution of an integral equation of the same form as (93). Numerical methods which are appropriate to the solution of such an equation, whether $G(x, \xi)$ is given analytically or by a table of values, are described in later sections of this chapter.

It is important to notice that the influence function itself incorporates the end conditions appropriate to the problem. Thus, as was mentioned previously, the integral equation generally serves to specify the problem completely.

Similar formulations of two- and three-dimensional problems are clearly possible. However, for the purpose of simplicity, attention will be restricted in most of what follows to problems involving only one independent variable. We consider next certain analytical procedures which are available for the *exact* solution of certain linear integral equations, after which we describe numerical methods of obtaining *approximate* solutions.

3.6. Fredholm equations with separable kernels. We shall speak of a kernel $K(x, \xi)$ as *separable* if it can be expressed as the sum of a finite number of terms, each of which is the product of a function of x alone and a function of ξ alone. Such a kernel is thus expressible in the form

$$K(x, \xi) = \sum_{n=1}^{N} f_n(x) g_n(\xi). \tag{95}$$

There is no loss in generality if we assume that the N functions $f_n(x)$ are *linearly independent* in the relevant interval.

Any *polynomial* in x and ξ is of this type. Further, we may notice, for example, that the kernel $\sin(x + \xi)$ is separable in this sense, since we can write

$$\sin(x + \xi) = \sin x \cos \xi + \cos x \sin \xi.$$

Integral equations with separable kernels do not occur frequently in practice. However, they are easily treated and, furthermore, the results of their consideration lead to a better understanding of integral equations of more general type. Also, it is often possible to apply the methods to be developed in this section to the *approximate* solution of Fredholm equations in which the kernel can be satisfactorily *approximated* by a polynomial in x and ξ, or by a separable kernel of more general form.

A Fredholm equation of the second kind, with (95) as its kernel, can be written in the form

$$y(x) = \lambda \int_a^b K(x, \xi) y(\xi)\, d\xi + F(x) \tag{96}$$

or

$$y(x) = \lambda \int_a^b \left[\sum_{n=1}^{N} f_n(x) g_n(\xi) \right] y(\xi)\, d\xi + F(x)$$

or

$$y(x) = \lambda \sum_{n=1}^{N} f_n(x) \left[\int_a^b g_n(\xi) y(\xi)\, d\xi \right] + F(x). \tag{97}$$

It is seen that the coefficients of $f_1(x), f_2(x), \cdots, f_N(x)$ in (97) are *constants*, although their values are unknown. If we introduce the abbreviation

$$c_n = \int_a^b g_n(x) y(x)\, dx \qquad (n = 1, 2, \cdots, N), \tag{98}$$

equation (97) takes the form

$$y(x) = F(x) + \lambda \sum_{n=1}^{N} c_n f_n(x). \tag{99}$$

This is thus the form of the required *solution* of the integral equation (96), and it remains only to determine the N constants $c_1, c_2, \cdots, c_N$.

For this purpose, we may obtain N equations involving only these N constants and calculable quantities by multiplying both members of (99) successively by $g_1(x), g_2(x), \cdots, g_N(x)$, and integrating the results over the interval (a, b). This procedure introduces calculable integrals which may be denoted as follows:

$$\alpha_{mn} = \int_a^b g_m(x) f_n(x)\, dx, \tag{100a}$$

$$\beta_m = \int_a^b g_m(x) F(x)\, dx. \tag{100b}$$

With these abbreviations, the N equations so obtained can be written in the form

$$\left.\begin{array}{r} (1 - \lambda\alpha_{11})c_1 - \lambda\alpha_{12}c_2 - \lambda\alpha_{13}c_3 + \cdots - \lambda\alpha_{1N}c_N = \beta_1, \\ -\lambda\alpha_{21}c_1 + (1 - \lambda\alpha_{22})c_2 - \lambda\alpha_{23}c_3 + \cdots - \lambda\alpha_{2N}c_N = \beta_2, \\ \cdots\cdots\cdots\cdots\cdots\cdots\cdots\cdots\cdots\cdots\cdots\cdots \\ -\lambda\alpha_{N1}c_1 - \lambda\alpha_{N2}c_2 - \lambda\alpha_{N3}c_3 + \cdots + (1 - \lambda\alpha_{NN})c_N = \beta_N \end{array}\right\}. \tag{101}$$

This set of equations possesses a *unique* solution for the c's if and only if the determinant Δ of the coefficients of the c's does not vanish. In the matrix notation of Chapter 1, the set may be written in the abbreviated form

$$(\mathbf{I} - \lambda\, \mathbf{A})\mathbf{c} = \boldsymbol{\beta}, \tag{102}$$

where $\mathbf{I}$ is the unit matrix of order N and $\mathbf{A}$ is the matrix $[\alpha_{ij}]$. Thus the results of Chapter 1, relevant to sets of linear equations, are immediately applicable to the present discussion of solutions of the integral equation (96), with a separable kernel.

If the function $F(x)$ is identically zero in (96), the integral equation is said to be *homogeneous*, and is obviously satisfied by the *trivial solution* $y(x) \equiv 0$, corresponding to the trivial solution $c_1 = c_2 = \cdots = c_N = 0$ of (101) when the right-hand members vanish. Unless the determinant $\Delta \equiv |\mathbf{I} - \lambda\, \mathbf{A}|$ vanishes, this is the *only* solution. However, if $\Delta = 0$, at least one of the c's can be assigned arbitrarily, and the remaining c's can be determined accordingly. Thus, in such cases, infinitely many solutions of the integral equation (96) exist.

Those values of λ for which $\Delta(\lambda) = 0$ are known as the *characteristic values* (or *eigenvalues*), and any nontrivial solution of the homogeneous integral equation (with a convenient choice of the arbitrary constant or constants) is then called a corresponding *characteristic function* (eigenfunction) of the integral equation. If k of the constants $c_1, c_2, \cdots, c_N$ can be assigned arbitrarily for a given characteristic value of λ, then k linearly independent corresponding characteristic functions are obtained.

If the function $F(x)$ is not identically zero, but is *orthogonal* to all the functions $g_1(x), g_2(x), \cdots, g_N(x)$, equation (100b) shows that the right-hand members of (101) again vanish. The preceding discussion again applies to this case, except for the fact that here the solution (99) of the integral equation involves also the function $F(x)$. The trivial values $c_1 = c_2 = \cdots = c_N = 0$ thus lead to the solution $y = F(x)$. Solutions corresponding to characteristic values of λ are now expressed as the sum of $F(x)$ and arbitrary multiples of characteristic functions.

Finally, if at least one right-hand member of (101) does not vanish, a unique nontrivial solution of (101) exists, leading to a unique nontrivial solution of the integral equation (96), *if the determinant* $\Delta(\lambda)$ *does not vanish.* However, in this case, *if* $\Delta(\lambda)$ *does vanish* equations (101) are either *incompatible*, and no solution exists, or they are *redundant*, and infinitely many solutions exist.

3.7. Illustrative example. The several possible cases just discussed may be illustrated by a consideration of the integral equation

$$y(x) = \lambda \int_0^1 (1 - 3x\xi) y(\xi)\, d\xi + F(x). \tag{103}$$

This equation can be rewritten in the form

$$y(x) = \lambda(c_1 - 3c_2 x) + F(x), \tag{104}$$

where

$$c_1 = \int_0^1 y(\xi)\, d\xi, \qquad c_2 = \int_0^1 \xi y(\xi)\, d\xi. \tag{105a,b}$$

To determine c_1 and c_2, we multiply both sides of (104) successively by 1 and x and integrate the results over (0, 1), to obtain the equations

$$c_1 = \lambda\left(c_1 - \frac{3}{2} c_2\right) + \int_0^1 F(x)\, dx,$$

$$c_2 = \lambda\left(\frac{1}{2} c_1 - c_2\right) + \int_0^1 xF(x)\, dx,$$

or

$$\left.\begin{aligned} (1 - \lambda)c_1 + \frac{3}{2}\lambda c_2 &= \int_0^1 F(x)\, dx, \\ -\frac{1}{2}\lambda c_1 + (1 + \lambda)c_2 &= \int_0^1 xF(x)\, dx \end{aligned}\right\}. \tag{106a,b}$$

The determinant of coefficients is given by

$$\Delta(\lambda) = \begin{vmatrix} 1-\lambda & \frac{3}{2}\lambda \\ -\frac{1}{2}\lambda & 1+\lambda \end{vmatrix} = \tfrac{1}{4}(4-\lambda^2). \tag{107}$$

It follows that a *unique* solution exists if and only if

$$\lambda \neq \pm 2, \tag{108}$$

and is obtained by solving (106a,b) for c_1 and c_2, and introducing the results into (104). In particular, if $F(x) = 0$ and $\lambda \neq \pm 2$, the only solution is the trivial one, $y(x) = 0$. The numbers $\lambda = \pm 2$ are the characteristic numbers of the problem.

If $\lambda = +2$, equations (106a,b) take the form

$$\left.\begin{aligned} -c_1 + 3c_2 &= \int_0^1 F(x)\,dx, \\ -c_1 + 3c_2 &= \int_0^1 xF(x)\,dx \end{aligned}\right\}, \tag{109a,b}$$

while if $\lambda = -2$ equations (106a,b) become

$$\left.\begin{aligned} c_1 - c_2 &= \frac{1}{3}\int_0^1 F(x)\,dx, \\ c_1 - c_2 &= \int_0^1 xF(x)\,dx \end{aligned}\right\}. \tag{110a,b}$$

Equations (109a,b) are incompatible unless the prescribed function $F(x)$ satisfies the condition

$$\int_0^1 F(x)\,dx = \int_0^1 xF(x)\,dx \quad or \quad \int_0^1 (1-x)F(x)\,dx = 0, \tag{111}$$

whereas (110a,b) are incompatible unless

$$\frac{1}{3}\int_0^1 F(x)\,dx = \int_0^1 xF(x)\,dx \quad or \quad \int_0^1 (1-3x)F(x)\,dx = 0, \tag{112}$$

in which cases the corresponding equation pairs (109) or (110) are redundant.

We consider first the case when

$$F(x) = 0, \tag{113}$$

so that (103) is *homogeneous*. Then, if $\lambda \neq \pm 2$, the only solution is the trivial one $y(x) = 0$, as was mentioned above. If $\lambda = 2$ (and $F = 0$) equations (109) are redundant, and either equation gives the single condition $c_1 = 3c_2$. Thus (104) then gives the solution

$$y(x) = A(1-x) \quad when \quad \lambda = 2, \tag{114}$$

where $A = 6c_2$ is an *arbitrary* constant. Thus the function $1 - x$ (or any convenient nonzero multiple of that function) is the *characteristic function* corresponding to the characteristic number $\lambda = +2$. In a similar way, we find the solution

$$y(x) = B(1 - 3x) \quad \textit{when} \quad \lambda = -2, \tag{115}$$

where $B = -2c_1 = -2c_2$ is an arbitrary constant, so that $1 - 3x$ is the characteristic function corresponding to $\lambda = -2$.

Equation (104) shows that any solution of (103) is expressible in the form

$$y(x) = F(x) + C_1(1 - x) + C_2(1 - 3x), \tag{116}$$

where we have written $C_1 = 3\lambda(c_1 - c_2)/2$ and $C_2 = \lambda(3c_2 - c_1)/2$. Thus it follows that *any solution of* (103) *can be expressed as the sum of* $F(x)$ *and some linear combination of the characteristic functions.* The fact that this statement can be applied to a wide class of integral equations is of basic importance, as will be seen.

In the *nonhomogeneous* case, $F(x) \not\equiv 0$, a *unique solution* exists if $\lambda \neq \pm 2$. If $\lambda = 2$, equation (111) shows that *no solution exists unless* $F(x)$ *is orthogonal to* $1 - x$ *over the relevant interval* (0, 1), that is, *unless* $F(x)$ *is orthogonal to the characteristic function corresponding to* $\lambda = 2$.* If F satisfies this restriction equations (109a,b) are again equivalent. Hence, if we use (109a), we may obtain $c_1 = 3c_2 - \int_0^1 F(x)\,dx$, so that (104) gives the solution as follows:

$$\lambda = 2: \quad y(x) = F(x) - 2\int_0^1 F(x)\,dx + A(1 - x),$$

when

$$\int_0^1 (1 - x)F(x)\,dx = 0. \tag{117}$$

Here $A = 6c_2$ is again an arbitrary constant. Thus, in this case, *infinitely many* solutions exist, *differing from each other by a multiple of the relevant characteristic function.*

Similarly, if $\lambda = -2$ there is no solution unless $F(x)$ is orthogonal to $1 - 3x$ over (0, 1), in which case infinitely many solutions exist as follows:

$$\lambda = -2: \quad y(x) = F(x) - \frac{2}{3}\int_0^1 F(x)\,dx + B(1 - 3x),$$

when

$$\int_0^1 (1 - 3x)F(x)\,dx = 0. \tag{118}$$

Here $B = -2c_2$ is an arbitrary constant.

* As will be seen in the following section, this situation is a consequence of the *symmetry* of the kernel $K(x, \xi) = 1 - 3x\xi$ in (103).

3.8. Hilbert-Schmidt theory. In those cases when the kernel $K(x, \xi)$ of a homogeneous Fredholm equation is not of the form (95), in particular, if $K(x, \xi)$ is given by different analytic expressions in the intervals for which $x < \xi$ and $x > \xi$, there are generally infinitely many characteristic numbers λ_n $(n = 1, 2, 3, \cdots)$, each corresponding to a characteristic function defined within an arbitrary multiplicative constant. In exceptional cases, a given characteristic number λ_k may correspond to two or more independent characteristic functions.* In this section we investigate certain properties of these characteristic functions.

Let $y_m(x)$ and $y_n(x)$ be characteristic functions corresponding respectively to two *different* characteristic numbers λ_m and λ_n of the homogeneous Fredholm equation

$$y(x) = \lambda \int_a^b K(x, \xi)y(\xi)\, d\xi, \tag{119}$$

and *suppose that the kernel* $K(x, \xi)$ *is symmetric*, so that

$$K(x, \xi) = K(\xi, x). \tag{120}$$

As has been indicated in preceding sections, such kernels are of frequent occurrence in the formulation of physically motivated problems. We may notice that $\lambda = 0$ *cannot* be a characteristic number since it leads necessarily to the trivial solution $y(x) \equiv 0$.

The functions y_m and y_n accordingly must satisfy the equations

$$\left.\begin{aligned} y_m(x) &= \lambda_m \int_a^b K(x, \xi)y_m(\xi)\, d\xi, \\ y_n(x) &= \lambda_n \int_a^b K(x, \xi)y_n(\xi)\, d\xi \end{aligned}\right\}. \tag{121a,b}$$

If we multiply both members of (121a) by $y_n(x)$, and integrate the results with respect to x over (a, b), there then follows

$$\int_a^b y_m(x)y_n(x)\, dx = \lambda_m \int_a^b y_n(x)\left[\int_a^b K(x, \xi)y_m(\xi)\, d\xi\right] dx. \tag{122}$$

If the order of integration is reversed in the right-hand member, equation (122) becomes

$$\int_a^b y_m(x)y_n(x)\, dx = \lambda_m \int_a^b y_m(\xi)\left[\int_a^b K(x, \xi)y_n(x)\, dx\right] d\xi. \tag{123}$$

* It is known, however, that when $K(x, \xi)$ is continuous and (a, b) is finite the characteristic numbers are of *finite multiplicity* and also are *isolated*, in the sense that they cannot cluster about a finite "limit point." In particular, when there are infinitely many characteristic numbers, successive magnitudes increase without limit.

We now make use of the assumed *symmetry* (120) to rewrite the inner integral on the right in the form

$$\int_a^b K(\xi, x) y_n(x)\, dx.$$

But this integral differs from the coefficient of λ_n in (121b) only in that x and ξ are interchanged, and hence it is equivalent to

$$\frac{1}{\lambda_n} y_n(\xi),$$

so that (123) becomes

$$\int_a^b y_m(x) y_n(x)\, dx = \frac{\lambda_m}{\lambda_n} \int_a^b y_m(\xi) y_n(\xi)\, d\xi. \tag{124}$$

Since the integrals in (124) are equivalent, (124) can be rewritten in the form

$$(\lambda_m - \lambda_n) \int_a^b y_m(x) y_n(x)\, dx = 0. \tag{125}$$

Thus we conclude that *if $y_m(x)$ and $y_n(x)$ are characteristic functions of* (119) *corresponding to distinct characteristic numbers, then $y_m(x)$ and $y_n(x)$ are orthogonal over the interval* (a, b).

If two or more linearly independent characteristic functions correspond to the same characteristic number, then an equal number of orthogonalized linear combinations can be formed by the Gram-Schmidt procedure (Section 1.13). In the remainder of this chapter it will be assumed that this has been done, when such exceptional cases arise.

It is important to notice that the preceding results apply only to a *symmetric* kernel.

We show next that *the characteristic numbers of a Fredholm equation with a real symmetric kernel are all real.** This result is established by noticing that if λ_m were an imaginary characteristic number, corresponding to a complex characteristic function $y_m(x)$, then the complex conjugate number $\bar{\lambda}_m$ would necessarily also be a characteristic number, corresponding to the characteristic function $\bar{y}_m(x)$ which is the complex conjugate of $y_m(x)$. Hence, by replacing λ_n by $\bar{\lambda}_m$ and y_n by $\bar{y}_m$ in (125), it would follow that

$$(\lambda_m - \bar{\lambda}_m) \int_a^b y_m(x) \bar{y}_m(x)\, dx = 0. \tag{126}$$

If we write $\lambda_m = \alpha_m + i\, \beta_m$ and $y_m(x) = f_m(x) + i\, g_m(x)$, this relation takes the form

$$2i\, \beta_m \int_a^b (f_m^{\,2} + g_m^{\,2})\, dx = 0. \tag{127}$$

* Proof that such an equation always possesses at least one characteristic number, when K is continuous, is omitted.

But, since $y_m(x) \not\equiv 0$, the integral cannot vanish, and we conclude that the imaginary part of λ_m must vanish, as was to be shown.

A Fredholm equation with a nonsymmetric kernel may possess characteristic numbers which are not real.

In more advanced works* the following basic theorem is established:

Any function $f(x)$ which can be generated from a continuous function $\Phi(x)$ by the operation $\int_a^b K(x, \xi)\Phi(\xi)\,d\xi$, where $K(x, \xi)$ is continuous, real, and symmetric, so that

$$f(x) = \int_a^b K(x, \xi)\Phi(\xi)\,d\xi$$

for *some* continuous function Φ, *can be represented over (a, b) by a linear combination of the characteristic functions $y_1(x), y_2(x), \cdots$, of the homogeneous Fredholm integral equation* (119) *with $K(x, \xi)$ as its kernel.*†

Because of the orthogonality, the coefficients in the representation

$$f(x) = \sum_n A_n y_n(x) \qquad (a \leqq x \leqq b) \tag{128a}$$

are then determined by the familiar formula

$$A_n \int_a^b [y_n(x)]^2\,dx = \int_a^b f(x)y_n(x)\,dx \qquad (n = 1, 2, \cdots). \tag{128b}$$

In those cases when only a finite number of characteristic functions exist, the functions generated by the operation

$$\int_a^b K(x, \xi)\Phi(\xi)\,d\xi$$

form a very restricted class. For example, if $K(x, \xi) = \sin(x + \xi)$ and $(a, b) = (0, 2\pi)$, there follows

$$\begin{aligned}\int_0^{2\pi} K(x, \xi)\Phi(\xi)\,d\xi &= \int_0^{2\pi} (\sin x \cos \xi + \cos x \sin \xi)\Phi(\xi)\,d\xi \\ &= \left[\int_0^{2\pi} \Phi(\xi)\cos\xi\,d\xi\right]\sin x + \left[\int_0^{2\pi} \Phi(\xi)\sin\xi\,d\xi\right]\cos x,\end{aligned} \tag{129}$$

and hence this operation can generate only functions of the form

$$f(x) = C_1 \sin x + C_2 \cos x, \tag{130}$$

* See, for example, Reference 2.

† When the set of characteristic functions is infinite, the resultant infinite series converges *absolutely* and *uniformly* in the interval (a, b). This theorem applies, more generally, when K is any continuous complex kernel with the property that $K(\xi, x) = \overline{K(x, \xi)}$, where $\overline{K(x, \xi)}$ is the complex conjugate of $K(x, \xi)$.

regardless of the form of Φ. The characteristic functions of the associated homogeneous Fredholm integral equation

$$y(x) = \lambda \int_0^{2\pi} \sin(x + \xi)\, y(\xi)\, d\xi \tag{131}$$

are readily found, by the methods of the preceding section, to be arbitrary multiples of the functions $y_1(x) = \sin x + \cos x$ and $y_2(x) = \sin x - \cos x$, corresponding respectively to $\lambda_1 = 1/\pi$ and $\lambda_2 = -1/\pi$. It is obvious that any function of the form (130), generated by $\int_0^{2\pi} \sin(x + \xi)\, \Phi(\xi)\, d\xi$, can indeed be expressed as a linear combination of $y_1(x)$ and $y_2(x)$.

Even though the number of relevant independent characteristic functions be infinite, it is not necessarily true that *any* continuous function $f(x)$ defined over (a, b) can be represented over that interval by a series of these functions; that is, the set of characteristic functions, even though infinite in number, may not comprise a *complete* set, in the sense defined in Section 1.28.

Suppose now that we have (in some manner) obtained all members of the set of characteristic functions $y_n(x)$, each corresponding to a characteristic number λ_n of the homogeneous equation

$$y(x) = \lambda \int_a^b K(x, \xi) y(\xi)\, d\xi, \tag{132}$$

where $K(x, \xi)$ is continuous, real, and symmetric. We next show that the knowledge of these functions and constants permits a simple determination of the continuous solution of the corresponding *nonhomogeneous* Fredholm equation of the second kind,

$$y(x) = F(x) + \lambda \int_a^b K(x, \xi) y(\xi)\, d\xi, \tag{133}$$

where $F(x)$ is a given continuous real function, when a continuous solution exists.

We suppose that the characteristic numbers have been ordered (say, with respect to magnitude), that characteristic numbers corresponding to k independent characteristic functions have been counted k times, and that such subsets of independent characteristic functions (corresponding to multiple characteristic numbers) have been orthogonalized.

In order to simplify the relations which follow, we suppose also that the arbitrary multiplicative constant associated with each characteristic function is so chosen that the function is *normalized* over the relevant interval (a, b). Thus we write

$$\phi_n = C_n y_n, \tag{134}$$

where the normalizing factor C_n is given by

$$C_n = \frac{1}{\sqrt{\int_a^b [y_n(x)]^2 \, dx}}, \tag{135}$$

so that there follows

$$\int_a^b [\phi_n(x)]^2 \, dx = 1. \tag{136}$$

The expansion (128) of a function $f(x)$ in a series of the normalized characteristic functions then takes the simpler form

$$f(x) = \sum_n a_n \phi_n(x) \quad \textit{where} \quad a_n = \int_a^b f(x)\phi_n(x) \, dx. \tag{137}$$

Since the series (128) and (137) are identical, it follows that

$$a_n \phi_n = A_n y_n, \qquad A_n = a_n C_n. \tag{138a,b}$$

These relations permit transition from the expressions to be obtained to corresponding expressions involving nonnormalized characteristic functions.

If the equation

$$y(x) = F(x) + \lambda \int_a^b K(x, \xi) y(\xi) \, d\xi \tag{139}$$

possesses a continuous solution $y(x)$, then the function $y(x) - F(x)$ is generated by the operation $\int_a^b K(x, \xi)[\lambda y(\xi)] \, d\xi$, and hence it can be represented by a series (or linear combination) of the normalized characteristic functions $\phi_n(x)$ $(n = 1, 2, \cdots)$, of the form

$$y(x) - F(x) = \sum_n a_n \phi_n(x) \qquad (a \leqq x \leqq b), \tag{140}$$

where the coefficients a_n are given, by virtue of (137), by

$$a_n = \int_a^b [y(x) - F(x)]\phi_n(x) \, dx. \tag{141}$$

With the convenient abbreviations

$$c_n = \int_a^b y(x)\phi_n(x) \, dx, \qquad f_n = \int_a^b F(x)\phi_n(x) \, dx, \tag{142a,b}$$

this relation takes the form

$$a_n = c_n - f_n. \tag{143}$$

In order to obtain a relation which permits the elimination of the unknown integral $c_n = \int_a^b y\phi_n \, dx$ from (143), we multiply both members of (139) by

$\phi_n(x)$ and integrate the results over (a, b), so that there follows

$$c_n = f_n + \lambda \int_a^b \phi_n(x) \left[\int_a^b K(x, \xi) y(\xi) \, d\xi \right] dx, \tag{144}$$

with the notation of (142). If the order of integration is reversed in the coefficient of λ, and use is made of the assumed *symmetry* in $K(x, \xi)$, that coefficient becomes

$$\int_a^b y(\xi) \left[\int_a^b K(\xi, x) \phi_n(x) \, dx \right] d\xi = \frac{1}{\lambda_n} \int_a^b y(\xi) \phi_n(\xi) \, d\xi = \frac{c_n}{\lambda_n},$$

and hence (144) is equivalent to the relation

$$c_n = f_n + \frac{\lambda}{\lambda_n} c_n. \tag{145}$$

The elimination of c_n between (143) and (145) then gives

$$a_n = \frac{\lambda}{\lambda_n - \lambda} f_n \tag{146}$$

if $\lambda \neq \lambda_n$. Hence the required solution (140) then takes the form

$$y(x) = F(x) + \lambda \sum_n \frac{f_n}{\lambda_n - \lambda} \phi_n(x) \qquad (\lambda \neq \lambda_n) \tag{147}$$

where, as previously defined,

$$f_n = \int_a^b F(x) \phi_n(x) \, dx \qquad (n = 1, 2, \cdots). \tag{148}$$

We may notice that the constants f_n would be the coefficients in the expansion $F(x) = \Sigma f_n \phi_n(x)$ if $F(x)$ were representable by such an expansion. It is of some importance to notice that in the preceding derivation it was not necessary to assume the validity of this representation.

The expansion (147) exists uniquely if and only if λ does not take on a characteristic value. Because of the presence of the term $f_k \phi_k(x)/(\lambda_k - \lambda)$, we see that if $\lambda = \lambda_k$, where λ_k is the kth characteristic number, the solution (147) is *nonexistent* unless also $f_k = 0$, that is, *unless* $F(x)$ *is orthogonal to the corresponding characteristic function or functions*. But if $\lambda = \lambda_k$ and $f_k = 0$ equation (145) reduces to the trivial identity when $n = k$, and hence imposes no restriction on c_k. From (143) it then follows that the coefficient of $\phi_k(x)$ in (147), which formally assumes the form 0/0, is truly *arbitrary*, so that in this case (139) possesses *infinitely many* solutions, differing from each other by arbitrary multiples of $\phi_k(x)$. If λ assumes a characteristic value and $F(x)$ is *not* orthogonal to the corresponding characteristic function or functions, *no solution exists*. These results are illustrated by the simple example of Section 3.7, which involves the symmetric kernel $K(x, \xi) = 1 - 3x\xi$.

By virtue of (138), the normalization of the characteristic functions is unnecessary, in the sense that (147) can be replaced by the expression

$$y(x) = F(x) + \lambda \sum_n \frac{F_n}{\lambda_n - \lambda} y_n(x) \qquad (\lambda \neq \lambda_n), \tag{147'}$$

where

$$F_n \int_a^b [y_n(x)]^2 \, dx = \int_a^b F(x) y_n(x) \, dx \qquad (n = 1, 2, \cdots). \tag{148'}$$

We consider next the Fredholm equation of the *first* kind,

$$F(x) = \int_a^b K(x, \xi) y(\xi) \, d\xi, \tag{149}$$

with a continuous, real, and symmetric kernel, where F is a prescribed continuous function and y is to be determined. It follows from the basic expansion theorem (page 253) that (149) has *no continuous solution* unless $F(x)$ can be expressed as a linear combination of the characteristic functions corresponding to the associated homogeneous equation of the second kind,

$$y(x) = \lambda \int_a^b K(x, \xi) y(\xi) \, d\xi. \tag{150}$$

For example, in the special case for which $K(x, \xi) = \sin (x + \xi)$ and $(a, b) = (0, 2\pi)$, equation (149) becomes

$$F(x) = \int_0^{2\pi} y(\xi) \sin (x + \xi) \, d\xi \tag{151a}$$

or

$$F(x) = \left[\int_0^{2\pi} y(\xi) \cos \xi \, d\xi \right] \sin x + \left[\int_0^{2\pi} y(\xi) \sin \xi \, d\xi \right] \cos x. \tag{151b}$$

This relation can be satisfied only if $F(x)$ is prescribed as a linear combination of $\sin x$ and $\cos x$ or, equivalently, as a related linear combination of the characteristic functions y_1 and y_2 of the associated homogeneous equation (131),

$$y_1 = \sin x + \cos x, \qquad y_2 = \sin x - \cos x, \tag{152}$$

corresponding to $\lambda_1 = 1/\pi$ and $\lambda_2 = -1/\pi$. If $F(x)$ is prescribed as

$$F(x) = A \sin x + B \cos x, \tag{153}$$

then (151b) is satisfied by any function y for which

$$\int_0^{2\pi} y(\xi) \cos \xi \, d\xi = A, \qquad \int_0^{2\pi} y(\xi) \sin \xi \, d\xi = B. \tag{154}$$

One such function is clearly

$$y(x) = \frac{1}{\pi} (A \cos x + B \sin x). \tag{155}$$

However, if we add to (155) *any function* which is orthogonal to both $\sin x$ and $\cos x$, and hence to the characteristic functions y_1 and y_2, over $(0, 2\pi)$, the conditions (154) will still be satisfied, so that the solution is by no means unique. Unless $F(x)$ is prescribed in the form (153), no solution exists.

Suppose now that (149) *does* possess a continuous solution. Then $F(x)$ is generated from $y(x)$ by the operation $\int_a^b K(x, \xi)y(\xi)\, d\xi$, and hence it can be expanded in a series

$$F(x) = \sum_n f_n\phi_n(x) \qquad (a \leqq x \leqq b), \tag{156}$$

where

$$f_n = \int_a^b F(x)\phi_n(x)\, dx, \tag{157}$$

and where ϕ_n is the nth characteristic function of (150). The series may be finite or infinite. But since ϕ_n satisfies the equation

$$\phi_n(x) = \lambda_n \int_a^b K(x, \xi)\phi_n(\xi)\, d\xi, \tag{158}$$

and since (149) must be satisfied, we may replace $F(x)$ and $\phi_n(x)$ by the right-hand members of (149) and (158), so that (156) takes the form

$$\int_a^b K(x, \xi)y(\xi)\, d\xi = \sum_n \lambda_n f_n \int_a^b K(x, \xi)\phi_n(\xi)\, d\xi$$

or

$$\int_a^b K(x, \xi)\Big[y(\xi) - \sum_n \lambda_n f_n \phi_n(\xi)\Big]\, d\xi = 0.* \tag{159}$$

This condition is satisfied if and only if $y(x)$ is of the form

$$y(x) = \sum_n \lambda_n f_n \phi_n(x) + \Phi(x), \tag{160}$$

where $\Phi(x)$ is a solution of the equation

$$\int_a^b K(x, \xi)\Phi(\xi)\, d\xi = 0. \tag{161}$$

We conclude that if (149) possesses a continuous solution, then that solution must be of the form (160), where Φ is any continuous function satisfying (161). From the homogeneity of (161) it is clear that either (161) is satisfied only by the trivial function $\Phi(x) \equiv 0$ or it possesses infinitely many solutions.

If we multiply both members of (161) by $\phi_n(x)$, integrate the results over (a, b) and make use of the assumed symmetry in K, we obtain the condition

$$\begin{aligned}\int_a^b \phi_n(x)\Big[\int_a^b K(x, \xi)\Phi(\xi)\, d\xi\Big]\, dx &= \int_a^b \Phi(\xi)\Big[\int_a^b K(\xi, x)\phi_n(x)\, dx\Big]\, d\xi \\ &= \frac{1}{\lambda_n}\int_a^b \Phi(\xi)\phi_n(\xi)\, d\xi = 0.\end{aligned} \tag{162}$$

* The validity of the interchange of order of integration and summation is a consequence of the uniformity of the convergence of (156) (see footnote on page 253).

Hence it follows that if (161) possesses a nontrivial solution, then that solution must be orthogonal to all the characteristic functions ϕ_n. If this set of functions is *finite*, then infinitely many linearly independent functions satisfying this condition exist. If the functions ϕ_n comprise an infinite *complete* set over (a, b), then *no* continuous nontrivial function can be simultaneously orthogonal to all functions of the set, so that in this case the function Φ in (160) must be identically zero.

In the preceding developments we made use of a known expansion theorem to show that if a Fredholm equation with a continuous, real, and symmetric kernel possesses a continuous solution, then that solution must be of a certain form. In particular, if the nonhomogeneous equation (133), of the second kind, possesses a continuous solution, then that solution is *unique* unless λ assumes a characteristic value, and is given by (147). If the equation (149), of the first kind, possesses a continuous solution, then it is given by (160) and it is or is not uniquely defined, according as (161) does not or does possess nontrivial solutions.

In the case of equation (133), it is known (see Section 3.11) that a continuous solution *does* in fact always exist when $F(x)$ and $K(x, \xi)$ are continuous and λ is noncharacteristic, so that the validity of (147) then is completely established. However, as has been seen, no such assurance exists in the case of a Fredholm equation of the first kind.*

In physically motivated problems, questions concerning existence and uniqueness of a solution usually can be resolved by physical considerations. Thus, for example, if the kernel in (161) is the Green's function (90) for a loaded string, a nontrivial solution of (161) clearly cannot exist, since it would represent a static loading which leads to no deflection at any point of the string. From the mathematical point of view, however, such questions are of considerable interest. Certain known results are included in the following sections.

It should be remarked that before the theory of this section can be applied it is necessary to determine the characteristic numbers and functions of the homogeneous equation. Except in special cases this determination must depend upon numerical (or graphical) procedures, certain of which are discussed in Section 3.14.

3.9. Iterative methods for solving equations of the second kind. In certain cases integral equations of the second kind can be solved by a method of successive approximations. In this section we describe the method and investigate its validity.

Suppose that in a Fredholm equation of the second kind,

$$y(x) = F(x) + \lambda \int_a^b K(x, \xi) y(\xi) \, d\xi, \tag{163}$$

* It is possible to prove, by direct substitution, that (160) does indeed satisfy (149) *when the series involved converges uniformly* (*or terminates*).

where F and K are continuous, we replace y under the integral sign by an initial approximation $y^{(0)}$. Then (163) determines an approximation $y^{(1)}$ in the form

$$y^{(1)}(x) = F(x) + \lambda \int_a^b K(x, \xi) y^{(0)}(\xi)\, d\xi. \tag{164}$$

By substituting this approximation into the right-hand member of (163), we may then obtain the next approximation $y^{(2)}$, and continue the process in such a way that successive approximations are determined by the formula

$$y^{(n)}(x) = F(x) + \lambda \int_a^b K(x, \xi) y^{(n-1)}(\xi)\, d\xi. \tag{165}$$

The same method is clearly applicable also when the upper limit b is replaced by the current variable x, so that the equation is of the *Volterra* type. It remains to determine under what conditions the successive approximations actually tend toward a continuous solution of (163).

In order to examine this procedure more closely, we write out explicitly the results of the indicated substitutions. Thus we first obtain the result of replacing y in the right-hand member of (163) by $y^{(1)}$, as given by (164). In this substitution, we must replace the current variable x in (164) by the dummy variable ξ appearing in (163). To avoid ambiguity, we must then replace ξ in (164) by another dummy variable, say ξ_1, so that (164) becomes

$$y^{(1)}(\xi) = F(\xi) + \lambda \int_a^b K(\xi, \xi_1) y^{(0)}(\xi_1)\, d\xi_1.$$

The result of the substitution then takes the form

$$y^{(2)}(x) = F(x) + \lambda \int_a^b K(x, \xi) \left[F(\xi) + \lambda \int_a^b K(\xi, \xi_1) y^{(0)}(\xi_1)\, d\xi_1 \right] d\xi$$

or

$$y^{(2)}(x) = F(x) + \lambda \int_a^b K(x, \xi) F(\xi)\, d\xi \\ + \lambda^2 \int_a^b K(x, \xi) \int_a^b K(\xi, \xi_1) y^{(0)}(\xi_1)\, d\xi_1\, d\xi. \tag{166}$$

If we now replace x by ξ, ξ by ξ_1, and ξ_1 by ξ_2, in (166), and substitute once more in the right-hand member of (163), there follows

$$y^{(3)}(x) = F(x) + \lambda \int_a^b K(x, \xi) F(\xi)\, d\xi \\ + \lambda^2 \int_a^b K(x, \xi) \int_a^b K(\xi, \xi_1) F(\xi_1)\, d\xi_1\, d\xi \\ + \lambda^3 \int_a^b K(x, \xi) \int_a^b K(\xi, \xi_1) \int_a^b K(\xi_1, \xi_2) y^{(0)}(\xi_2)\, d\xi_2\, d\xi_1\, d\xi. \tag{167}$$

The analysis is abbreviated considerably if we introduce an *integral operator* $\mathscr{K}$, defined by the equation

$$\mathscr{K}f(x) \equiv \int_a^b K(x, \xi)f(\xi)\, d\xi. \tag{168}$$

The integral equation (163) then takes the symbolic form

$$y(x) = F(x) + \lambda\, \mathscr{K}y(x), \tag{169}$$

while (165) becomes

$$y^{(n)}(x) = F(x) + \lambda\, \mathscr{K}y^{(n-1)}(x). \tag{170}$$

Further, equations (164), (166), and (167) take the forms

$$\left.\begin{aligned} y^{(1)}(x) &= F(x) + \lambda\, \mathscr{K}y^{(0)}(x), \\ y^{(2)}(x) &= F(x) + \lambda\, \mathscr{K}F(x) + \lambda^2\mathscr{K}^2y^{(0)}(x), \\ y^{(3)}(x) &= F(x) + \lambda\, \mathscr{K}F(x) + \lambda^2\mathscr{K}^2F(x) + \lambda^3\mathscr{K}^3y^{(0)}(x) \end{aligned}\right\}. \tag{171}$$

More generally, after the nth substitution we have

$$y^{(n)}(x) = F(x) + \lambda\, \mathscr{K}F(x) + \lambda^2\mathscr{K}^2F(x) + \lambda^3\mathscr{K}^3F(x) \\ + \cdots + \lambda^{n-1}\mathscr{K}^{n-1}F(x) + R_n(x), \tag{172}$$

where $R_n(x)$ is defined by

$$R_n(x) = \lambda^n\mathscr{K}^ny^{(0)}(x). \tag{173}$$

Hence, as $n \to \infty$, we are led to the possibility that the desired solution of (163) can be expressed as the infinite series

$$y(x) = F(x) + \sum_{n=1}^{\infty} \lambda^n\mathscr{K}^nF(x), \tag{174}$$

It remains to determine conditions under which the expression $R_n(x)$ tends to zero and under which the formal series (174) actually converges, and represents a continuous solution of (163).

Since $K(x, \xi)$ and $F(x)$ are assumed to be continuous for all values of x and ξ in (a, b), their magnitudes certainly are bounded in (a, b), so that there exist positive constants M and m such that

$$|K(x, \xi)| \leqq M, \qquad |F(x)| \leqq m \tag{175a,b}$$

in (a, b). We suppose also that the magnitude of the initial approximation $y^{(0)}(x)$ is bounded in (a, b), so that

$$|y^{(0)}(x)| \leqq C \tag{176}$$

in that interval.

With the understanding that $b > a$, there then follows

$$|\mathscr{K}y^{(0)}(x)| = \left| \int_a^b K(x, \xi)y^{(0)}(\xi)\, d\xi \right| \leqq \int_a^b MC\, d\xi = M(b - a)C.$$

More generally, we find by iteration that

$$|\mathscr{K}^n y^{(0)}(x)| \leqq M^n(b-a)^n C \tag{177}$$

and, similarly,

$$|\mathscr{K}^n F(x)| \leqq M^n(b-a)^n m. \tag{178}$$

Hence, according to (173) and (177), there follows

$$|R_n(x)| \leqq |\lambda|^n M^n(b-a)^n C, \tag{179}$$

and we may deduce that $R_n(x)$ tends to zero with increasing n if

$$|\lambda| < \frac{1}{M(b-a)}. \tag{180}$$

Further, from (178) it follows that the series

$$|F(x)| + \sum_{n=1}^{\infty} |\lambda|^n \, |\mathscr{K}^n F(x)|$$

is dominated by the constant series

$$m\left[1 + \sum_{n=1}^{\infty} |\lambda|^n M^n(b-a)^n\right].$$

Since this geometric series converges when λ satisfies the inequality (180), we may deduce that the series (174) converges absolutely and uniformly* in (a, b) when (180) is satisfied.

It is then easily seen, by direct substitution and term-by-term integration, that the series (174) actually satisfies the integral equation (163) and hence represents the continuous solution of (163) when (180) is satisfied and K is continuous.

The series (174) is a power series in λ. If we recall that the solution to (163) generally fails to exist when λ takes on a characteristic value, we are led to expect that the series solution (174) will cease to converge at least as soon as $|\lambda|$ becomes equal to the absolute value of the *smallest* characteristic number λ_1. It can be shown that this is the case and, indeed, that *the series* (174) *converges when* $|\lambda| < |\lambda_1|$, *and only then.*

Since the condition (180) may be conservative, in the sense that the series (174) *may* perhaps converge even though (180) is *not* satisfied, we thus obtain as a by-product of this derivation the useful relation

$$|\lambda_1| \geqq \frac{1}{M(b-a)}, \tag{181}$$

* A series of functions of x which is dominated by a convergent series of positive *constants* independent of x, for all values of x in an interval (a, b), is *uniformly convergent* in (a, b). Such a series of *continuous* functions *represents* a continuous function, and it can be integrated term by term in (a, b).

which gives a lower bound for the magnitude of the smallest characteristic number λ_1. When K is *real and symmetric*, a somewhat more involved analysis (see Problem 91) leads to the inequality

$$|\lambda_1| \geqq \frac{1}{\sqrt{\int_a^b \int_a^b [K(x, \xi)]^2 \, dx \, d\xi}} \tag{182}$$

which gives a sharper lower bound in that case.

It is of some interest to notice that the relation (174) can also be obtained by writing (169) in the operational form

$$(\mathscr{I} - \lambda \mathscr{K})y(x) = F(x), \tag{183}$$

where $\mathscr{I}$ is the identity operator, deducing that

$$y(x) = (\mathscr{I} - \lambda \mathscr{K})^{-1} F(x), \tag{184}$$

where $(\mathscr{I} - \lambda \mathscr{K})^{-1}$ represents an operator *inverse* to $\mathscr{I} - \lambda \mathscr{K}$, and *formally* applying the binomial theorem to deduce that

$$(\mathscr{I} - \lambda \mathscr{K})^{-1} = \mathscr{I} + \lambda \mathscr{K} + \lambda^2 \mathscr{K}^2 + \cdots. \tag{185}$$

Only when $|\lambda| < |\lambda_1|$ is this formal result valid.

In the case of the *Volterra* equation,

$$y(x) = F(x) + \lambda \int_a^x K(x, \xi) y(\xi) \, d\xi, \tag{186}$$

with a *variable* upper limit, we define $\mathscr{K}_x$ as the integral operator such that

$$\mathscr{K}_x f(x) \equiv \int_a^x K(x, \xi) f(\xi) \, d\xi, \tag{187}$$

and a procedure completely analogous to that used above leads to the formal solution

$$y(x) = F(x) + \sum_{n=1}^{\infty} \lambda^n \mathscr{K}_x^{\,n} F(x), \tag{188}$$

if the expression

$$R_n(x) = \lambda^n \mathscr{K}_x^{\,n} y^{(0)}(x) \tag{189}$$

tends to zero as $n \to \infty$.

In this case, if we consider any interval (a, b), where b is *any* number larger than a, and again assume the bounds (175) and (176) over (a, b), we have

$$|\mathscr{K}_x y^{(0)}(x)| = \left| \int_a^x K(x, \xi) y^{(0)}(\xi) \, d\xi \right| \leqq MC \int_a^x d\xi = MC(x - a),$$

when $a \leqq x \leqq b$, and hence also

$$|\mathscr{K}_x^2 y^{(0)}(x)| \leqq \int_a^x |K(x, \xi)|\, MC(\xi - a)\, d\xi$$
$$\leqq M^2 C \int_a^x (\xi - a)\, d\xi = \frac{M^2(x - a)^2}{2 \cdot 1} C.$$

Inductive reasoning then leads to the result

$$|R_n(x)| = |\lambda^n \mathscr{K}_x^n y^{(0)}(x)| \leqq |\lambda|^n \frac{M^n(x - a)^n}{n!} C$$
$$\leqq |\lambda|^n \frac{M^n(b - a)^n}{n!} C, \tag{190}$$

for $a \leqq x \leqq b$. In a similar way, it is found that

$$|\lambda^n \mathscr{K}_x^n F(x)| \leqq |\lambda|^n \frac{M^n(b - a)^n}{n!} m \tag{191}$$

for $a \leqq x \leqq b$. But the last member of (190) tends to zero as $n \to \infty$, and also the series whose nth term is given by the right-hand member of (191) converges, for *any* finite value of λ. Thus the method of successive substitutions converges to the series (188) and that series converges absolutely and uniformly, for any finite value of λ, in any interval (a, b) for which $b > a$. A similar result follows for any $b < a$. It then follows, by direct substitution, that *the series* (188) *converges to the unique continuous solution of the Volterra equation* (186) *for all values of* λ, *in any interval* (a, b) *in which* $F(x)$ *and* $K(x, \xi)$ *are continuous.*

We may notice that, as was to be expected, the final solution in each case is independent of the initial approximation $y^{(0)}(x)$. In practice, it is often desirable to merely evaluate successive terms in the series (174) or (188) by iteration, as is illustrated by an example which follows, rather than to actually pursue the method of successive *substitutions* which motivated (174) and (188). The latter method reduces to the former when the initial approximation

$$y^{(0)}(x) = F(x)$$

is used. However, when advance information as to the nature of the required solution is available, a better choice of $y^{(0)}(x)$ may suggest itself.

As a very simple illustration of these results, we consider the Fredholm equation

$$y(x) = 1 + \lambda \int_0^1 (1 - 3x\xi) y(\xi)\, d\xi, \tag{192}$$

the solution of which can be obtained from the results of Section 3.7 in the form

$$y(x) = \frac{4 + 2\lambda(2 - 3x)}{4 - \lambda^2} \qquad (\lambda \neq \pm 2). \tag{193}$$

The operation $\mathscr{K}f(x)$ is then of the form

$$\mathscr{K}f(x) = \int_0^1 (1 - 3x\xi)f(\xi)\,d\xi.$$

To obtain the series solution (174), we make the calculations

$$\mathscr{K}F = \int_0^1 (1 - 3x\xi)\,d\xi = 1 - \frac{3}{2}x,$$

$$\mathscr{K}^2F = \int_0^1 (1 - 3x\xi)\left(1 - \frac{3}{2}\xi\right)d\xi = \frac{1}{4},$$

$$\mathscr{K}^3F = \int_0^1 (1 - 3x\xi)\frac{1}{4}\,d\xi = \frac{1}{4}\left(1 - \frac{3}{2}x\right),$$

and so forth. From the form of these results the form of the general result is obvious, and (174) becomes

$$y(x) = 1 + \lambda\left(1 - \frac{3}{2}x\right) + \frac{\lambda^2}{4} + \frac{\lambda^3}{4}\left(1 - \frac{3}{2}x\right) + \frac{\lambda^4}{16} + \frac{\lambda^5}{16}\left(1 - \frac{3}{2}x\right) + \cdots. \tag{194a}$$

This result can be expressed in the form

$$y(x) = \left(1 + \frac{\lambda^2}{4} + \frac{\lambda^4}{16} + \cdots\right)\left[1 + \lambda\left(1 - \frac{3}{2}x\right)\right]. \tag{194b}$$

The power series in (194b) is a geometric series, convergent when $|\lambda| < 2$, with the sum $1/(1 - \lambda^2/4)$. Hence (194) *is the power series expansion, valid when and only when* $|\lambda| < 2$, *of the solution*

$$y(x) = \frac{1 + \lambda(1 - \frac{3}{2}x)}{1 - \frac{1}{4}\lambda^2}, \tag{195}$$

which is identical with (193), *and which itself is valid for all values of* λ *except* $\lambda = \pm 2$.

We may notice that consequently the method of successive substitutions *would not converge*, for example, if it were applied to the integral equation

$$y(x) = 1 + 4\int_0^1 (1 - 3x\xi)y(\xi)\,d\xi,$$

while it *would* converge if applied to the same equation with the factor 4 replaced by any number smaller than two in absolute value.

Obviously the method of Section 3.6 is generally to be preferred when the kernel $K(x, \xi)$ is separable, as in the preceding case. It is important to notice that the methods of Section 3.6 express the solution as the ratio of a function of x and λ to a polynomial $\Delta(\lambda)$, valid for all values of λ except the characteristic values for which $\Delta(\lambda) = 0$. The method of successive substitutions, in such cases, leads to the power series expansion of this *ratio*, valid only when $|\lambda|$ is smaller than the magnitude of the smallest characteristic number.

In the more important cases when the kernel is not separable, there exists a method, due to *Fredholm*, which generalizes the procedure of Section 3.6. This method is discussed in Section 3.11.

In the following section the preceding developments are considered from a slightly different viewpoint.

3.10. The Neumann series. With the notation of equation (168),

$$\mathscr{K}f(x) = \int_a^b K(x, \xi)f(\xi)\, d\xi, \tag{196}$$

there follows also

$$\begin{aligned}\mathscr{K}^2f(x) &= \int_a^b K(x, \xi_1)\mathscr{K}f(\xi_1)\, d\xi_1 \\ &= \int_a^b K(x, \xi_1)\left[\int_a^b K(\xi_1, \xi)f(\xi)\, d\xi\right] d\xi_1 \\ &= \int_a^b \left[\int_a^b K(x, \xi_1)K(\xi_1, \xi)\, d\xi_1\right] f(\xi)\, d\xi. \end{aligned}\tag{197}$$

If we define the *iterated kernel* $K_2(x, \xi)$ by the relation

$$K_2(x, \xi) = \int_a^b K(x, \xi_1)K(\xi_1, \xi)\, d\xi_1, \tag{198}$$

equation (197) takes the form

$$\mathscr{K}^2f(x) = \int_a^b K_2(x, \xi)f(\xi)\, d\xi. \tag{199}$$

By repeating this process, it is easily seen that one can write

$$\mathscr{K}^nf(x) = \int_a^b K_n(x, \xi)f(\xi)\, d\xi, \tag{200}$$

where $K_n(x, \xi)$ is the nth iterated kernel, defined by the recurrence formula

$$K_n(x, \xi) = \int_a^b K(x, \xi_1)K_{n-1}(\xi_1, \xi)\, d\xi_1, \tag{201a}$$

for $n = 2, 3, 4, \cdots$, and where we write

$$K_1(x, \xi) \equiv K(x, \xi). \tag{201b}$$

It is not difficult to establish the consequent validity of the relation

$$K_{p+q}(x, \xi) = \int_a^b K_p(x, \xi_1)K_q(\xi_1, \xi)\, d\xi_1, \tag{202}$$

for any positive integers p and q. Further, if $K(x, \xi)$ is bounded in (a, b), in such a way that

$$|K(x, \xi)| \leqq M \tag{203}$$

in (a, b), then it follows easily that also

$$|K_n(x, \xi)| \leqq M^n(b - a)^{n-1} \tag{204}$$

for values of x and ξ in (a, b).

With the notation of (200), the series (174), representing the solution of the equation

$$y(x) = F(x) + \lambda \int_a^b K(x, \xi)y(\xi)\, d\xi \tag{205}$$

for sufficiently small values of $|\lambda|$, takes the form

$$\begin{aligned} y(x) &= F(x) + \sum_{n=1}^{\infty} \lambda^n \int_a^b K_n(x, \xi)F(\xi)\, d\xi \\ &= F(x) + \lambda \int_a^b \left[\sum_{n=0}^{\infty} \lambda^n K_{n+1}(x, \xi)\right] F(\xi)\, d\xi, \end{aligned} \tag{206}$$

assuming the legitimacy of interchange of summation and integration. If we introduce the abbreviation

$$\begin{aligned} \Gamma(x, \xi; \lambda) &= \sum_{n=0}^{\infty} \lambda^n K_{n+1}(x, \xi) \\ &= K(x, \xi) + \lambda K_2(x, \xi) + \lambda^2 K_3(x, \xi) + \cdots, \end{aligned} \tag{207}$$

equation (206) takes the form

$$y(x) = F(x) + \lambda \int_a^b \Gamma(x, \xi; \lambda)F(\xi)\, d\xi. \tag{208}$$

The function $\Gamma(x, \xi; \lambda)$ is known as the *reciprocal* or *resolvent kernel* associated with the kernel $K(x, \xi)$ in the interval (a, b). Further, the series (207) [or, in some references, the series (206)] is known as the *Neumann series*. If use is made of (204), it is found that this series converges (absolutely and uniformly) when $|\lambda| < 1/M(b - a)$. A more precise analysis shows indeed that the series converges when $|\lambda| < |\lambda_1|$, where λ_1 is the smallest characteristic number, as was the case in the analogous expansions of the preceding section. In fact, equation (208) is merely an abbreviation for (206), which is equivalent to (174) by virtue of (200).

In practice, unless the solution of (205) is required for several choices of $F(x)$, it may be more convenient to obtain leading terms of the series (206) or (174) by the iterative methods of the preceding section, than actually to evaluate terms of (207) and insert the result into (208); the net results are, of course, the same. However, the Neumann series and the resolvent kernel are of importance in theoretical developments. An interesting result in this connection is obtained if we rewrite (207) in the form

$$\Gamma(x, \xi; \lambda) = K(x, \xi) + \lambda \sum_{n=0}^{\infty} \lambda^n K_{n+2}(x, \xi)$$

$$= K(x, \xi) + \lambda \sum_{n=0}^{\infty} \lambda^n \int_a^b K(x, \xi_1) K_{n+1}(\xi_1, \xi)\, d\xi_1$$

and hence, again referring to (207), deduce the relation

$$\Gamma(x, \xi; \lambda) = K(x, \xi) + \lambda \int_a^b K(x, \xi_1)\Gamma(\xi_1, \xi; \lambda)\, d\xi_1,$$

or, with a change in notation,

$$\Gamma(x, t; \lambda) = K(x, t) + \lambda \int_a^b K(x, \xi)\Gamma(\xi, t; \lambda)\, d\xi. \qquad (209)$$

Thus it follows that *the resolvent kernel* Γ, *considered as a function of the two variables* x *and* t *and the parameter* λ, *is the solution of equation* (205) *when the prescribed function* F *is replaced by the kernel* K, *considered as a function of* x *and* t.

In order to illustrate the actual determination of the resolvent kernel in a very simple case, we again consider equation (192). With

$$K(x, \xi) = 1 - 3x\xi,$$

there follows

$$K_2(x, \xi) = \int_0^1 (1 - 3x\xi_1)(1 - 3\xi_1\xi)\, d\xi_1 = 1 - \frac{3}{2}(x + \xi) + 3x\xi$$

and, similarly,

$$K_3(x, \xi) = \int_0^1 K(x, \xi_1)K_2(\xi_1, \xi)\, d\xi_1 = \frac{1}{4}(1 - 3x\xi).$$

Since, in this special case, we therefore have $K_3 = K_1/4$, it follows easily that $K_n = K_{n-2}/4$ for $n \geqq 3$, and hence we have

$$\Gamma = K_1 + \lambda K_2 + \lambda^2 K_3 + \cdots$$

$$= \left(1 + \frac{\lambda^2}{4} + \frac{\lambda^4}{16} + \cdots\right) K_1 + \lambda \left(1 + \frac{\lambda^2}{4} + \frac{\lambda^4}{16} + \cdots\right) K_2$$

or

$$\Gamma(x, \xi; \lambda) = \frac{1}{1 - \lambda^2/4}\left[(1 + \lambda) - \frac{3}{2}\lambda(x + \xi) - 3(1 - \lambda)x\xi\right]$$

$$(|\lambda| < 2). \qquad (210)$$

The introduction of this function into (208), with $F(x) = 1$, leads again to the solution (195).

It is important to notice that the result obtained is correct for *all* values of λ except $\lambda = \pm 2$. That is, the resolvent kernel is correctly given by (210) for all such values of λ. However, the series involved in the equation preceding (210) converges only when $|\lambda| < 2$. It happens that we are able to sum that series explicitly in the present example, and that the resultant function correctly represents the resolvent kernel for *all* values of λ other than characteristic values.

3.11. Fredholm theory. In the general case of a continuous kernel $K(x, \xi)$, which need not be real, it is possible to express the resolvent kernel $\Gamma(x, \xi; \lambda)$ as the *ratio* of *two* infinite series of powers of λ, in such a way that *both series converge for all values of* λ. The *derivation* of the basic equations, due originally to Fredholm, involves considerable algebraic manipulation and is not considered here.

If the resolvent kernel is expressed as the ratio

$$\Gamma(x, \xi; \lambda) = \frac{D(x, \xi; \lambda)}{\Delta(\lambda)}, \tag{211}$$

where

$$D(x, \xi; \lambda) = K(x, \xi) + \lambda D_1(x, \xi) + \lambda^2 D_2(x, \xi) + \cdots \tag{212}$$

and

$$\Delta(\lambda) = 1 + \lambda C_1 + \lambda^2 C_2 + \cdots, \tag{213}$$

it is found that the coefficients C_n and the functions $D_n(x, \xi)$ can be determined successively by the following sequence of calculations:

$$C_1 = -\int_a^b K(x, x)\, dx, \quad D_1(x, \xi) = C_1 K(x, \xi) + \int_a^b K(x, \xi_1) K(\xi_1, \xi)\, d\xi_1;$$

$$2C_2 = -\int_a^b D_1(x, x)\, dx, \quad D_2(x, \xi) = C_2 K(x, \xi) + \int_a^b K(x, \xi_1) D_1(\xi_1, \xi)\, d\xi_1;$$

. . . .

$$nC_n = -\int_a^b D_{n-1}(x, x)\, dx, \; D_n(x, \xi) = C_n K(x, \xi) + \int_a^b K(x, \xi_1) D_{n-1}(\xi_1, \xi)\, d\xi_1. \tag{214}$$

The solution of the equation

$$y(x) = F(x) + \lambda \int_a^b K(x, \xi) y(\xi)\, d\xi \tag{215}$$

is then obtained by introducing (211) into (208), in the form

$$y(x) = F(x) + \lambda \frac{\int_a^b D(x, \xi; \lambda) F(\xi)\, d\xi}{\Delta(\lambda)}. \tag{216}$$

In those cases when $K(x, \xi)$ is *separable*, this result is identical in form with the solution obtained by the methods of Section 3.6. The series (212) and (213) then each involve only a finite number of terms.

More generally, if the ratio of the two power series involved in (216) were expressed as a single power series in λ (by division or otherwise) the result would reduce to the series (174). However, the result of this operation would converge only for small values of $|\lambda|$ (when $|\lambda| < |\lambda_1|$), whereas the separate series expansions of the numerator and denominator in the last term of (216) each converge for *all* values of λ.

The denominator $\Delta(\lambda)$ vanishes only when λ takes on a characteristic value, in which case either no solution or infinitely many solutions of (215) exist, and (216) is no longer valid.

Despite the generality of the solution just described, the practical usefulness of the result is limited by the fact that the relevant calculations usually involve a prohibitive amount of labor unless $K(x, \xi)$ is separable (and hence the simpler methods of Section 3.6 are usually preferable). Nevertheless, the rigorous development of the underlying theory has led to valuable information concerning existence and uniqueness of solutions of (215). In the following paragraph we summarize certain known facts which generalize results already obtained in the special cases when the kernel is either separable or real and symmetric (see Reference 2).

The equation

$$y(x) = F(x) + \lambda \int_a^b K(x, \xi) y(\xi)\, d\xi, \tag{217}$$

where (a, b) *is a finite interval and where* $F(x)$ *and* $K(x, \xi)$ *are continuous in* (a, b), *possesses one and only one continuous solution for any fixed value of* λ *which is not a characteristic value. If* λ_c *is a characteristic number of multiplicity* r, *that is, if the associated homogeneous equation*

$$y(x) = \lambda_c \int_a^b K(x, \xi) y(\xi)\, d\xi \tag{218}$$

possesses r *linearly independent nontrivial solutions* $\phi_1, \phi_2, \cdots, \phi_r$, *then* r *is finite and the associated transposed homogeneous equation*

$$z(x) = \lambda_c \int_a^b K(\xi, x) z(\xi)\, d\xi \tag{219}$$

also possesses r *linearly independent nontrivial solutions* $\psi_1, \psi_2, \cdots, \psi_r$. *In this exceptional case,* (217) *possesses no solution unless* $F(x)$ *is orthogonal to each of the characteristic functions* $\psi_1, \psi_2, \cdots, \psi_r$,

$$\int_a^b F(x) \psi_k(x)\, dx = 0 \qquad (k = 1, 2, \cdots, r). \tag{220}$$

Finally, if $\lambda = \lambda_c$ *and* (220) *is satisfied, then the solution of* (217) *is determinate only within an additive linear combination* $c_1\phi_1 + c_2\phi_2 + \cdots + c_r\phi_r$, *where the* r *constants* c_n *are arbitrary.**

* See also Problem 67.

When $K(x, \xi)$ is real and *symmetric*, equations (218) and (219) are identical and the preceding results reduce to those given by the Hilbert-Schmidt theory of Section 3.8.

It is useful to notice the complete analogy between the preceding results and the corresponding results relevant to existence and uniqueness of solutions of sets of n linear algebraic equations in n unknowns (see Section 1.11). Indeed, the plausibility of these statements was first suggested by the possibility of considering a Fredholm integral equation as the limit of such a set of equations as the number n of equations and unknowns becomes infinite.

A *Volterra* integral equation, of the form

$$y(x) = F(x) + \lambda \int_a^x K(x, \xi) y(\xi)\, d\xi, \tag{221}$$

can be considered as a special form of a Fredholm equation, with a kernel given by the expressions

$$\tilde{K}(x, \xi) = \begin{cases} 0 & \text{when} \quad x < \xi, \\ K(x, \xi) & \text{when} \quad x \geqq \xi. \end{cases} \tag{222}$$

However, unless $K(x, x) = 0$, the modified kernel $\tilde{K}(x, \xi)$ is *discontinuous* when $x = \xi$. The results of Section 3.9 show that *if $F(x)$ and $K(x, \xi)$ are continuous the Volterra equation* (221) *possesses one and only one continuous solution, and that solution is given by the series* (188) *for any value of λ.* In particular, when $F(x) \equiv 0$ the only possible continuous solution of (221) is then the *trivial* solution $y(x) \equiv 0$.

3.12. Singular integral equations. An integral equation in which the range of integration is infinite, or in which the kernel $K(x, \xi)$ is discontinuous, is called a *singular* integral equation. Thus, in illustration, the equations

$$F(x) = \int_0^\infty \sin(x\xi)\, y(\xi)\, d\xi, \tag{223}$$

$$F(x) = \int_0^\infty e^{-x\xi} y(\xi)\, d\xi, \tag{224}$$

and

$$F(x) = \int_0^x \frac{y(\xi)}{\sqrt{x - \xi}}\, d\xi \tag{225}$$

are all singular integral equations of the first kind. As will be seen, such equations may possess very unusual properties. The three preceding examples were chosen here because of the fact that they have been studied rather extensively.

The function $F(x)$ defined by the right-hand member of (223) may be recognized as the *Fourier sine transform* of $y(x)$. If $F(x)$ is piecewise differentiable when $x > 0$, and if $\int_0^\infty |F(x)|\, dx$ exists, then it is known that equation (223) can be inverted uniquely in the form

$$y(x) = \frac{2}{\pi}\int_0^\infty \sin(x\xi)\, F(\xi)\, d\xi \qquad (x > 0). \tag{226}$$

This result leads to an interesting property of the homogeneous integral equation of the *second* kind,

$$y(x) = \lambda\int_0^\infty \sin(x\xi)\, y(\xi)\, d\xi, \tag{227}$$

associated with (223) and obtained from (223) by replacing $F(x)$ by $y(x)/\lambda$, since the corresponding inversion of (227) is then of the form

$$y(x) = \frac{2}{\pi\lambda}\int_0^\infty \sin(x\xi)\, y(\xi)\, d\xi. \tag{228}$$

Unless $y(x) \equiv 0$, equations (227) and (228) are compatible only if

$$\lambda = \pm\sqrt{\frac{2}{\pi}}. \tag{229}$$

Thus we conclude that if (227) possesses characteristic numbers, those numbers can only be $\lambda = \sqrt{2/\pi}$ and $\lambda = -\sqrt{2/\pi}$.

That these values of λ *are* actually characteristic values follows from the verifiable relation

$$\sqrt{\frac{\pi}{2}}\, e^{-ax} \pm \frac{x}{a^2 + x^2} = \pm\sqrt{\frac{2}{\pi}}\int_0^\infty \sin(x\xi)\left[\sqrt{\frac{\pi}{2}}\, e^{-a\xi} \pm \frac{\xi}{a^2 + \xi^2}\right] d\xi, \tag{230}$$

when $x > 0$ and $a > 0$. This equation states that when $\lambda = \sqrt{2/\pi}$ equation (227) is satisfied by the function

$$y_1(x) = \sqrt{\frac{\pi}{2}}\, e^{-ax} + \frac{x}{a^2 + x^2} \qquad (x > 0), \tag{231}$$

for any positive constant value of a, whereas when $\lambda = -\sqrt{2/\pi}$ the function

$$y_2(x) = \sqrt{\frac{\pi}{2}}\, e^{-ax} - \frac{x}{a^2 + x^2} \qquad (x > 0) \tag{232}$$

is a solution for any positive value of a. Thus, the two characteristic values of λ here are of *infinite multiplicity;* that is, each value corresponds to infinitely many independent characteristic functions. This situation is in contrast with the fact that any characteristic number of a *nonsingular* Fredholm equation corresponds only to a *finite* number of independent characteristic functions.

The function $F(x)$ defined by the right-hand member of equation (224) is the *Laplace transform* of the function $y(x)$. It is known that, while not all functions can be Laplace transforms of other functions, there cannot be two distinct continuous functions with the same transform. Thus for a prescribed function $F(x)$, if (224) possesses a continuous solution then that solution is unique, and it can be determined by known methods. In order to establish an unusual property of the associated homogeneous equation of the *second* kind,

$$y(x) = \lambda \int_0^\infty e^{-x\xi} y(\xi)\, d\xi \qquad (x > 0), \tag{233}$$

we notice that, in accordance with the definition of the *Gamma function*, we have the relation

$$\int_0^\infty e^{-x\xi} \xi^{a-1}\, d\xi = \Gamma(a) x^{-a} \qquad (a > 0). \tag{234}$$

The result of replacing a by $1 - a$ is then of the form

$$\int_0^\infty e^{-x\xi} \xi^{-a}\, d\xi = \Gamma(1 - a) x^{a-1} \qquad (a < 1). \tag{235}$$

If (234) is divided by $\sqrt{\Gamma(a)}$, and (235) is divided by $\sqrt{\Gamma(1 - a)}$, and if the resultant equations are added to each other, the truth of the equation

$$\int_0^\infty e^{-x\xi} \left[\sqrt{\Gamma(1-a)}\, \xi^{a-1} + \sqrt{\Gamma(a)}\, \xi^{-a}\right] d\xi$$
$$= \sqrt{\Gamma(a)\Gamma(1-a)} \left[\sqrt{\Gamma(1-a)}\, x^{a-1} + \sqrt{\Gamma(a)}\, x^{-a}\right] \quad (0 < a < 1) \tag{236}$$

is established. This equation is identified with (233) by writing

$$\lambda = \frac{1}{\sqrt{\Gamma(a)\Gamma(1-a)}} \qquad (0 < a < 1) \tag{237}$$

and

$$y(x) = \sqrt{\Gamma(1-a)}\, x^{a-1} + \sqrt{\Gamma(a)}\, x^{-a} \qquad (x > 0). \tag{238}$$

It thus follows that for any value of the parameter a such that $0 < a < 1$ a value of λ is determined by (237), corresponding to which (233) possesses a nontrivial solution specified by (238). In consequence of the identity

$$\Gamma(a)\Gamma(1-a) = \frac{\pi}{\sin \pi a} \qquad (0 < a < 1), \tag{239}$$

equation (237) can be written in the form

$$\lambda = \sqrt{\frac{\sin \pi a}{\pi}} \qquad (0 < a < 1), \tag{240}$$

from which it follows that *all values of λ in the interval* $0 < \lambda \leqq 1/\sqrt{\pi}$ *are characteristic values for the singular integral equation* (233).

This situation is in contrast with the fact that the characteristic values of λ for a *nonsingular* equation are *discretely* distributed, and cannot constitute a *continuous* "spectrum."

It can be shown further that all values of λ in the interval

$$-1/\sqrt{\pi} \leqq \lambda < 0$$

are also characteristic values for equation (233).

Other singular Fredholm equations may possess only discretely distributed characteristic numbers, or they may possess both a discrete and a continuous spectrum of characteristic numbers.

Equation (225), in which the range of integration is finite but the kernel is unbounded, is considered in the following section.

3.13. Special devices. In this section we present techniques which are useful in dealing with certain special types of integral equations.

1. *Transforms.* If a relationship of the form

$$y(x) = \int_a^b \int_a^b \Gamma(x, \xi_1) K(\xi_1, \xi) y(\xi) \, d\xi \, d\xi_1 \tag{241}$$

is known to be valid (for a suitably restricted class of functions y) and if the double integral can be evaluated as an iterated integral, then it follows that if

$$F(x) = \int_a^b K(x, \xi) y(\xi) \, d\xi \tag{242}$$

we have also

$$y(x) = \int_a^b \Gamma(x, \xi) F(\xi) \, d\xi. \tag{243}$$

Thus, if (242) is considered as an integral equation in y, a solution is given by (243), whereas if (243) is considered as an integral equation in F a solution is given by (242). It is conventional to refer to one of the functions as the *transform* of the second function, and to the second function as an *inverse transform* of the first. The correspondence may or may not be unique. Thus, for example, the Fourier sine-integral formula

$$y(x) = \frac{2}{\pi} \int_0^\infty \int_0^\infty \sin(x\xi_1) \sin(\xi_1 \xi) \, y(\xi) \, d\xi \, d\xi_1$$

leads to the reciprocal relations (223) and (226).

2. *The convolution.* The function defined by the integral

$$\int_0^x u(x - \xi) v(\xi) \, d\xi \tag{244}$$

is known as the *convolution* of $u(x)$ and $v(x)$. The known fact that *the Laplace transform of the convolution of u and v is equal to the product of the transforms of u and v* permits the reduction of the problem of solving the special Volterra equation

$$y(x) = F(x) + \int_0^x K(x - \xi)y(\xi)\,d\xi \tag{245}$$

to the problem of determining an inverse Laplace transform. If we denote the Laplace transform of a function f by $\mathscr{L}f$, the result of taking the transforms of the equal members of (245) takes the form

$$\mathscr{L}y(x) = \mathscr{L}F(x) + \mathscr{L}K(x)\mathscr{L}y(x),$$

and hence there follows

$$\mathscr{L}y(x) = \frac{\mathscr{L}F(x)}{1 - \mathscr{L}K(x)}. \tag{246}$$

The right-hand member of (246) is calculable, and it remains only to determine (by use of tables or otherwise) its inverse transform. Equations of the form (245) occur rather frequently in practice.

3. *Volterra equations of the first kind.* It is often possible to reduce an integral equation of the form

$$F(x) = \int_0^x K(x, \xi)y(\xi)\,d\xi \tag{247}$$

to an equation of the *second* kind. Such a reduction is desirable since the method of successive substitutions is then applicable. Under the assumption that the kernel is continuously differentiable when $\xi \leqq x$, two different procedures are available. First, if $F(x)$ is differentiable, we may obtain the relation

$$F'(x) = K(x, x)y(x) + \int_0^x \frac{\partial K(x, \xi)}{\partial x}\,y(\xi)\,d\xi$$

by differentiating the equal members of (247). If $K(x, x)$ is never zero, this equation can be put in the form

$$y(x) = \tilde{F}(x) + \int_0^x \tilde{K}(x, \xi)y(\xi)\,d\xi, \tag{248}$$

where

$$\tilde{F}(x) = \frac{F'(x)}{K(x, x)}, \qquad \tilde{K}(x, \xi) = -\frac{1}{K(x, x)}\frac{\partial K(x, \xi)}{\partial x}. \tag{249}$$

Alternatively, if we define the function

$$Y(x) = \int_0^x y(\xi)\,d\xi, \tag{250}$$

equation (247) takes the form

$$F(x) = \int_0^x K(x, \xi)Y'(\xi)\,d\xi$$

and an integration by parts leads to the relation

$$F(x) = K(x, x)Y(x) - \int_0^x \frac{\partial K(x, \xi)}{\partial \xi}\, Y(\xi)\, d\xi.$$

If $K(x, x)$ is never zero, we may rewrite this equation in the form

$$Y(x) = \tilde{F}(x) + \int_0^x \tilde{K}(x, \xi) Y(\xi)\, d\xi, \tag{251}$$

where

$$\tilde{F}(x) = \frac{F(x)}{K(x, x)}, \quad \tilde{K}(x, \xi) = \frac{1}{K(x, x)} \frac{\partial K(x, \xi)}{\partial \xi}. \tag{252}$$

The solution of (247) is then related to the solution of (251) by the equation $y(x) = Y'(x)$.

The preceding reductions are both unsatisfactory when $K(x, x)$ is zero for one or more isolated values of x in (a, b), since then $\tilde{F}(x)$ and $\tilde{K}(x, \xi)$ generally will not remain finite. However, if $K(x, x)$ is *identically* zero, so that the new equation is still of the first kind, it may be possible to obtain a nonsingular equation of the second kind by *repeating* the reduction process.

4. *Abel's equation.* The Volterra equation

$$F(x) = \int_0^x \frac{y(\xi)}{\sqrt{x - \xi}}\, d\xi \tag{253}$$

is known as *Abel's integral equation.* Because of the singularity in the kernel, the preceding reductions cannot be used. However, the equation can be solved, under appropriate restrictions on the prescribed function F, by an indirect method in which we divide both sides of (253) by $\sqrt{s - x}$, where s is a parameter, and integrate the results with respect to x over $(0, s)$. This procedure leads to the equation

$$\int_0^s \frac{F(x)}{\sqrt{s - x}}\, dx = \int_0^s \left\{ \int_0^x \frac{y(\xi)}{\sqrt{x - \xi}}\, d\xi \right\} \frac{dx}{\sqrt{s - x}}. \tag{254a}$$

If the order of integration in the right-hand member is inverted, and the limits of integration are modified accordingly, this equation becomes

$$\int_0^s \frac{F(x)}{\sqrt{s - x}}\, dx = \int_0^s \left\{ \int_\xi^s \frac{dx}{\sqrt{(x - \xi)(s - x)}} \right\} y(\xi)\, d\xi. \tag{254b}$$

The success of this special method depends upon the fact that the inner integral on the right can be evaluated by elementary methods* to give the

* With $x = (s - \xi)t + \xi$, this integral takes the form $\int_0^1 dt/\sqrt{t(1 - t)}$.

constant value

$$\int_{\xi}^{s} \frac{dx}{\sqrt{(x-\xi)(s-x)}} = \pi.$$

Hence (254b) is equivalent to the relation

$$\int_0^s y(\xi)\, d\xi = \frac{1}{\pi} \int_0^s \frac{F(x)}{\sqrt{s-x}}\, dx$$

or, with a more convenient notation,

$$\int_0^x y(\xi)\, d\xi = \frac{1}{\pi} \int_0^x \frac{F(\xi)}{\sqrt{x-\xi}}\, d\xi. \tag{255}$$

By differentiating this relation, we then obtain the desired solution

$$y(x) = \frac{1}{\pi} \frac{d}{dx} \int_0^x \frac{F(\xi)}{\sqrt{x-\xi}}\, d\xi. \tag{256}$$

Unless F is prescribed in such a way that the right-hand member of (256) exists and is continuous, the equation (253) does not possess a continuous solution. A more direct derivation of (256) can be accomplished by the use of Laplace transforms (see Problem 76).

It is of some interest to consider the mechanical problem which led Abel to consider this equation. Suppose that a particle of mass m starts from rest at the time $t = 0$, and slides to the ground along a smooth curve in a vertical plane under the action of gravity. If the initial point is at height x above the ground and if the height is ξ at time t, then the speed at time t is given by $\sqrt{2g(x-\xi)}$, regardless of the shape of the curve. However, the *time of descent* will depend upon this shape. If distance along the curve from the terminal point at time t is denoted by $s(\xi)$, there follows

$$\frac{ds}{dt} = -\sqrt{2g(x-\xi)}$$

and hence the time of descent is given by

$$T = \frac{1}{\sqrt{2g}} \int_0^x \frac{s'(\xi)}{\sqrt{x-\xi}}\, d\xi. \tag{257}$$

For a specified curve, this relation permits the calculation of T as a function of the initial height x. Abel considered the converse problem, in which the time of descent is specified as a function of x, and the curve is to be determined. Equation (257) reduces to (253) if we write $F(x) = \sqrt{2g}\, T(x)$ and $y(x) = s'(x)$.

The more general equation

$$F(x) = \int_0^x \frac{y(\xi)}{(x-\xi)^\alpha}\, d\xi \qquad (0 < \alpha < 1), \tag{258}$$

which was also considered by Abel, can be solved in a similar way (see Problem 74).

3.14. Iterative approximations to characteristic functions. Methods analogous to those given in Section 1.23, for the approximate determination of characteristic numbers and functions, can be applied to the homogeneous equation

$$y(x) = \lambda \int_a^b K(x, \xi) y(\xi) \, d\xi, \tag{259}$$

where $K(x, \xi)$ is continuous, real, and symmetric. If this equation is written in the operational form $y = \lambda \mathcal{K} y$, it appears that here the parameter λ is analogous to the inverse parameter $1/\lambda$ in the matrix equation $\mathbf{A}\,\mathbf{x} = \lambda\,\mathbf{x}$ of Section 1.23. Correspondingly, whereas the methods of that section tend to determine the *largest* characteristic value of λ, the analogous procedures in the present case tend to determine the characteristic number with *smallest* absolute value. Except in those cases when the kernel is separable, the integral equation (259), with a real symmetric kernel, possesses an infinite set of characteristic numbers (see Problem 49), and it is known that this set does not *possess* a largest member, in terms of absolute value.

In order to approximate the fundamental characteristic function we choose an initial approximation $y^{(1)}(x)$ and calculate a corresponding approximation from the equation

$$y(x) = \lambda \int_a^b K(x, \xi) y^{(1)}(\xi) \, d\xi \equiv \lambda f^{(1)}(x). \tag{260}$$

A convenient multiple of $f^{(1)}(x)$ is then taken as a new approximation $y^{(2)}(x)$, and the process is repeated until satisfactory convergence is indicated. In those cases when the kernel $K(x, \xi)$ is continuous, real, and symmetric, it can be shown that the successive approximations $y^{(n)}(x)$ tend to a characteristic function $y_1(x)$ corresponding to the characteristic number λ_1 with smallest absolute value unless it happens that the initial approximation is orthogonal to that function. Further, the ratio of the input $y^{(n)}(x)$ to the output $f^{(n)}(x)$ in the nth cycle tends to λ_1 as n increases. The proof is completely analogous to that given in Section 1.24.

Estimates of the value of λ_1 in the nth cycle are afforded by use of any of the following formulas (see Problem 95):

$$\lambda_1 \approx \frac{\int_a^b y^{(n)}(x)\,dx}{\int_a^b f^{(n)}(x)\,dx} \quad \text{or} \quad \frac{\int_a^b [y^{(n)}(x)]^2\,dx}{\int_a^b y^{(n)}(x) f^{(n)}(x)\,dx} \quad \text{or} \quad \frac{\int_a^b y^{(n)}(x) f^{(n)}(x)\,dx}{\int_a^b [f^{(n)}(x)]^2\,dx}. \tag{261a,b,c}$$

The approximation given by the ratio (261b) is in general more nearly accurate than that given by (261a), whereas (261c) is in general still more efficient. It is easily shown that (when K is real and symmetric) the formula (261b) always yields an approximation which is *too large* in absolute value, and that the same is true of (261c) if the characteristic numbers are all *positive*. [Compare equations (250a,b) of Section 1.24.]

If the characteristic function $y_1(x)$ were known *exactly*, and the next higher characteristic quantities were required, a sequence of approximations tending toward $y_2(x)$ would be obtained by starting with an initial approximation which is orthogonal to $y_1(x)$ over (a, b). That is, we would choose a convenient function $F(x)$ and take

$$y^{(1)}(x) = F(x) - cy_1(x), \tag{262}$$

where c is determined by the equation

$$c\int_a^b [y_1(x)]^2\,dx = \int_a^b F(x)y_1(x)\,dx, \tag{263}$$

so that "the y_1 component of F is subtracted from F." Since $y_1(x)$ is not known exactly, its approximation must be substituted for it in (262) and (263). Convergence to λ_2 and $y_2(x)$ will generally then obtain if before each cycle the initial approximation is (approximately) "cleared" of $y_1(x)$ before substitution into (259). Once λ_2 and $y_2(x)$ are satisfactorily approximated, the successive initial approximations in the next stage must be cleared of both $y_1(x)$ and $y_2(x)$, and the process may be continued indefinitely. However, it is found that unless the fundamental characteristic functions are determined to a high degree of accuracy, the accuracy and rate of convergence of following calculations may be seriously impaired.

Modified procedures can be devised for the purpose of dealing with more general classes of kernels, in analogy with the modifications of the corresponding matrix method which are suggested in Sections 1.25 and 1.26. In particular, in the case of the equation

$$y(x) = \lambda\int_a^b G(x, \xi)r(\xi)y(\xi)\,d\xi, \tag{264}$$

where G is continuous, real, and symmetric and $r(x)$ is continuous, real, and *positive* in (a, b), the basic theory differs from that associated with (259) only in that the characteristic functions corresponding to distinct characteristic numbers are orthogonal *with respect to the weighting function* $r(x)$ (see Problem 45). The iterative procedure accordingly is modified only to the extent that the weighting function r is to be introduced into the integrands appearing in equations (259), (260), (261), and (263). [Compare equations (282a,b) of Section 1.25.]

3.15. Approximation of Fredholm equations by sets of algebraic equations. It has already been pointed out that a Fredholm integral equation can be considered as the limit of a set of n algebraic equations, as the number of equations increases without limit. Use can be made of this fact to obtain approximate solutions of such integral equations.

For this purpose, we recall first that a definite integral of the form

$$I = \int_a^b f(\xi)\, d\xi \tag{265}$$

is defined as a limit of the form

$$I = \lim_{n\to\infty} \sum_{k=1}^{n} f(x_k)(\Delta x)_k, \tag{266}$$

where the interval (a, b) is divided into n subintervals of lengths $(\Delta x)_1, \cdots, (\Delta x)_n$, and x_k is a point of the kth subinterval. An approximate evaluation can be obtained by not proceeding to the limit, and hence by expressing I approximately as the weighted sum of the ordinates $f(x_k)$ at n conveniently chosen points $x_1, x_2, \cdots, x_n$ of the interval (a, b):

$$I \approx \sum_{k=1}^{n} D_k f(x_k), \tag{267}$$

where D_k is the "weighting coefficient" associated with the point x_k.

The coefficient D_k may be identified with the length $(\Delta x)_k$ of the subinterval associated with the point x_k, as is suggested by (266). However, when the points $x_1, x_2, \cdots, x_n$ are equally spaced, more nearly accurate approximations are generally obtained by choosing these coefficients in accordance with a formula such as the *trapezoidal rule* or *Simpson's rule.* More elaborate formulas are also available in the literature. If the points x_1 and x_n are identified with the end points $x = a$ and $x = b$, respectively, and a uniform spacing h is chosen, so that

$$(n - 1)h = b - a, \tag{268}$$

we recall that the *trapezoidal rule* gives

$$\{D_1, D_2, D_3, D_4, \cdots, D_{n-2}, D_{n-1}, D_n\} = h\{\tfrac{1}{2}, 1, 1, 1, \cdots, 1, 1, \tfrac{1}{2}\}. \tag{269}$$

According to *Simpson's rule,* which is applicable only if n is *odd,* the weighting coefficients are of the form

$$\{D_1, D_2, D_3, D_4, \cdots, D_{n-2}, D_{n-1}, D_n\} = \frac{h}{3}\{1, 4, 2, 4, \cdots, 2, 4, 1\}, \tag{270a}$$

when $n = 5, 7, 9, \cdots$, and are of the form

$$\{D_1, D_2, D_3\} = \frac{h}{3}\{1, 4, 1\} \tag{270b}$$

in the special case when $n = 3$.*

* It may be recalled that the trapezoidal rule results from approximating the integrand by joining the ordinates at successive division points by straight lines; Simpson's rule results from passing *parabolas* through successive sets of *three* ordinates, and is frequently more nearly accurate.

In the same way, the integral equation

$$y(x) = F(x) + \lambda \int_a^b K(x, \xi) y(\xi)\, d\xi \tag{271}$$

can be approximated in the form

$$y(x) \approx F(x) + \lambda \sum_{k=1}^{n} D_k K(x, x_k) y(x_k), \tag{272}$$

where the points x_k are n conveniently chosen points in the interval (a, b), and the constants D_k are corresponding weighting coefficients. If we now require that the two members of (272) be equal *at each of the n chosen points*, we obtain the n linear equations

$$y(x_i) = F(x_i) + \lambda \sum_{k=1}^{n} D_k K(x_i, x_k) y(x_k) \qquad (i = 1, 2, \cdots, n), \tag{273}$$

in the n unknowns $y(x_1), \cdots, y(x_n)$ which specify approximate values of the unknown function $y(x)$ at the n points.

If we introduce the abbreviations

$$y_i = y(x_i), \qquad F_i = F(x_i), \qquad K_{ij} = K(x_i, x_j), \tag{274}$$

where K_{ij} is hence the value of $K(x, \xi)$ when $x = x_i$ and $\xi = x_j$, this set of equations can be written in the form

$$y_i = F_i + \lambda \sum_{k=1}^{n} K_{ik} D_k y_k \qquad (i = 1, 2, \cdots, n). \tag{275}$$

Thus, if we consider the numbers y_i and F_i as components of the vectors $\mathbf{y}$ and $\mathbf{f}$, and define the matrix $\mathbf{K} = [K_{ij}]$, the set of equations (273) can be written concisely in the form

$$\mathbf{y} = \mathbf{f} + \lambda\, \mathbf{K}\, \mathbf{D}\, \mathbf{y}.$$

Here $\mathbf{D} = [D_i \delta_{ij}]$ is a diagonal matrix, and the product $\mathbf{K}\,\mathbf{D}$ is the matrix obtained by multiplying successive *columns* of $\mathbf{K}$ by successive weighting coefficients. Hence the required set of equations is of the form

$$(\mathbf{I} - \lambda\, \mathbf{K}\, \mathbf{D})\mathbf{y} = \mathbf{f} \tag{276}$$

where $\mathbf{I}$ is the unit matrix of order n.

To illustrate the use of this approximate procedure, we apply it to the solution of the integral equation

$$y(x) = x + \int_0^1 K(x, \xi) y(\xi)\, d\xi, \tag{277}$$

where the kernel is of the form defined by (23),

$$K(x, \xi) = \begin{cases} x(1 - \xi) & \textit{when} \quad x < \xi, \\ \xi(1 - x) & \textit{when} \quad x > \xi. \end{cases} \tag{278}$$

In this particular example, the integral equation can be reduced to the differential equation $d^2y/dx^2 + y = 0$ with the end conditions $y(0) = 0$, $y(1) = 1$, so that the *exact* solution is obtainable in the form

$$y(x) = \frac{\sin x}{\sin 1}. \tag{279}$$

For simplicity, we take $n = 5$ equally spaced points, so that

$$x_1 = 0, \qquad x_2 = \tfrac{1}{4}, \qquad x_3 = \tfrac{1}{2}, \qquad x_4 = \tfrac{3}{4}, \qquad x_5 = 1. \tag{280}$$

The corresponding matrix $\mathbf{K}$ is then easily determined in the form

$$\mathbf{K} = \begin{bmatrix} 0 & 0 & 0 & 0 & 0 \\ 0 & \frac{3}{16} & \frac{1}{8} & \frac{1}{16} & 0 \\ 0 & \frac{1}{8} & \frac{1}{4} & \frac{1}{8} & 0 \\ 0 & \frac{1}{16} & \frac{1}{8} & \frac{3}{16} & 0 \\ 0 & 0 & 0 & 0 & 0 \end{bmatrix}. \tag{281}$$

If the weighting coefficients of the trapezoidal rule are used with $h = \frac{1}{4}$, the matrix of coefficients of the linear equations corresponding to (276) with $\lambda = 1$ is then obtained, in the present special case, in the form $\mathbf{I} - \frac{1}{4}\mathbf{K}$. The required equations then follow:

$$\left.\begin{aligned} y_1 &= 0, \\ \tfrac{61}{64}y_2 - \tfrac{1}{32}y_3 - \tfrac{1}{64}y_4 &= \tfrac{1}{4}, \\ -\tfrac{1}{32}y_2 + \tfrac{15}{16}y_3 - \tfrac{1}{32}y_4 &= \tfrac{1}{2}, \\ -\tfrac{1}{64}y_2 - \tfrac{1}{32}y_3 + \tfrac{61}{64}y_4 &= \tfrac{3}{4}, \\ y_5 &= 1 \end{aligned}\right\}. \tag{282}$$

The solution of this set of equations, when the results are rounded to four decimal places, is obtained as follows:

$$y_1 = 0, \quad y_2 = 0.2943, \quad y_3 = 0.5702, \quad y_4 = 0.8104, \quad y_5 = 1. \tag{283}$$

These approximate values of the solution $y(x)$ at the points $x = 0, \frac{1}{4}, \frac{1}{2}, \frac{3}{4}$, and 1 may be compared with the rounded true values which are obtained from (279) as follows:

$$y_1 = 0, \quad y_2 = 0.2940, \quad y_3 = 0.5697, \quad y_4 = 0.8100, \quad y_5 = 1. \tag{284}$$

In correspondence with the presence of a *corner* in the graph of the integrand in (277), when $x = \xi$, it happens that in this case the use of Simpson's rule is found to give less nearly accurate results.

The preceding method clearly can be applied equally well to the approximate solution of integral equations of the *first* kind, and to the treatment of *characteristic-value* problems. In the latter case, a corresponding problem of the type considered in Chapter 1 is obtained, and the iterative methods developed in that chapter are applicable.

Once approximate values of the solution are determined at the n points $x_1, \cdots, x_n$, additional values may be obtained from them by interpolation. Alternatively, the values of the n calculated approximate ordinates may be introduced into the right-hand member of (272), which then may serve to define an approximation to $y(x)$ over all of (a, b).

It should be noticed that the present method is particularly useful when the kernel $K(x, \xi)$ is not given analytically, but is specified by empirical data. In this case, the matrix $\mathbf{K}$ is immediately available as a *table of values* of the empirical influence function.

3.16. Approximate methods of undetermined coefficients. Other numerical methods for obtaining approximate solutions of integral equations also generally consist of reducing the problem to the consideration of a finite set of algebraic equations. In particular, the solution of the equation

$$y(x) = F(x) + \lambda \int_a^b K(x, \xi) y(\xi)\, d\xi \tag{285}$$

may be approximated by a linear combination of n suitably chosen functions $\phi_1, \phi_2, \cdots, \phi_n$, of the form

$$y(x) \approx \sum_{k=1}^{n} c_k \phi_k(x), \tag{286}$$

where the n constants of combination are to be determined in such a way that (285) is satisfied as nearly as possible (in some sense) by (286) over the interval (a, b).

The requirement that (286) approximately satisfy (285) takes the form

$$\sum_{k=1}^{n} c_k \phi_k(x) \approx F(x) + \lambda \sum_{k=1}^{n} c_k \int_a^b K(x, \xi) \phi_k(\xi)\, d\xi \quad (a \leqq x \leqq b). \tag{287}$$

With the convenient abbreviation

$$\Phi_k(x) = \int_a^b K(x, \xi) \phi_k(\xi)\, d\xi, \tag{288}$$

this condition becomes merely

$$\sum_{k=1}^{n} c_k [\phi_k(x) - \lambda \Phi_k(x)] \approx F(x) \qquad (a \leqq x \leqq b). \tag{289}$$

The coefficients $c_1, \cdots, c_n$ are then to be determined by a set of n conditions which tends to reduce the two members of (289) to an equality over the interval (a, b). Several procedures which lead to such sets of conditions are outlined and illustrated in the sections which follow.

In practical cases, advance information concerning the nature of the behavior of the unknown function $y(x)$ is frequently at hand, and the choice of the approximating functions is motivated by this knowledge. It is frequently convenient to take the approximation in the form of a polynomial of degree n, so that $\phi_k(x)$ may be identified with x^k. However, if it happens that one or more of the end values $y(a)$ and $y(b)$ is known in advance (or obtainable by inspection from the integral equation, as indeed was the case in the example of the preceding section), it may be desirable to take the assumed approximation in the form

$$y(x) \approx \phi_0(x) + \sum_{k=1}^{n} c_k \phi_k(x), \tag{290}$$

where ϕ_0 is chosen in such a way that it assumes the known end values, and the remaining ϕ's are made to *vanish* at the corresponding end or ends of the interval.

The dependability of the approximation obtained can be judged to some extent by comparing the resultant left-hand member of (289) with the right-hand member. It should be pointed out, however, that situations unfortunately exist in which a *large* change in the function $y(x)$ may correspond to a *small* change in the function

$$y(x) - \lambda \int_a^b K(x, \xi) y(\xi)\, d\xi.$$

In such cases, it may happen that the integral equation is very nearly satisfied over the interval (a, b) by an "approximation" $\tilde{y}(x)$, in the sense that the difference between the two members of (289) then is everywhere small relative to either of those members, but nevertheless $\tilde{y}(x)$ may differ appreciably from the exact solution $y(x)$ over that interval.

A somewhat more satisfactory estimate of dependability is obtained by comparing the result of an n-term approximation with the result of an $(n + 1)$-term approximation, where the constants of combination are determined by the same technique in both cases.

3.17. The method of collocation. If we introduce the abbreviation

$$s_k(x) = \phi_k(x) - \lambda \Phi_k(x) \equiv \phi_k(x) - \lambda \int_a^b K(x, \xi) \phi_k(\xi)\, d\xi, \tag{291}$$

the requirement that (286) approximately satisfy (285) can be expressed in the form

$$\sum_{k=1}^{n} c_k s_k(x) \approx F(x) \qquad (a \leq x \leq b). \tag{292}$$

A set of n conditions for the determination of the n constants of combination can be obtained most simply by requiring that (292) be an *equality* at n

distinct points in the interval (a, b). If we denote these points by x_i $(i = 1, 2, \cdots, n)$, the resultant conditions are then of the form

$$\sum_{k=1}^{n} c_k s_k(x_i) = F(x_i) \qquad (i = 1, 2, \cdots, n). \tag{293}$$

The matrix of the coefficients of the c's in this set of equations is then given by

$$\mathbf{S} = [s_{ij}] \quad \textit{where} \quad s_{ij} = s_j(x_i), \tag{294}$$

that is, the set of equations can be expressed in the matrix form

$$\mathbf{S}\,\mathbf{c} = \mathbf{f} \tag{295}$$

where $\mathbf{c} = \{c_i\}$ and $\mathbf{f} = \{F_i\}$.

To illustrate the method, we again consider the integral equation (277), where $K(x, \xi)$ is defined by (278). For simplicity, we assume a three-term approximation of the polynomial form

$$y(x) \approx c_1 + c_2 x + c_3 x^2. \tag{296}$$

With $\phi_1 = 1$, $\phi_2 = x$, and $\phi_3 = x^2$, and with the notation of (288), there follows by direct integration

$$\Phi_1 = \tfrac{1}{2}x(1 - x), \quad \Phi_2 = \tfrac{1}{6}x(1 - x^2), \quad \Phi_3 = \tfrac{1}{12}x(1 - x^3). \tag{297}$$

Thus, with the notation of (291), equation (292) here takes the form

$$c_1\left[1 - \frac{x}{2}(1 - x)\right] + c_2\left[x - \frac{x}{6}(1 - x^2)\right] + c_3\left[x^2 - \frac{x}{12}(1 - x^3)\right] \approx x \qquad (0 \leqq x \leqq 1). \tag{298}$$

If we require that this relation be an equality at the three points $x = 0$, $x = \frac{1}{2}$, and $x = 1$, we obtain the conditions

$$\left.\begin{aligned} c_1 &= 0, \\ \tfrac{7}{8}c_1 + \tfrac{7}{16}c_2 + \tfrac{41}{192}c_3 &= \tfrac{1}{2}, \\ c_1 + c_2 + c_3 &= 1 \end{aligned}\right\}. \tag{299}$$

The solution of this set of equations, with the results rounded to four decimal places, is then given by

$$c_1 = 0, \qquad c_2 = 1.2791, \qquad c_3 = -0.2791, \tag{300}$$

so that the desired approximate solution is of the form

$$y(x) \approx 1.2791x - 0.2791x^2. \tag{301}$$

A comparison with the exact solution is postponed until Section 3.19 (page 291).

In those cases where the integrals defining the Φ's in (288) are not readily evaluated, or where $K(x, \xi)$ is defined empirically, the integrals may be evaluated approximately as weighted sums of ordinates (see Problem 106).

3.18 The method of weighting functions. A second method of obtaining n conditions for the determination of the constants, which is often associated with the name of Galerkin, consists of requiring that the difference between the two members of (292) be *orthogonal* to n linearly independent functions $\psi_i(x)$ $(i = 1, 2, \cdots, n)$ over the interval (a, b).

Thus, the conditions obtained in this way are of the form

$$\sum_{k=1}^{n} c_k \int_a^b \psi_i s_k \, dx = \int_a^a \psi_i F \, dx \qquad (i = 1, 2, \cdots, n), \tag{302}$$

or, equivalently,

$$\mathbf{M}\,\mathbf{c} = \mathbf{b} \quad \textit{where} \quad m_{ij} = \int_a^b \psi_i s_j \, dx \quad \textit{and} \quad b_i = \int_a^b \psi_i F \, dx. \tag{303}$$

The procedure can be interpreted as weighting the two members of (292) by each of the functions $\psi_i(x)$, and requiring that the integrals of the weighted members be equal. A particularly convenient choice of the n weighting functions is the set $1, x, x^2, \cdots, x^{n-1}$. In this case the graphical representations of the two members of (292) are required to determine areas with the x axis which are equal, and whose first $n - 1$ *moments* are equal. It is desirable to choose the functions ψ_i as n members of a *complete* set of functions (see Section 1.28), since then the relation (292) must tend to an equality over (a, b) as n increases without limit. It is often convenient to identify the *weighting* functions ψ_i with the *approximating* functions ϕ_i.

In illustration, the application of this procedure to the example considered previously, with the weighting functions 1, x, and x^2, leads to the conditions

$$\left.\begin{aligned} \tfrac{11}{12}c_1 + \tfrac{11}{24}c_2 + \tfrac{37}{120}c_3 &= \tfrac{1}{2}, \\ \tfrac{11}{24}c_1 + \tfrac{14}{45}c_2 + \tfrac{17}{72}c_3 &= \tfrac{1}{3}, \\ \tfrac{37}{120}c_1 + \tfrac{17}{72}c_2 + \tfrac{321}{1680}c_3 &= \tfrac{1}{4} \end{aligned}\right\} . \tag{304}$$

These equations are obtained by multiplying the two members of (298) successively by 1, x, and x^2, and equating the integrals of the results over (0, 1). The solution is found to be

$$c_1 = -0.0088, \qquad c_2 = 1.2968, \qquad c_3 = -0.2798, \tag{305}$$

leading to the approximation

$$y(x) \approx -0.0088 + 1.2968x - 0.2798x^2. \tag{306}$$

A comparison with the exact solution, and with the approximation (301), is presented in Section 3.19 (page 291).

3.19. The method of least squares. The accuracy obtained by the procedures of the two preceding sections will in general depend upon the choice of appropriate points of collocation or weighting functions. A method which avoids this dependence upon the judgment of the computer is next presented.

In place of requiring that the integral equation be satisfied *exactly* at a number of points equal to the number of undetermined coefficients (Section 3.17), we may require that the integral of the square of the difference between the two members, over (a, b), be as small as possible. Thus the basic condition is of the form

$$\int_a^b \left[\sum_{k=1}^{n} c_k s_k(x) - F(x) \right]^2 dx = \text{minimum}, \tag{307}$$

with the notation of (291) and (292). In order that (307) be satisfied, the derivative of the left-hand member with respect to each parameter c_i must vanish, so that we must have

$$\int_a^b s_i(x) \left[\sum_{k=1}^{n} c_k s_k(x) - F(x) \right] dx = 0 \quad (i = 1, 2, \cdots, n). \tag{308}$$

These conditions take the form

$$\sum_{k=1}^{n} c_k \int_a^b s_i s_k \, dx = \int_a^b s_i F \, dx \quad (i = 1, 2, \cdots, n), \tag{309}$$

and hence are equivalent to the conditions (302) where the weighting functions $\psi_i(x)$ are identified with the functions $s_i(x)$. Thus it follows that *if the integral equation is to be satisfied as well as possible over (a, b) in the least-squares sense, the weighting functions of the preceding section must be identified with the functions $s_i(x)$.*

In many practical cases, the functions s_i are such that the integrations involved in (309) are not feasible. Therefore, a modification which incorporates most of the advantages of this method over the collocation procedure, with only a small increase in the amount of calculation involved, is now formulated.

If the integrals in (307) and (308) are approximated by weighted sums of the relevant ordinates at N conveniently chosen points, the resultant minimal conditions (308) take the form

$$\sum_{r=1}^{N} D_r s_i(x_r) \left[\sum_{k=1}^{n} c_k s_k(x_r) - F(x_r) \right] = 0 \quad (i = 1, 2, \cdots, n), \tag{310}$$

where the numbers D_r $(r = 1, 2, \cdots, N)$ are appropriate weighting coefficients, associated with the points $x_1, x_2, \cdots, x_N$ involved in the approximate integration. These conditions can also be expressed in the form

$$\sum_{k=1}^{n} c_k \left[\sum_{r=1}^{N} D_r s_i(x_r) s_k(x_r) \right] = \sum_{r=1}^{N} D_r s_i(x_r) F(x_r) \quad (i = 1, 2, \cdots, n). \tag{311}$$

In spite of the rather formidable appearance of this set of conditions, the coefficients in the set of linear algebraic equations which it represents can be obtained very simply by matrix multiplication, as is next shown.

Equation (311) can be written in the abbreviated form

$$\sum_{k=1}^{n} c_k p_{ik} = q_i \qquad (i = 1, 2, \cdots, n), \tag{312}$$

where

$$p_{ik} = \sum_{r=1}^{N} D_r s_{ri} s_{rk} \tag{313a}$$

and

$$q_i = \sum_{r=1}^{N} D_r s_{ri} F_r, \tag{313b}$$

and where we have written

$$s_{ij} = s_j(x_i) \tag{314}$$

in accordance with the notation of (294). With a change of indices, equation (313a) can be written in the form

$$p_{ij} = \sum_{k=1}^{N} s_{ki} D_k s_{kj} \qquad (i, j = 1, 2, \cdots, n), \tag{315}$$

and hence can be expressed in the matrix form

$$\mathbf{P} = \mathbf{S}^T \mathbf{D} \mathbf{S} \tag{316}$$

where $\mathbf{P} = [p_{ij}]$, $\mathbf{S} = [s_{ij}] \equiv [s_j(x_i)]$, and $\mathbf{D} = [D_i \delta_{ij}]$. In the same way, we obtain also

$$\mathbf{q} = \mathbf{S}^T \mathbf{D} \mathbf{f} \tag{317}$$

where $\mathbf{q} = \{q_i\}$ and $\mathbf{f} = \{F_i\}$. Further, since $\mathbf{D}$ is a diagonal matrix, there follows also

$$\mathbf{S}^T \mathbf{D} = (\mathbf{D} \mathbf{S})^T,$$

so that (316) and (317) become

$$\mathbf{P} = (\mathbf{D} \mathbf{S})^T \mathbf{S}, \qquad \mathbf{q} = (\mathbf{D} \mathbf{S})^T \mathbf{f}. \tag{318}$$

If we notice that $\mathbf{S}$ is the matrix of the coefficients of the c's in the equations

$$\sum_{k=1}^{n} c_k s_k(x_i) = F(x_i) \qquad (i = 1, 2, \cdots, N), \tag{319}$$

we see that the matrix of coefficients $\mathbf{P}$ in (311) can be obtained by premultiplying the matrix of coefficients $\mathbf{S}$ in (318) by the transpose of the matrix $\mathbf{D} \mathbf{S}$, and the column of right-hand members $\mathbf{q}$ in (311) can be obtained similarly by premultiplying the corresponding column $\mathbf{f}$ in (319) by the same matrix.

This result leads to the following procedure for determining the n linear equations represented by (311):

1. Choose N points $x_1, x_2, \cdots, x_N$ in the interval (a, b) and write down the N equations (319) which would require that the integral equation be satisfied at those points when $y(x)$ is replaced by the approximation (286).

2. Denote the $N \times n$ matrix of coefficients in this set of equations by $\mathbf{S}$, and form an associated "weighting matrix"

$$\mathbf{S}^* = \mathbf{D} \mathbf{S}$$

by multiplying the ith *row* of $\mathbf{S}$ by the weighting coefficient D_i associated with the point x_i in an approximate integration scheme involving the N points.

3. Premultiply the *augmented matrix* of (319) by the *transpose* of the weighting matrix $\mathbf{S}^*$. The resultant matrix is the augmented matrix of the required set of n linear equations which determines the constants $c_1, c_2, \cdots, c_n$.

We may notice that, since (311) is homogeneous in the D's, these weighting coefficients may be multiplied by any convenient common factor in the formation of $\mathbf{S}^*$. Thus if the formula of the trapezoidal rule is used, the successive coefficients are conveniently taken to be merely $\frac{1}{2}, 1, 1, \cdots, 1, 1, \frac{1}{2}$, so that the elements of the first and last rows of $\mathbf{S}$ are divided by two and the remaining entries are unchanged. Similarly, if Simpson's rule is used, the coefficients can be taken as $\frac{1}{2}, 2, 1, 2, 1, \cdots, 1, 2, \frac{1}{2}$.

As may be expected, this method leads to a set of equations equivalent to the original set when $N = n$, that is, when the number of chosen points x_i is equal to the number of c's to be determined. However, when $N > n$, it permits us to choose a number of points greater than n and to require that the integral equation be satisfied as nearly as possible at those points, rather than to require that it be satisfied *exactly* at n points. The weighting coefficients D_i weight the squared errors committed in failing to satisfy the equation at the respective points x_i in proportion to the influence of the ordinate at x_i in the integration of the squared error over (a, b). Whereas the N equations (319) are in general incompatible, this procedure affords the "best possible" solution in a least-squares sense.

It should be noticed that, by substitution of the calculated c's into the left-hand members of the equations (319), the difference between the two members of the integral equation can be readily calculated at the N chosen points, to give an indication of the dependability of the solution. In physically motivated problems, it is often clear from the nature of the relevant physical phenomenon that small errors in the satisfaction of the integral equation necessarily imply also small errors in the unknown function. However, as was mentioned earlier, this situation does not *always* exist.

The present procedure differs from the collocation procedure of Section 3.17 in that, first, more than n equations are formed initially; second, a weighting matrix must be determined; and, third, an additional matrix multiplication is involved. Since additional equations would be needed in any case for the purpose of investigating the degree of satisfaction of the integral equation, this feature involves no additional calculation. As was shown, the formation of the weighting matrix need involve only multiplication or division of certain elements in the original coefficient matrix by a factor of two if the formulas of the trapezoidal rule or of Simpson's rule are used. The principal source of increased labor is involved in the matrix multiplication. However, the relevant operations are particularly well adapted (for example) to the use of automatic desk calculators, each element of the product matrix being determined by a single continuous sequence of machine operations.

In those cases when N is large (so that the matrix $\mathbf{S}$ possesses a large number of *rows*) it is often inconvenient to actually write the weighting matrix $\mathbf{S}^*$ in transposed form. In such cases it may be preferable to merely write the matrix $\mathbf{S}^*$ to the left of the original augmented matrix, without transposing its rows and columns, and to determine the product by column-into-column (rather than row-into-column) multiplication. The element in the ith row and jth column of the product matrix is then formed from the ith column of the first factor and the jth column of the second factor.

It is useful to notice that it follows from (315) that

$$p_{ji} = p_{ij}, \tag{320}$$

so that the coefficient matrix of the final set of equations is *symmetric*. This means that all elements below the principal diagonal of $\mathbf{P}$ need not be calculated directly, but may be written down by symmetry once the remaining elements have been determined. The symmetry of the coefficient matrix also permits an appreciable reduction of labor in the actual solution of the corresponding set of equations (see Appendix, page 340).

To illustrate the procedure just considered, we again deal with the example of the preceding sections. If we choose the five points $x = 0, \frac{1}{4}, \frac{1}{2}, \frac{3}{4}$, and 1 as the points x_i, the five equations corresponding to (319) are obtained by equating the two members of (298) for those five values of x, in the form

$$\left.\begin{aligned} c_1 &= 0, \\ \tfrac{29}{32}c_1 + \tfrac{27}{128}c_2 + \tfrac{43}{1024}c_3 &= \tfrac{1}{4}, \\ \tfrac{7}{8}c_1 + \tfrac{7}{16}c_2 + \tfrac{41}{192}c_3 &= \tfrac{1}{2}, \\ \tfrac{29}{32}c_1 + \tfrac{89}{128}c_2 + \tfrac{539}{1024}c_3 &= \tfrac{3}{4}, \\ c_1 + c_2 + c_3 &= 1 \end{aligned}\right\} . \tag{321}$$

If five decimal places are retained in the calculations, the augmented matrix of the required set of three equations is then obtained as follows:

$$\begin{bmatrix} 0.50000 & 0 & 0 \\ 1.81250 & 0.42188 & 0.08398 \\ 0.87500 & 0.43750 & 0.21354 \\ 1.81250 & 1.39062 & 1.05274 \\ 0.50000 & 0.50000 & 0.50000 \end{bmatrix}^T \begin{bmatrix} 1 & 0 & 0 & 0 \\ 0.90625 & 0.21094 & 0.04199 & 0.25000 \\ 0.87500 & 0.43750 & 0.21354 & 0.50000 \\ 0.90625 & 0.69531 & 0.52637 & 0.75000 \\ 1 & 1 & 1 & 1 \end{bmatrix}$$

$$= \begin{bmatrix} 5.05078 & 2.52539 & 1.71700 & 2.75000 \\ 2.52539 & 1.74731 & 1.34312 & 1.86718 \\ 1.71700 & 1.34312 & 1.10326 & 1.41732 \end{bmatrix} . \tag{322}$$

The second factor in the product is merely the augmented matrix of (321). In forming the weighting matrix which precedes it, we have used the weighting coefficients $\frac{1}{2}$, 2, 1, 2, $\frac{1}{2}$ corresponding to the formula of Simpson's rule. The corresponding set of equations,

$$\left.\begin{aligned} 5.05078c_1 + 2.52539c_2 + 1.71700c_3 &= 2.75000, \\ 2.52539c_1 + 1.74731c_2 + 1.34312c_3 &= 1.86718, \\ 1.71700c_1 + 1.34312c_2 + 1.10326c_3 &= 1.41732 \end{aligned}\right\}, \tag{323}$$

then is found to possess the solution

$$c_1 = -0.0079, \qquad c_2 = 1.2939, \qquad c_3 = -0.2783, \tag{324}$$

leading to the approximation

$$y(x) \approx -0.0079 + 1.2939x - 0.2783x^2. \tag{325}$$

In the following table we compare the results of (A) three-point collocation [equation (301)], (B) use of weighting functions 1, x, and x^2 [equation (306)], and (C) five-point least squares with Simpson's rule [equation (325)], with the exact solution given by equation (279):

		Approximate Solutions			
x	$y(x)$	(A)	(B)	(C)	$\tilde{y}(x)$
0	0	0	−0.0088	−0.0079	−0.0090
0.1	0.1186	0.1251	0.1181	0.1187	0.1180
0.2	0.2361	0.2446	0.2394	0.2398	0.2393
0.3	0.3512	0.3586	0.3550	0.3552	0.3550
0.4	0.4628	0.4670	0.4651	0.4652	0.4652
0.5	0.5697	0.5698	0.5696	0.5695	0.5697
0.6	0.6710	0.6670	0.6685	0.6683	0.6686
0.7	0.7656	0.7586	0.7618	0.7615	0.7619
0.8	0.8526	0.8446	0.8495	0.8491	0.8496
0.9	0.9309	0.9251	0.9316	0.9312	0.9316
1.0	1.0000	1.0000	1.0081	1.0077	1.0081

For the purpose of further comparison, there are included in the last column of the table the values of the parabola $\tilde{y} = c_1 + c_2x + c_3x^2$ which gives the best least-squares approximation to the exact solution itself, over the interval (0, 1). The coefficients were determined in such a way that the integral

$$\int_0^1 \left[\frac{\sin x}{\sin 1} - (c_1 + c_2x + c_3x^2)\right]^2 dx$$

takes on a minimum value, and are defined by the equation

$$\tilde{y}(x) = -0.0090 + 1.2976x - 0.2805x^2. \tag{326}$$

The example considered was chosen for the purpose of simplicity, and also for the reason that the exact solution is known and can be reasonably well approximated by a parabola over the relevant interval. It may be noticed that approximation (B) agrees very closely with the "best possible" parabolic approximation $\tilde{y}$, that it affords a better approximation to the exact solution y than does the collocation approximation (A), and that the five-point least-squares approximation (C) is only slightly less accurate than (B).

Because of the simplicity of the coefficient functions $s_i(x)$ appearing in (298), in the present case, the formulation of equations (304) was very easily accomplished. Indeed, the amount of relevant calculation was less than that involved in the formation of the approximate least-squares equations (323). In more involved problems, in which the integrals involved in (302) frequently must be evaluated by approximate methods, the modified least-squares procedure of the present section is usually preferable because of the fact that the relevant numerical calculations are carried out in a systematic way.

In view of the fact that the integral equation specified by (277) and (278) implies the obvious end conditions $y(0) = 0$, $y(1) = 1$, it is to be expected that a three-parameter approximation of the form

$$y(x) \approx x + x(1 - x)(d_1 + d_2x + d_3x^2)$$

would lead to much more nearly accurate results.

When polynomial approximation is used in connection with the method of collocation, or the modified least-squares procedure, the calculations involved can be further systematized if the approximating polynomial is expressed in the so-called *Lagrangian form*. Details may be found in References 3 and 5.

Certain difficulties which arise when least-squares methods are applied to *characteristic-value* problems, and methods of circumventing these difficulties, are treated in Reference 4.

3.20. Approximation of the kernel. As was mentioned in Section 3.6, it is sometimes convenient to approximate the *kernel* of a Fredholm integral equation by a polynomial in x and ξ, or by a *separable kernel* of more general form, and to solve the resultant approximate equation by the methods of that section.

Thus, for example, the kernel (278) could be approximated by a three-parameter polynomial, of the form $A_1 + A_2x + A_3x^2$ or of the more appropriate form $x(1 - x)(B_1 + B_2x + B_3x^2)$, where the A's or B's are determined as functions of ξ by three-point collocation, the use of appropriate weighting functions, or the use of least-square techniques.

To illustrate the procedure, we assume a crude approximation of the form

$$K(x, \xi) \approx Bx(1 - x). \tag{327}$$

Noticing that the approximation is exact at the end points $x = 0$ and $x = 1$, we determine the coefficient B in such a way that the integral of the kernel over (0, 1) is equal to the integral of its approximation over that interval:

$$\int_0^1 K(x, \xi)\, dx = B\int_0^1 x(1 - x)\, dx. \tag{328}$$

Direct calculation then gives the determination

$$B = 3\xi(1 - \xi), \tag{329}$$

and the introduction of the corresponding approximate kernel into (277) leads to the approximating integral equation

$$y(x) = x + 3x(1 - x)\int_0^1 \xi(1 - \xi)y(\xi)\, d\xi. \tag{330}$$

Following the method of Section 3.6, we introduce the abbreviation

$$c = \int_0^1 x(1 - x)y(x)\, dx, \tag{331}$$

and rewrite (330) in the form

$$y(x) = x + 3cx(1 - x). \tag{332}$$

In order to determine c, we multiply the equal members of (332) by $x(1 - x)$ and integrate the results over (0, 1), to obtain the condition

$$c = \int_0^1 x^2(1 - x)\, dx + 3c\int_0^1 x^2(1 - x)^2\, dx,$$

and the evaluation

$$c = \tfrac{5}{54}. \tag{333}$$

Hence the desired approximate solution (332) is obtained in the form

$$y(x) \approx x + \tfrac{5}{18}x(1 - x) \doteq 1.2778x - 0.2778x^2. \tag{334}$$

The approximation very nearly coincides with that of (301).

More generally, it is easily seen that a kernel approximation of the special form

$$K(x, \xi) \approx x\xi(1 - x)(1 - \xi)(a_1 + a_2x\xi + a_3x^2\xi^2 + \cdots) \tag{335}$$

would lead to an approximate solution of the form

$$y(x) \approx x + x(1 - x)(c_1 + c_2x + c_3x^2 + \cdots) \tag{336}$$

in the case of the present example.

Related procedures involving kernel approximation are outlined in Problems 117 and 118.

REFERENCES

1. Bückner, H.: *Die praktische Behandlung von Integralgleichungen*, Ergebnisse der angewandten Mathematik, Bd. 1, Springer-Verlag, Berlin, 1952.
2. Courant, R., and D. Hilbert: *Methods of Mathematical Physics*, Interscience Publishers, Inc., New York, 1953.
3. Crout, P. D.: "An Application of Polynomial Approximation to the Solution of Integral Equations Arising in Physical Problems," *J. Math. Phys.*, Vol. 19, No. 1 (1940).
4. Crout, P. D.: "A Least Square Procedure for Solving Homogeneous Integral Equations," *J. Math. Phys.*, Vol. 40, No. 2 (1961).
5. Crout, P. D., and F. B. Hildebrand: "A Least Square Procedure for Solving Integral Equations by Polynomial Approximation," *J. Math. Phys.*, Vol. 20, No. 3 (1941).
6. Lovitt, W. V.: *Linear Integral Equations*, McGraw-Hill Book Company, New York, 1924.
7. Mikhlin, S. G.: *Integral Equations*, Pergamon Press, London, 1957.
8. Smithies, F.: *Integral Equations*, Cambridge University Press, London, 1958.
9. Tricomi, F. G.: *Integral Equations*, Interscience Publishers, Inc., New York, 1957.
10. Whittaker, E. T., and G. N. Watson: *Modern Analysis*, Cambridge University Press, New York, 1958.

PROBLEMS

Section 3.1.

1. (a) If $y''(x) = F(x)$, and y satisfies the initial conditions $y(0) = y_0$ and $y'(0) = y_0'$, show that

$$y(x) = \int_0^x (x - \xi)F(\xi)\,d\xi + y_0'x + y_0.$$

[Notice that $y'(x) = \int_0^x F(x_1)\,dx_1 + y_0'$, and use (10).]

(b) Verify that this expression satisfies the prescribed differential equation and initial conditions.

2. (a) If $y''(x) = F(x)$, and y satisfies the end conditions $y(0) = 0$ and $y(1) = 0$, show that

$$y(x) = \int_0^x (x - \xi)F(\xi)\,d\xi - x\int_0^1 (1 - \xi)F(\xi)\,d\xi.$$

[Set $y_0 = 0$ in the result of Problem 1, and determine y_0' so that $y(1) = 0$.]

(b) Show that the result of part (a) can be written in the form

$$y(x) = \int_0^1 K(x, \xi)F(\xi)\, d\xi,$$

where $K(x, \xi)$ is defined by the relations

$$K(x, \xi) = \begin{cases} \xi(x - 1) & \textit{when} \quad \xi < x, \\ x(\xi - 1) & \textit{when} \quad \xi > x. \end{cases}$$

(c) Verify directly that the expression obtained satisfies the prescribed differential equation and end conditions.

Section 3.2.

3. (a) Show that, if $y(x)$ satisfies the differential equation

$$\frac{d^2y}{dx^2} + xy = 1$$

and the conditions $y(0) = y'(0) = 0$, then y also satisfies the Volterra equation

$$y(x) = \int_0^x \xi(\xi - x)y(\xi)\, d\xi + \tfrac{1}{2}x^2.$$

(b) Prove that the converse of the preceding statement is also true.

4. Suppose that a sequence of approximate solutions is obtained for the integral equation of Problem 3, the $(n + 1)$th approximation $y^{(n+1)}(x)$ being defined by substitution of the nth approximation into the right-hand member:

$$y^{(n+1)}(x) = \int_0^x \xi(\xi - x)y^{(n)}(\xi)\, d\xi + \tfrac{1}{2}x^2.$$

(a) Taking $y^{(0)}(x) = 0$, obtain the functions

$$y^{(1)}(x) = \tfrac{1}{2}x^2 \quad \text{and} \quad y^{(2)}(x) = \tfrac{1}{2}x^2 - \tfrac{1}{40}x^5$$

as the two succeeding approximations.

(b) Obtain the first two nonvanishing terms in the power-series solution of the problem considered in Problem 3(a), in the form

$$y(x) = a_0 + a_1x + a_2x^2 + \cdots,$$

and compare the result with that of part (a). [Notice that we must have $a_0 = a_1 = 0$, to satisfy the initial conditions, and determine the remaining a's by substitution in the differential equation.]

5. (a) Show that, if $y(x)$ satisfies the differential equation

$$x\frac{d^2y}{dx^2} + \frac{dy}{dx} + xy = x$$

when $x \geqq 0$, and if y, y', and y'' are finite at $x = 0$, then there must follow $y'(0) = 0$.

[This conclusion follows directly from the differential equation. Further, the most general solution which is finite at $x = 0$ is found to be $y(x) = 1 + cJ_0(x)$, where $J_0(x)$ is the Bessel function of first kind, of order zero. By virtue of the fact that $J_0(0) = 1$ and $J_0'(0) = 0$, the value of $y(0)$ can be arbitrarily prescribed, whereas $y'(0)$ *cannot* be prescribed. This situation is a consequence of the fact that $x = 0$ is a *singular point* of the differential equation.]

(b) Show that the integral-equation formulation of equation (13) is not applicable to the problem of part (a), if the initial conditions are prescribed at point $x = 0$. [When the equation is written in the form of equation (11), the function $A(x)$ is not finite at $x = 0$, and hence the right-hand member of (14b) is undefined. The integration by parts which led to (13) was not legitimate in this case.]

6. By integrating the equal members of the differential equation of Problem 5(a) twice over the interval $(0, x)$, and simplifying the integrals $\int xy''\,dx$ and $\int xy'\,dx$ by integration by parts in successive steps, show that $y(x)$ must satisfy the integral equation

$$xy(x) = \int_0^x [\xi(\xi - x) + 1]y(\xi)\,d\xi + \tfrac{1}{6}x^3,$$

of the "third kind" regardless of the prescribed initial condition. [Notice that this equation hence must possess *infinitely many* solutions, each of the form $y(x) = 1 + c\,J_0(x)$, where c is an arbitrary constant. This situation is a consequence of the fact that, when this equation is written in the form (13), the kernel $K(x, \xi)$ becomes infinite when $x = 0$.]

7. Obtain an alternative integral-equation formulation of the problem described by equations (11) and (12), by first setting $y'' = u$, and showing that

$$y(x) = \int_a^x (x - \xi)u(\xi)\,d\xi + y_0'(x - a) + y_0$$

where $u(x)$ satisfies an integral equation of the form

$$u(x) = \int_a^x [(\xi - x)B(x) - A(x)]u(\xi)\,d\xi + F(x).$$

8. Show that the application of the method of Section 3.2 to the problem $y'' + Ay' + By = 0$, $y(0) = y(1) = 0$, where A and B are *constants*, leads to the integral equation $y(x) = \int_0^1 K(x, \xi)y(\xi)\,d\xi$, where

$$K(x, \xi) = \begin{cases} B\xi(1 - x) + Ax - A & \textit{when} \quad \xi < x, \\ Bx(1 - \xi) + Ax & \textit{when} \quad \xi > x. \end{cases}$$

[Notice that the kernel obtained in this way is nonsymmetric, and discontinuous at $\xi = x$, unless $A = 0$.]

Section 3.3.

9. Transform the problem

$$\frac{d^2y}{dx^2} + xy = 1, \qquad y(0) = y(1) = 0$$

to the integral equation

$$y(x) = \int_0^1 G(x, \xi)\xi y(\xi)\, d\xi - \tfrac{1}{2}x(1 - x),$$

where $G(x, \xi) = x(1 - \xi)$ when $x < \xi$ and $G(x, \xi) = \xi(1 - x)$ when $x > \xi$.

10. Transform the problem

$$\frac{d^2y}{dx^2} + y + \epsilon y^2 = f(x), \qquad y(0) = 0, \quad y(1) = 0$$

to a nonlinear Fredholm integral equation in each of the two following ways:

(a) Use the method of Section 3.3 with $\mathscr{L}y = y''$.

(b) Proceed as in part (a) with $\mathscr{L}y = y'' + y$.

11. Transform the problem

$$\frac{d^2y}{dx^2} + y = x, \qquad y(0) = 0, \quad y'(1) = 0$$

to a Fredholm integral equation.

12. Transform the problem

$$\frac{d^2y}{dx^2} + y = x, \qquad y(0) = 1, \quad y'(1) = 0$$

to a Fredholm integral equation.

13. (a) Determine $p(x)$ and $q(x)$ in such a way that the equation

$$x^2\frac{d^2y}{dx^2} - 2x\frac{dy}{dx} + 2y = 0$$

is equivalent to the equation $\dfrac{d}{dx}\left(p\dfrac{dy}{dx}\right) + qy = 0$, thus showing that the equation can be written in the "self-adjoint" form

$$\frac{d}{dx}\left(\frac{1}{x^2}\frac{dy}{dx}\right) + \frac{2}{x^4}y = 0.$$

(b) Verify that $u(x) = x$ and $v(x) = x^2$ are linearly independent solutions of the equation of part (a), and verify Abel's formula in this case, showing that here $A = 1$ in equation (31).

(c) Show that, if we take $u(x) = a_1x + a_2x^2$ and $v(x) = b_1x + b_2x^2$, there follows $A = a_1b_2 - a_2b_1$. [Notice that hence $A \neq 0$ unless u and v are in a constant ratio, and hence are not linearly independent.]

14. (a) Show that the Green's function for the Bessel operator of order n,

$$\mathscr{L}y = \frac{d}{dx}\left(x\frac{dy}{dx}\right) - \frac{n^2}{x}y,$$

relevant to the end conditions $y(0) = y(1) = 0$, is of the form

$$G(x, \xi) = \begin{cases} \dfrac{x^n/\xi^n}{2n}(1 - \xi^{2n}) & \textit{when} \quad x < \xi, \\[2ex] \dfrac{\xi^n/x^n}{2n}(1 - x^{2n}) & \textit{when} \quad x > \xi, \end{cases}$$

when $n \neq 0$.

(b) Use the result of part (a) to reduce the problem

$$x^2\frac{d^2y}{dx^2} + x\frac{dy}{dx} + (\lambda x^2 - n^2)y = 0, \qquad y(0) = y(1) = 0$$

to an integral equation, when $n \neq 0$.

15. Transform the problem

$$\frac{d^4y}{dx^4} + \Phi(x) = 0, \qquad y(0) = y'(0) = y(1) = y'(1) = 0$$

to the relation

$$y(x) = \int_0^1 G(x, \xi)\Phi(\xi)\, d\xi,$$

where

$$G(x,\xi) = \begin{cases} \frac{1}{6}x^2(1 - \xi)^2(2x\xi + x - 3\xi) & \textit{when} \quad x < \xi, \\ \frac{1}{6}\xi^2(1 - x)^2(2x\xi + \xi - 3x) & \textit{when} \quad x > \xi. \end{cases}$$

16. (a) If $\mathscr{L}$ is the self-adjoint operator $\mathscr{L} \equiv \dfrac{d}{dx}\left(p\dfrac{dy}{dx}\right) + q$, show that

$$\int_A^B f(x)[\mathscr{L}g(x)]\, dx = \int_A^B g(x)[\mathscr{L}f(x)]\, dx + \Big[p(x)\{f(x)g'(x) - f'(x)g(x)\}\Big]_A^B$$

for any functions $f(x)$ and $g(x)$ which possess a continuous first derivative in (A, B). [Write the left-hand member in the form $\int_A^B f(pg')'\, dx + \int_A^B fqg\, dx$, and integrate the first term by parts. Then integrate the resultant term again by parts, in an appropriate way. This formula is known as *Green's formula* for the operator $\mathscr{L}$.]

(b) If $f(x)$ and $g(x)$ satisfy identical homogeneous conditions at the ends $x = A$ and $x = B$ of the interval, show that the integrated terms vanish, and there follows

$$\int_A^B f(x)[\mathscr{L}g(x)]\, dx = \int_A^B g(x)[\mathscr{L}f(x)]\, dx.$$

[Notice that this situation exists if at each end we require the *vanishing* of each function, or of the derivative of each function, or of a linear combination of each function and its derivative. Further, if $p(x)$ vanishes at an end, we may require at that end only that each function and its derivative remain finite. Also, if $p(B) = p(A)$, we may require merely that each function and its derivative take on the same values at $x = B$ as at $x = A$.]

17. Prove by the following steps, that the differential equation (34), with associated homogeneous conditions at the ends of the interval (a, b), implies the relation (33).

(a) Assuming that $y(x)$ satisfies (34), show that y also must satisfy the equation

$$-\int_a^b G(x, \xi)\, \mathscr{L}y(x)\, dx = \int_a^b G(x, \xi)\Phi(x)\, dx,$$

where $G(x, \xi) = G(\xi, x)$ is the relevant Green's function.

(b) Divide the interval of integration in the left-hand member into the subintervals (a, ξ) and (ξ, b) [so that Green's formula of Problem 16(a) applies over each subinterval], and show that the resultant expression

$$-\int_a^\xi G_1(x, \xi)\mathscr{L}y(x)\, dx - \int_\xi^b G_2(x, \xi)\mathscr{L}y(x)\, dx,$$

can be written in the form

$$-p(\xi)\left[G_1(\xi, \xi)y'(\xi) - \frac{\partial G_1(x, \xi)}{\partial x}\bigg|_{x=\xi} y(\xi)\right] + p(\xi)\left[G_2(\xi, \xi)y'(\xi) - \frac{\partial G_2(x, \xi)}{\partial x}\bigg|_{x=\xi} y(\xi)\right] = y(\xi).$$

(c) By appropriately changing variables, deduce that $y(x)$ therefore must satisfy (33).

18. Suppose that a function $U(x)$ satisfies the equation $\mathscr{L}y = 0$ in an interval (a, b), where $\mathscr{L}$ is the self-adjoint operator defined in Problem 16, and also satisfies certain homogeneous conditions at the ends of the interval. Prove that the equation $\mathscr{L}y + \Phi = 0$ cannot possess a solution, valid everywhere in (a, b) and satisfying the *same* homogeneous conditions at the end points, unless $\Phi(x)$ is "orthogonal" to $U(x)$, that is, unless the condition

$$\int_a^b U(x)\Phi(x)\, dx = 0$$

is satisfied. [Assume that such a solution $y(x)$ exists. Multiply the equal members of the relation $\Phi = -\mathscr{L}y$ by $U(x)$, integrate the results over (a, b), and use Green's formula of Problem 16(b).] Also, verify this result in the case of the problem $y'' + \Phi = 0$, $y'(0) = y'(1) = 0$, with $\Phi = 1$ and with $\Phi = 2x - 1$, noticing that here $U(x) =$ constant.

19. *The generalized Green's function.* Suppose that a problem, consisting of the differential equation $\mathscr{L}y \equiv (py')' + qy = 0$ and homogeneous conditions prescribed at the ends of an interval (a, b), is satisfied by a nontrivial function $y = U(x)$, so that the Green's function defined by properties 1 to 4 of page 235 does not exist. The *generalized* Green's function is then defined as a function H which, when considered as a function of x for a fixed number ξ, posesses the following properties:

1. H satisfies the differential equation

$$\mathscr{L}H = CU(x)U(\xi)$$

in the subintervals (a, ξ) and (ξ, b), where C is any nonzero constant.

2. H satisfies the prescribed end conditions.
3. H is continuous at $x = \xi$.
4. The x derivative of H possesses a jump of magnitude $-1/p(\xi)$ as the point $x = \xi$ is crossed in the positive x direction.
5. H satisfies the condition

$$\int_a^b H(x, \xi)U(x)\,dx = 0.$$

Show, by the following steps, that (if such a function exists) the function H has the property that, if Φ is any function such that $\int_a^b U\Phi\,dx = 0$, then the relation

$$y(x) = \int_a^b H(x, \xi)\Phi(\xi)\,d\xi + AU(x),$$

where A is any constant, implies the differential equation

$$\mathscr{L}y + \Phi = 0,$$

together with the homogeneous end conditions associated with H.

(a) By writing $H = H_1$ when $x < \xi$ and $H = H_2$ when $x > \xi$, show that there follows

$$y(x) = \int_a^x H_2(x, \xi)\Phi(\xi)\,d\xi + \int_x^b H_1(x, \xi)\Phi(\xi)\,d\xi + AU(x),$$

$$y'(x) = \int_a^x \frac{\partial H_2(x, \xi)}{\partial x}\Phi(\xi)\,d\xi + \int_x^b \frac{\partial H_1(x, \xi)}{\partial x}\Phi(\xi)\,d\xi + AU'(x),$$

and

$$y''(x) = \int_a^x \frac{\partial^2 H_2(x, \xi)}{\partial x^2}\Phi(\xi)\,d\xi + \int_x^b \frac{\partial^2 H_1(x, \xi)}{\partial x^2}\Phi(\xi)\,d\xi - \frac{1}{p(x)}\Phi(x) + AU''(x).$$

[Make use of properties 3 and 4.]

(b) Verify that

$$\begin{aligned}\mathscr{L}y(x) &= \int_a^x [\mathscr{L}H_2(x, \xi)]\Phi(\xi)\,d\xi + \int_x^b [\mathscr{L}H_1(x, \xi)]\Phi(\xi)\,d\xi - \Phi(x)\\ &= \int_a^b [CU(x)U(\xi)]\Phi(\xi)\,d\xi - \Phi(x)\\ &= -\Phi(x).\end{aligned}$$

[Make use of property 1 and the restriction on Φ. *Notice that satisfaction of property 5 is not necessary.* (See, however, Problem 22.)]

20. Suppose that $U(x)$ is a nontrivial solution of the equation $\mathscr{L}y = 0$ in the interval (a, b), and that $U(x)$ satisfies certain prescribed homogeneous conditions at the ends of that interval. Let $u(x)$ denote a function such that $\mathscr{L}u(x) = CU(x)$ when $a \leqq x \leqq \xi$, and let $v(x)$ denote a function such that $\mathscr{L}v(x) = CU(x)$ when $\xi \leqq x \leqq b$. Finally, suppose that $u(x)$ satisfies the prescribed condition at $x = a$, whereas $v(x)$ satisfies the prescribed condition at $x = b$.

(a) By setting $f = U$ and $g = u$ in Green's formula [Problem 16(a)], show that

$$C\int_a^\xi [U(x)]^2\,dx = p(\xi)[U(\xi)u'(\xi) - U'(\xi)u(\xi)].$$

(b) In a similar way, show that

$$C\int_\xi^b [U(x)]^2\,dx = -p(\xi)[U(\xi)v'(\xi) - U'(\xi)v(\xi)].$$

(c) Deduce that

$$p(\xi)\{[u'(\xi) - v'(\xi)]U(\xi) - [u(\xi) - v(\xi)]U'(\xi)\} = C\int_a^b [U(x)]^2\,dx.$$

21. With the terminology of Problems 19 and 20, verify that, if C and the function $U(x)$ are chosen in such a way that

$$C\int_a^b [U(x)]^2\,dx = 1,$$

then the function

$$H(x, \xi) = \alpha(\xi)U(x) + \begin{cases} v(\xi)U(x) + u(x)U(\xi) & \text{when} \quad x < \xi, \\ u(\xi)U(x) + v(x)U(\xi) & \text{when} \quad x > \xi \end{cases}$$

satisfies properties 1, 2, 3, and 4 of Problem 19, regardless of the form of the function $\alpha(\xi)$, so that the required generalized Green's function is obtained by determining $\alpha(\xi)$ by condition 5, in such a way that

$$\int_a^b H(x, \xi)U(x)\,dx = 0.$$

[Use the result of Problem 20 in investigating the satisfaction of property 4.]

22. (a) By writing $f(x) = H(x, r)$ and $g(x) = H(x, s)$ in Green's formula [Problem 16(a)], where $H(x, \xi)$ is the generalized Green's function relevant to the operator $\mathscr{L}$ in (a, b), show that

$$\int_a^b [H(x, r)\mathscr{L}H(x, s) - H(x, s)\mathscr{L}H(x, r)]\,dx = H(s, r) - H(r, s).$$

[Write $\int_a^b = \int_a^{r-} + \int_{r+}^{s-} + \int_{s+}^b$ and recall that $\partial H(x, r)/\partial x$ has a jump of magnitude $-1/p(r)$ at $x = r$, whereas $\partial H(x, s)/\partial x$ has a jump of magnitude $-1/p(s)$ at $x = s$.]

(b) Show that the left-hand member of the preceding equation can be written in the form

$$U(s)\int_a^b H(x, r)U(x)\,dx - U(r)\int_a^b H(x, s)U(x)\,dx,$$

and hence vanishes in consequence of the satisfaction of Property 5. Thus deduce that the generalized Green's function is symmetric:

$$H(x, \xi) = H(\xi, x).$$

[Notice that this proof applies also to the conventional Green's function (with $U=0$), although symmetry in this case is obvious from the explicit form (32).]

23. Prove (by methods similar to those used in Problem 17) that the differential equation $\mathscr{L}y + \Phi = 0$, with associated homogeneous conditions imposed at the ends of the interval (a, b), implies the relation

$$y(x) = \int_a^b H(x, \xi)\Phi(\xi)\,d\xi + AU(x),$$

where H is the corresponding generalized Green's function, U is any nontrivial function satisfying $\mathscr{L}U = 0$ and the prescribed end conditions, and A is a constant. Show also that

$$A\int_a^b U^2\,dx = \int_a^b yU\,dx,$$

so that A is determined when also the value of $\int_a^b yU\,dx$ is known.

24. Determine the generalized Green's function, relevant to the end conditions $y'(0) = y'(1) = 0$, for the expression

$$\mathscr{L}y = \frac{d^2y}{dx^2},$$

in the form

$$H(x, \xi) = \frac{1}{3} + \frac{1}{2}(x^2 + \xi^2) - \begin{cases} \xi & \text{when} \quad x < \xi, \\ x & \text{when} \quad x > \xi. \end{cases}$$

[Use either the basic properties of Problem 19 or the formula of Problem 21, with $U(x) = 1$.] Deduce that the problem

$$\frac{d^2y}{dx^2} + \lambda y = f, \qquad y'(0) = y'(1) = 0$$

is transformable to the integral equation

$$y(x) = \lambda\int_0^1 H(x, \xi)y(\xi)\,d\xi - \int_0^1 H(x, \xi)f(\xi)\,d\xi + \int_0^1 y(\xi)\,d\xi.$$

Also, by integrating the equal members of the differential equation over (0, 1), show that

$$\int_0^1 y\,d\xi = \frac{1}{\lambda}\int_0^1 f\,d\xi.$$

25. For the Legendre operator $\mathscr{L} = \dfrac{d}{dx}\left[(1 - x^2)\dfrac{d}{dx}\right]$ there follows $p(x) = 1 - x^2$, and hence $p(\pm 1) = 0$, with the notation of equation (26). Here appropriate end conditions for the expression $\mathscr{L}y$ in the interval $(-1, 1)$ consists of the requirements that $y(-1)$ and $y(1)$ be *finite*. Noticing that the function $U(x) =$ constant satisfies the equation $\mathscr{L}y = 0$ and these finiteness conditions (so that the conventional Green's function does not exist), obtain the generalized Green's function in the form

$$H(x, \xi) = \log 2 - \frac{1}{2} - \begin{cases} \dfrac{1}{2}\log\left[(1 + \xi)(1 - x)\right] & \textit{when} \quad x < \xi, \\ \dfrac{1}{2}\log\left[(1 - \xi)(1 + x)\right] & \textit{when} \quad x > \xi. \end{cases}$$

Deduce that the problem

$$(1 - x^2)\frac{d^2y}{dx^2} - 2x\frac{dy}{dx} + \lambda y = f, \qquad y(\pm 1)\ \textit{finite}$$

transforms into the integral equation

$$y(x) = \lambda\int_{-1}^{1} H(x, \xi)y(\xi)\, d\xi - \int_{-1}^{1} H(x, \xi)f(\xi)\, d\xi + \frac{1}{2}\int_{-1}^{1} y(\xi)\, d\xi.$$

Also show that

$$\int_{-1}^{1} y\, d\xi = \frac{1}{\lambda}\int_{-1}^{1} f\, d\xi.$$

Section 3.4.

26. Obtain an explicit solution of the problem $y''(x) + h_\epsilon(x) = 0$, in the interval $(0, 1)$, where $h_\epsilon(x)$ vanishes outside the interval $(\xi - \epsilon, \xi + \epsilon)$ and is given by $1/(2\epsilon)$ inside that interval, and where $y(0) = 0$ and $y(1) = 0$. [Determine the general solution in each of the subintervals $(0, \xi - \epsilon)$, $(\xi - \epsilon, \xi + \epsilon)$, and $(\xi + \epsilon, 1)$, and determine the six constants of integration by satisfying the end conditions and requiring that y and y' be continuous at the transition points.] Show that the solution is of the form

$$y(x) = \begin{cases} x(1 - \xi) & \textit{when} \quad 0 < x < \xi - \epsilon, \\ \dfrac{x + \xi}{2} - x\xi - \dfrac{\epsilon}{4} - \dfrac{(x - \xi)^2}{4\epsilon} & \textit{when} \quad \xi - \epsilon < x < \xi + \epsilon, \\ (1 - x)\xi & \textit{when} \quad \xi + \epsilon < x < 1, \end{cases}$$

and notice that this form tends to the relevant Green's function of $\mathscr{L}y = d^2y/dx^2$, subject to the prescribed end conditions, as $\epsilon \to 0$.

27. Suppose that $G(x, y; \xi, \eta)$ is the Green's function for the Laplacian expression $\nabla^2 w$ in a simple region $\mathscr{R}$ of the xy plane, relevant to the requirement that w vanish along the boundary $\mathscr{C}$, so that the solution of Poisson's equation

$\nabla^2 w + \Phi(x, y) = 0$, subject to that boundary condition, is of the form

$$w(x, y) = \iint_{\mathscr{R}} G(x, y; \xi, \eta)\, \Phi(\xi, \eta)\, d\xi\, d\eta, \tag{A}$$

and the corresponding solution of the equation $\nabla^2 w + \lambda q w = 0$ also satisfies the integral equation

$$w(x, y) = \lambda \iint_{\mathscr{R}} G(x, y; \xi, \eta)\, q(\xi, \eta) w(\xi, \eta)\, d\xi\, d\eta. \tag{B}$$

It can be shown (by a method analogous to that used in Problem 22) that G is *symmetric* in (x, y) and (ξ, η), so that $G(x, y; \xi, \eta) = G(\xi, \eta; x, y)$. Assuming this fact, obtain a further useful property of the Green's function, by the following steps:

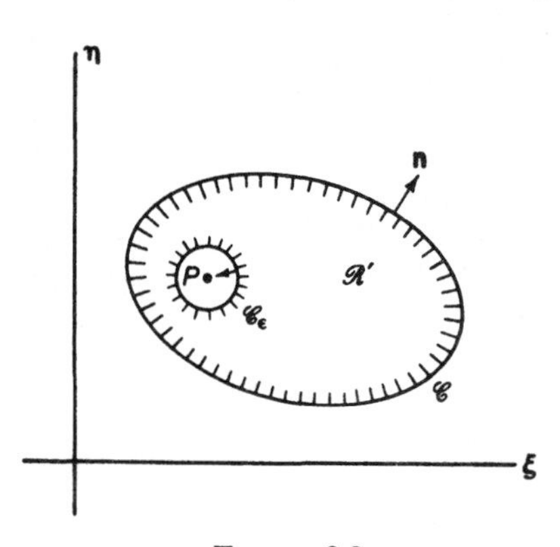

FIGURE 3.2

(a) By applying Green's theorem of vector analysis, in the form

$$\iint_{\mathscr{R}} (\phi_1 \nabla^2 \phi_2 - \phi_2 \nabla^2 \phi_1)\, dA = \oint_{\mathscr{C}} \left(\phi_1 \frac{\partial \phi_2}{\partial n} - \phi_2 \frac{\partial \phi_1}{\partial n}\right) ds,$$

to the region $\mathscr{R}'$ of the $\xi\eta$ plane in Figure 3.2, with $\phi_1 = G(x, y; \xi, \eta)$ and $\phi_2 = w(\xi, \eta)$, where the interior of a small circle $\mathscr{C}_\epsilon$ of radius ϵ about the point $P(x, y)$ is deleted from $\mathscr{R}$, show that

$$\iint_{\mathscr{R}'} (G \nabla^2 w - w \nabla^2 G)\, d\xi\, d\eta = \oint_{\mathscr{C}} \left(G \frac{\partial w}{\partial n} - w \frac{\partial G}{\partial n}\right) ds + \oint_{\mathscr{C}_\epsilon} \left(G \frac{\partial w}{\partial n} - w \frac{\partial G}{\partial n}\right) ds,$$

where here $\nabla^2 = \dfrac{\partial^2}{\partial \xi^2} + \dfrac{\partial^2}{\partial \eta^2}$, and where the normal differentiation is with respect to the coordinates ξ and η.

(b) Suppose that w satisfies Laplace's equation, $\nabla^2 w = 0$, everywhere inside $\mathscr{R}$. Noticing that also $\nabla^2 G = 0$ except at the point P, and that the normal derivatives calculated along $\mathscr{C}_\epsilon$ are along the *inward* normal relative to $\mathscr{C}_\epsilon$, show that, as $\epsilon \to 0$, there follows

$$\oint_{\mathscr{C}} \left(G \frac{\partial w}{\partial n} - w \frac{\partial G}{\partial n}\right) ds = \lim_{r \to 0} \int_0^{2\pi} \left(G \frac{\partial w}{\partial r} - w \frac{\partial G}{\partial r}\right) r\, d\theta,$$

where r denotes radial distance outward from the point P (so that $\partial/\partial n = -\partial/\partial r$ on $\mathscr{C}_\epsilon$), and θ denotes angular position along $\mathscr{C}_\epsilon$. Show also that the limit on the right is given formally by $w(x, y)$, in consequence of the properties of the Green's function. [See equation (75′).] Deduce that, since G vanishes along the boundary $\mathscr{C}$, there follows

$$w(x, y) = -\oint_{\mathscr{C}} w \frac{\partial G}{\partial n}\, ds, \tag{C}$$

where the normal differentiation is with respect to the coordinates ξ and η, so that the value of w at an interior point of $\mathscr{R}$ is thus expressed in terms of *prescribed* values of w along the boundary $\mathscr{C}$ with the help of the normal derivative of the Green's function along $\mathscr{C}$, and the solution of the *Dirichlet problem* for the interior of the region $\mathscr{R}$ is obtained.

(c) If instead, w satisfies the equation $\nabla^2 w + \Phi = 0$ in $\mathscr{R}$ and is arbitrarily prescribed along $\mathscr{C}$, show that there then follows

$$w(x, y) = \iint_{\mathscr{R}} G(x, y; \xi, \eta)\Phi(\xi, \eta)\, d\xi\, d\eta - \oint_{\mathscr{C}} w \frac{\partial G}{\partial n}\, ds. \qquad \text{(D)}$$

[When $w \equiv 0$ on $\mathscr{C}$, this relation reduces to (A) if Φ is a given function and to (B) if Φ is replaced by $\lambda q w$. When $\Phi \equiv 0$, the relation (D) reduces to (C).]

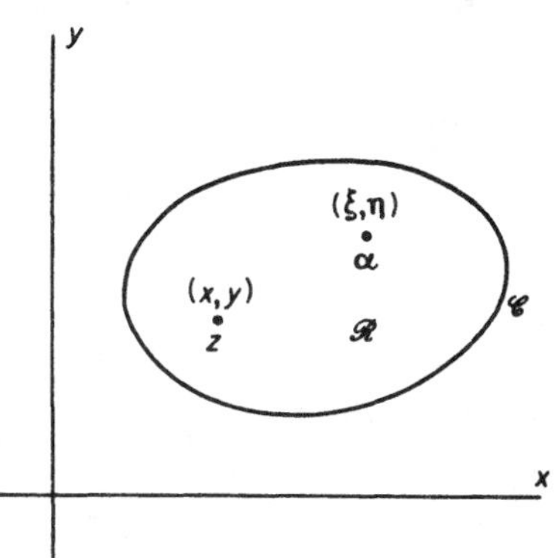

FIGURE 3.3

28. Suppose that a function $f(z)$ of the complex variable $z = x + iy$ can be found with the following properties (Figure 3.3):

1. $f(z)$ is analytic everywhere inside a region $\mathscr{R}$ of the xy plane and on the boundary $\mathscr{C}$.
2. $|f(z)| = 1$ at all points of $\mathscr{C}$ and $|f(z)| < 1$ inside $\mathscr{C}$.
3. $f(z)$ possesses a simple zero at the point (ξ, η), that is, at the complex point $\alpha = \xi + i\eta$, and differs from zero elsewhere in $\mathscr{R}$ and on $\mathscr{C}$.

(a) Show that the function

$$G(x, y; \xi, \eta) = -\frac{1}{2\pi} \log |f(z)| = -\frac{1}{2\pi} \operatorname{Re}\,[\log f(z)]$$

is the Green's function of the Laplacian expression $\nabla^2 w$ relevant to the requirement that w vanish along the boundary of $\mathscr{R}$. [Recall that the real and imaginary parts of $f(z)$ satisfy Laplace's equation at points where $f(z)$ is analytic, and that $\log f(z)$ is analytic when $f(z)$ is analytic and $f(z) \neq 0$. Also notice that here we may write $f(z) = (z - \alpha)\phi(z)$, where $\phi(z)$ is analytic *and nonzero* everywhere in $\mathscr{R}$ and on $\mathscr{C}$.]

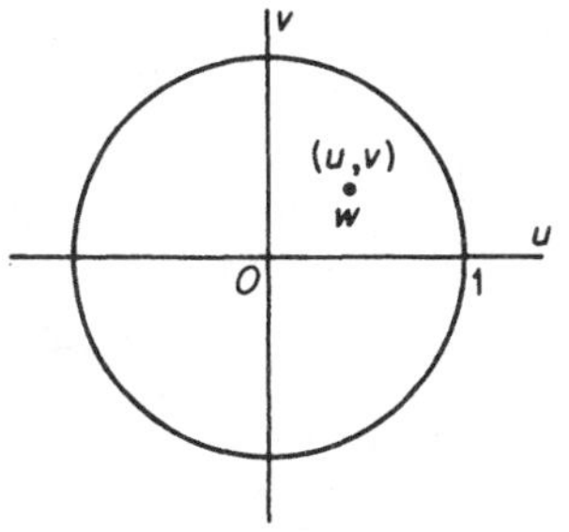

FIGURE 3.4

(b) Show that the function $f(z)$ defined in part (a) has the property that the relation $w = f(z)$ maps the interior of $\mathscr{R}$ onto the interior of the unit circle $|w| = 1$ of the w plane (Figure 3.4), with the point $\alpha = \xi + i\eta$ being mapped into the origin.

[Notice that the requirement that $f(z)$ possess only a *simple* zero at $z = \alpha$ insures that $f'(z) \neq 0$, so that the mapping is indeed one to one.]

29. (a) Verify that the mapping

$$w = e^{ia}\frac{z-\alpha}{1-\bar{\alpha}z} \equiv -\frac{e^{ia}}{\bar{\alpha}}\frac{z-\alpha}{z-1/\bar{\alpha}},$$

where $\alpha = \xi + i\eta$ and $\bar{\alpha} = \xi - i\eta$ and a is any real constant, maps the boundary and interior of the unit circle in the z plane onto the boundary and interior of the unit circle in the w plane if α is inside the unit circle (that is, if $|\alpha| < 1$), and that the point α maps into the origin. [Calculate $w\bar{w}$, and show that $w\bar{w} \equiv |w|^2$ is unity when $z\bar{z} \equiv |z|^2 = 1$. Also, show that the point $1/\bar{\alpha}$ is the image of the point α in an inversion relative to the unit circle (see Figure 3.5) and hence deduce from the second form of w that $|w| < 1$ when z and α are inside the unit circle, recalling that $|z - \alpha|$ is the *distance* between the points z and α.]

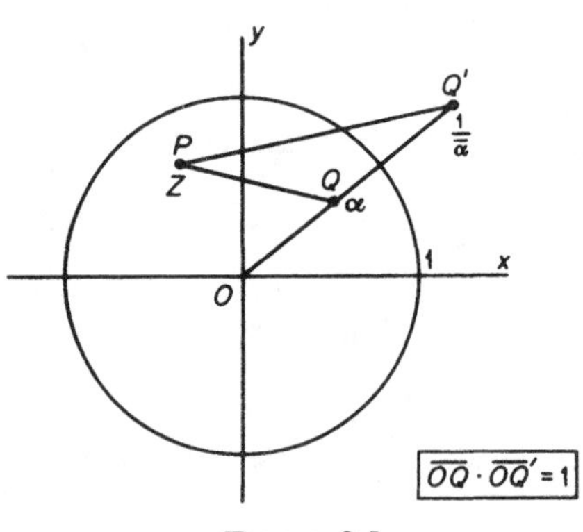

FIGURE 3.5

(b) Deduce that the Green's function of $\nabla^2 w$ for the interior of the unit circle, relevant to the requirement that w vanish along the boundary, is of the form

$$G(x, y; \xi, \eta) = -\frac{1}{2\pi}\log\left|\frac{z-\alpha}{1-\bar{\alpha}z}\right|,$$

where $\alpha = \xi + i\eta$, $\bar{\alpha} = \xi - i\eta$, and $z = x + iy$.

30. (a) With the introduction of the polar representations

$$z = \rho\, e^{i\theta}, \qquad \alpha = \beta\, e^{i\phi},$$

so that $x = \rho\cos\theta$, $y = \rho\sin\theta$, $\xi = \beta\cos\phi$, and $\eta = \beta\sin\phi$, show that the Green's function of Problem 29 takes the form

$$G(\rho, \theta; \beta, \phi) = \frac{1}{4\pi}\log\frac{1 - 2\beta\rho\cos(\theta-\phi) + \beta^2\rho^2}{\beta^2 - 2\beta\rho\cos(\theta-\phi) + \rho^2}.$$

(b) Deduce that the formal solution of the problem $\nabla^2 w + \Phi = 0$ inside the unit circle $\rho = 1$, where $w(1, \theta) = 0$, is of the form

$$w(\rho, \theta) = \int_0^1 d\beta \int_0^{2\pi} G(\rho, \theta; \beta, \phi)\Phi(\beta, \phi)\, d\phi,$$

and that the formal solution of the Dirichlet problem $\nabla^2 w = 0$ in the same region, when $w(1, \theta)$ is prescribed, is of the form

$$w(\rho, \theta) = -\int_0^{2\pi} \left[\frac{\partial G}{\partial \beta}\right]_{\beta=1} w(1, \phi)\, d\phi$$

$$= \frac{1}{2\pi}\int_0^{2\pi} \frac{1-\rho^2}{1 - 2\rho\cos(\theta - \phi) + \rho^2}\, w(1, \phi)\, d\phi.$$

[See Problem 27(b). The last result is the well-known *Poisson integral formula*, relevant to the unit circle.]

31. Suppose that an analytic function $F(z)$ maps the interior of a region $\mathscr{R}$ onto the interior of the unit circle, but does not necessarily map the point $z = \alpha = \xi + i\eta$ into the origin. Use the result of Problem 29 to show that the Green's function described in that problem is of the form

$$G(x, y; \xi, \eta) = -\frac{1}{2\pi}\log\left|\frac{F(z) - F(\alpha)}{1 - \overline{F(\alpha)}F(z)}\right|.$$

[Let the relation $t = F(z)$ map $\mathscr{R}$ onto the interior of the unit circle of a t plane, so that $z = \alpha$ maps into $t = F(\alpha)$, and map the t plane onto the w plane by the mapping of Problem 29(a).]

32. (a) Verify that the function

$$f(z) = e^{ia}\frac{z-\alpha}{z-\bar{\alpha}},$$

where a is a real constant, maps the upper half-plane ($y > 0$) onto the interior of the unit circle when α is in the upper half-plane, and deduce from Problem 28 that the Green's function for the upper half-plane is of the form

$$G(x, y; \xi, \eta) = -\frac{1}{2\pi}\log\left|\frac{z-\alpha}{z-\bar{\alpha}}\right| = -\frac{1}{2\pi}\log\left|\frac{(x-\xi) + i(y-\eta)}{(x-\xi)+i(y+\eta)}\right|$$

$$= \frac{1}{4\pi}\log\frac{(x-\xi)^2 + (y+\eta)^2}{(x-\xi)^2 + (y-\eta)^2}.$$

(b) Deduce that the formal solution of the problem $\nabla^2 w + \Phi = 0$ in the upper half-plane, where $w(x, 0) = 0$, is of the form

$$w(x, y) = \int_{-\infty}^{\infty} d\xi \int_0^{\infty} G(x, y; \xi, \eta)\Phi(\xi, \eta)\, d\eta,$$

and that the formal solution of the problem $\nabla^2 w = 0$ in the same region, where $w(x, 0) = \phi(x)$, is of the form

$$w(x, y) = -\int_{-\infty}^{\infty}\left[\frac{\partial G}{\partial(-\eta)}\right]_{\eta=0}\phi(\xi)\, d\xi$$

$$= \frac{y}{\pi}\int_{-\infty}^{\infty}\frac{\phi(\xi)\, d\xi}{(\xi - x)^2 + y^2} \qquad (y > 0).$$

[See Problem 27(b). Notice that the *outward* normal, relative to the upper half-plane, is in the *negative* η direction.]

33. Suppose that the analytic function $F(z)$ maps the interior of a region $\mathscr{R}$ onto the upper half-plane. Use the result of Problem 32(a) to show that the Green's function for $\nabla^2 w$ in the region $\mathscr{R}$, relevant to the requirement that w vanish on the boundary of $\mathscr{R}$, is of the form

$$G(x, y; \xi, \eta) = -\frac{1}{2\pi} \log \left| \frac{F(z) - F(\alpha)}{F(z) - \overline{F(\alpha)}} \right|,$$

where $z = x + iy$ and $\alpha = \xi + i\eta$. [When the boundary of $\mathscr{R}$ is *polygonal*, the function $F(z)$ can be determined by the *Schwarz-Christoffel* transformation.]

Section 3.5.

34. (a) In a certain linear system, the effect e at time t, due to a unit cause at a time τ, is a function only of the elapsed time $t - \tau$. If the system is inactive when $t < 0$, show that the cause-effect relation is a Volterra equation of the first kind, of the form

$$e(t) = \int_0^t K(t - \tau)c(\tau)\, d\tau.$$

[Equations of this form are considered in Section 3.13.]

(b) Show that the equation of part (a) can also be written in the form

$$e(t) = \int_0^t K(\tau)c(t - \tau)\, d\tau,$$

by replacing τ by $t - \tau$. [Notice that $K(\tau)$ can be considered as the effect at time τ due to a unit cause at time $t = 0$.]

(c) Noticing that the corresponding homogeneous Volterra equation of the second kind,

$$c(t) = \lambda \int_0^t K(t - \tau)c(\tau)\, d\tau,$$

expresses the requirement that the effect instantaneously reproduce the cause at all times, within a constant multiplicative factor, consider the possibility of existence of continuous nontrivial solutions of such an equation.

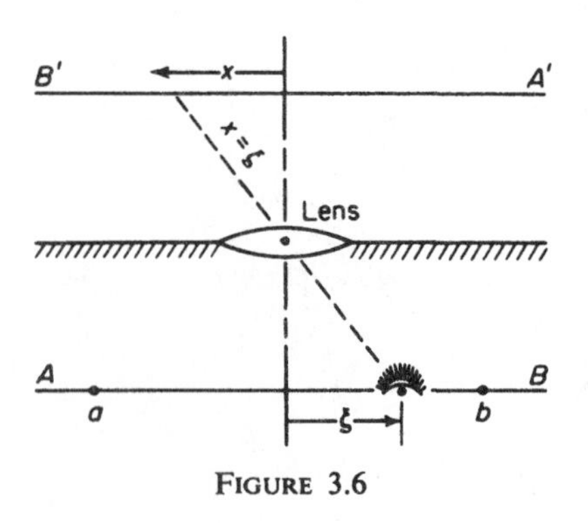

FIGURE 3.6

35. Figure 3.6 is a schematic representation of an optical system in which a distribution of illumination emanating from a one-dimensional object along the line AB passes through a refracting lens and is projected into a one-dimensional (reversed) image along the line $A'B'$. With the notation of that figure, the light intensity at a point x, due to a unit source at ξ, is found to be a certain symmetrical function of the difference $x - \xi$, with a maximum at the point $x = \xi$, for a given lens.

(a) Formulate the problem of determining the object illumination distribution $I_o(\xi)$ over an interval $a \leqq \xi \leqq b$, corresponding to a prescribed image distribution $I_i(x)$ over the interval $a \leqq x \leqq b$, as an integral equation.

(b) Formulate the problem of determining those object distributions over $a \leqq \xi \leqq b$ which are magnified (and reversed), but not distorted, when projected on the line $A'B'$.

36. (a) If heat radiating from a unit point source is constrained to flow in a plane (as in a thin plate with insulated faces), show that the temperature T at a distance r from that point is given by

$$T = -\frac{1}{2\pi K}\log r + \text{constant},$$

where K is the thermal conductivity of the medium, in the absence of other sources or boundaries.

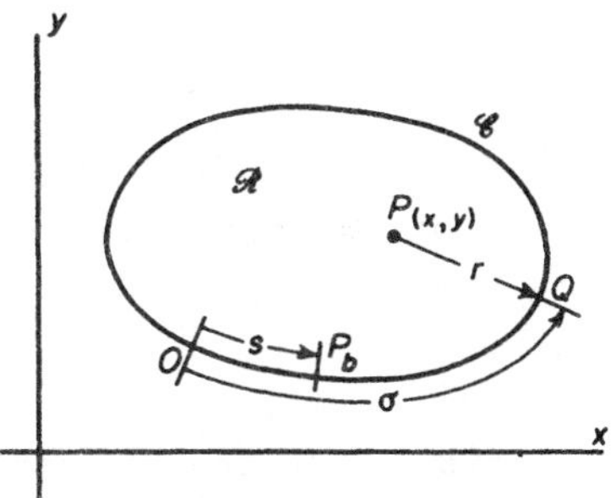

FIGURE 3.7

(b) Suppose that heat sources (and sinks) are continuously distributed along the closed boundary $\mathscr{C}$ of a region $\mathscr{R}$ in the xy plane. Suppose also that the *net* heat supplied per second along the entire boundary is *zero*, so that temperatures at interior points of $\mathscr{R}$ do not change with time. Show that the steady-state temperature $T(x, y)$ at an interior point $P(x, y)$ can be expressed in the form

$$T(x, y) = -\frac{1}{2\pi K}\oint_{\mathscr{C}} q(\sigma)\log r(x, y; \sigma)\, d\sigma + A,$$

where σ represents distance along $\mathscr{C}$, from a reference point O to a point Q, at which point heat is being generated at the rate $q(\sigma)$ calories per second per unit length of arc, where $r(x, y; \sigma)$ represents the distance from P to Q (Figure 3.7), and where A is an undetermined constant. (Notice that the presence of A is in accordance with the fact that a *uniform* temperature distribution in $\mathscr{R}$ can be maintained without supplying heat along the boundary $\mathscr{C}$.)

(c) Suppose that $q(\sigma)$ is not prescribed, but that the *temperature* of each boundary point P_b, at a distance s from O along $\mathscr{C}$, is prescribed as $f(s)$. Denoting by $r(s; \sigma)$ the length of the chord $\overline{P_bQ}$, show that $q(\sigma)$ must satisfy the integral equation

$$f(s) = -\frac{1}{2\pi K}\oint_{\mathscr{C}} q(\sigma)\log r(s; \sigma)\, d\sigma + A,$$

where the constant A is to be determined in such a way that $\oint_{\mathscr{C}} q\, d\sigma = 0$.

37. (a) By appropriately specializing the results of Problem 36, show that the steady-state temperature $T(\rho, \theta)$ at a point $x = \rho\cos\theta$, $y = \rho\sin\theta$, inside the boundary of a circular plate of radius a, with center at the origin (Figure 3.8), can be expressed in the form

$$T(\rho, \theta) = -\frac{1}{2\pi K}\int_0^{2\pi} q(\phi) \log \sqrt{\rho^2 - 2a\rho \cos(\theta - \phi) + a^2}\, d\phi + A,$$

where, if the temperature along the boundary is prescribed as $T(a, \theta) = f(\theta)$, the function $q(\phi)$ satisfies the integral equation

$$f(\theta) = -\frac{1}{4\pi K}\int_0^{2\pi} q(\phi) \log \sin^2 \frac{\theta - \phi}{2}\, d\phi + A,$$

in which A is to be determined in such a way that the condition

$$\int_0^{2\pi} q(\phi)\, d\phi = 0$$

is satisfied.

(b) By considering the result of integrating the equal members of the integral equation over $(0, 2\pi)$ with respect to θ [and using symmetry to show that the integral of the coefficient of $q(\phi)$ is independent of ϕ], show that A must then be taken as the mean value of the prescribed function $f(\theta)$ along the boundary,

$$A = \frac{1}{2\pi}\int_0^{2\pi} f(\theta)\, d\theta,$$

and verify further that A is also identified with the temperature $T(0, \theta)$ at the center of the circle. [In the case of a *circular* boundary, the Dirichlet problem can also be solved *directly* by use of the Poisson integral formula (see Problem 30). However, in the case of a general boundary, a formulation analogous to the preceding one (based on the results of Problem 36) is well adapted to numerical calculation.]

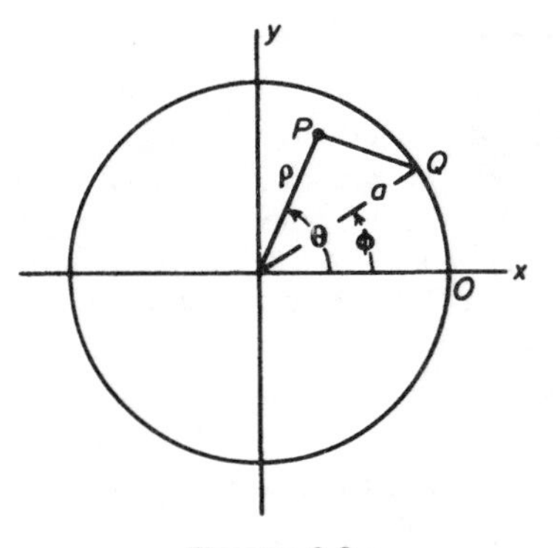

FIGURE 3.8

38. (a) If a unit heat source is present at a point Q, show that the rate of heat flow at a point P, per unit length of arc, across a curve $\mathscr{C}$ passing through P, is given by

$$-K\frac{\partial T}{\partial n}\bigg|_P = \frac{1}{2\pi}\frac{\partial}{\partial n}\log r\bigg|_P = \frac{1}{2\pi}\frac{\cos(n, r)}{r}\bigg|_P$$

where the differentiation is in the direction of the normal to $\mathscr{C}$ at the point P, where r denotes the distance from Q to P, and where (n, r) denotes the angle between QP and the normal.

(b) In the formulation of Problem 36, deduce that the net rate of heat flow *into* $\mathscr{R}$ across $\mathscr{C}$ at a boundary point P_b is given by

$$K\frac{\partial T}{\partial n}\bigg|_{P_b} = \frac{1}{2}q(s) - \frac{1}{2\pi}\oint_{\mathscr{C}} q(\sigma)\frac{\cos[n(\sigma), r(s; \sigma)]}{r(s; \sigma)}\, d\sigma,$$

where the angle involved in the integrand is that between the vector from the point Q to the point P_b and the *outward* normal at P_b. [The first term on the right corresponds to the fact that only one half of the heat generated by the source at P_b enters $\mathscr{R}$; the second term subtracts the net rate of flow outward at P_b from the remaining sources along $\mathscr{C}$. Notice that, in this formulation, when $K\,\partial T/\partial n$ is *prescribed* along $\mathscr{C}$, the source distribution q is hence obtained as the solution of an integral equation of the *second* kind. Notice also that the right-hand member of the preceding equation is *not* the result of differentiating the right-hand member of the equation of Problem 36 (b) under the integral sign, and evaluating the result at the boundary, but that the additional term $q(s)/2$ is present. *Inside* $\mathscr{C}$, however, the integrand does not become infinite, and differentiation under the integral sign can then be justified, in general.]

(c) In the case of a *circular* boundary (see Problem 37), verify that the integral involved in part (b) vanishes, so that $q(\phi)/K$ is identified with *twice* the boundary value of $\partial T/\partial n$ in this case, and the first equation of Problem 37 then gives the solution of the *Neumann* problem for the circle explicitly, with A an arbitrary constant. [Show that the coefficient of $q(\sigma)$ in the integral of part (b) is then given by the value of

$$\frac{\partial}{\partial \rho} \log \sqrt{\rho^2 - 2a\rho\cos(\theta - \phi) + a^2}$$

when $\rho = a$, and that this expression has the *constant* value $1/(2a)$.]

Sections 3.6, 3.7.

39. Express the solution of equation (105) in the form

$$y(x) = F(x) + \lambda \int_0^1 \Gamma(x, \xi; \lambda) F(\xi)\, d\xi$$

when $\lambda \neq \pm 2$.

40. (a) Show that the characteristic values of λ for the equation

$$y(x) = \lambda \int_0^{2\pi} \sin(x + \xi)\, y(\xi)\, d\xi$$

are $\lambda_1 = 1/\pi$ and $\lambda_2 = -1/\pi$, with corresponding characteristic functions of the form $y_1(x) = \sin x + \cos x$ and $y_2(x) = \sin x - \cos x$.

(b) Obtain the most general solution of the equation

$$y(x) = \lambda \int_0^{2\pi} \sin(x + \xi)\, y(\xi)\, d\xi + F(x)$$

when $F(x) = x$ and when $F(x) = 1$, under the assumption that $\lambda \neq \pm 1/\pi$.

(c) Prove that the equation

$$y(x) = \frac{1}{\pi} \int_0^{2\pi} \sin(x + \xi)\, y(\xi)\, d\xi + F(x)$$

possesses no solution when $F(x) = x$, but that it possesses infinitely many solutions when $F(x) = 1$. Determine all such solutions.

(d) Determine the most general form of the prescribed function $F(x)$, for which the integral equation

$$\int_0^{2\pi} \sin(x + \xi)\, y(\xi)\, d\xi = F(x),$$

of the first kind, possesses a solution.

41. Consider the equation

$$y(x) = F(x) + \lambda \int_0^{2\pi} \cos(x + \xi)\, y(\xi)\, d\xi.$$

(a) Determine the characteristic values of λ and the corresponding characteristic functions.

(b) Express the solution in the form

$$y(x) = F(x) + \lambda \int_0^{2\pi} \Gamma(x, \xi; \lambda) F(\xi)\, d\xi$$

when λ is not characteristic.

(c) Obtain the general solution (when it exists) if $F(x) = \sin x$, considering all possible cases.

42. Solve the equation

$$y(x) = 1 + \lambda \int_{-\pi}^{\pi} e^{i\omega(x-\xi)}\, y(\xi)\, d\xi,$$

considering separately all exceptional cases.

43. Obtain an approximate solution of the integral equation

$$y(x) = \int_0^1 \sin(x\xi)\, y(\xi)\, d\xi + x^2,$$

by replacing $\sin(x\xi)$ by the first two terms of its power-series development

$$\sin(x\xi) = (x\xi) - \frac{(x\xi)^3}{3!} + \cdots.$$

Section 3.8.

44. (a) Show that the kernel $K(x, \xi) = 1 + 3x\xi$ has a double characteristic number associated with $(-1, 1)$, with two independent characteristic functions.

(b) Show that the kernel $K(x, \xi) = 1 + \xi + 3x\xi$ has a double characteristic number associated with $(-1, 1)$, with only one characteristic function.

(c) Show that the kernel $K(x, \xi) = \sin x \cos \xi$ has *no* characteristic numbers associated with $(0, 2\pi)$.

45. Suppose that the kernel $K(x, \xi)$ of Section 3.8 is not necessarily symmetric, but is expressible in the form

$$K(x, \xi) = r(\xi) G(x, \xi),$$

where $r(\xi)$ is real and continuous in (a, b) and does not change sign in (a, b), and where $G(x, \xi)$ is real, continuous, and symmetric. By appropriately modifying the treatments of that section, establish the following results:

(a) Two characteristic functions $y_m(x)$ and $y_n(x)$, corresponding to distinct characteristic numbers λ_m and λ_n, are orthogonal over (a, b) with respect to the weighting function $r(x)$,

$$\int_a^b r(x)y_m(x)y_n(x)\,dx = 0.$$

(b) The characteristic numbers of $K(x, \xi)$ are all real.

(c) If equation (139) possesses a continuous solution, then that solution is given by (147) or (147′), where the weighting function $r(x)$ is to be inserted in the integrands involved in (136) and in (148) or (148′).

(d) If equation (149) possesses a continuous solution, then that solution is given by (160), where the weighting function $r(x)$ is to be inserted in the integrands involved in (136) and (157).

46. A complex kernel $K(x, \xi)$ [such as $e^{i(x-\xi)}$] which has the property

$$K(\xi, x) = \overline{K(x, \xi)},$$

where $\overline{K(x, \xi)}$ represents the complex conjugate of $K(x, \xi)$, is called a *Hermitian kernel*. Assuming that $K(x, \xi)$ is *continuous*, and making use of the known fact that the expansion theorem of page 253 applies also to Hermitian kernels, establish the following results by appropriately modifying the treatments of Section 3.8:

(a) The characteristic numbers associated with a Hermitian kernel over (a, b) are all real.

(b) Two characteristic functions $y_m(x)$ and $y_n(x)$, corresponding to distinct characteristic numbers λ_m and λ_n, are orthogonal over (a, b) in the Hermitian sense,

$$\int_a^b \overline{y_m(x)}y_n(x)\,dx = 0.$$

(c) Let the characteristic functions be normalized in the Hermitian sense:

$$\int_a^b \overline{\phi_n(x)}\phi_n(x)\,dx = 1.$$

Then, if equation (139) possesses a continuous solution, that solution is given by (147), where (148) is replaced by the definition

$$f_n = \int_a^b \overline{\phi_n(x)}F(x)\,dx,$$

or by (147′), where (148′) is replaced by the definition

$$F_n \int_a^b \overline{y_n(x)}y_n(x)\,dx = \int_a^b \overline{y_n(x)}F(x)\,dx.$$

(d) If (149) possesses a continuous solution, then that solution is given by (160), where f_n is defined in part (c).

47. If $u(x)$ is a characteristic function of a complex kernel $K(x, \xi)$ (which need not be Hermitian), corresponding to a characteristic number λ over (a, b), and $v(x)$ is a characteristic function of the transposed complex conjugate kernel $\overline{K(\xi, x)}$, corresponding to a characteristic number μ, show that

$$\int_a^b u(x)\overline{v(x)}\, dx = 0,$$

if $\lambda \neq \bar{\mu}$. [The kernel $\overline{K(\xi, x)}$ is sometimes called the *adjoint* of $K(x, \xi)$. Notice that real symmetric kernels and Hermitian kernels are *self-adjoint*.]

48. When λ is *nearly* equal to a characteristic number λ_m, to which there corresponds only one characteristic function, show that the solution of the equation

$$y(x) = F(x) + \lambda \int_a^b K(x, \xi) y(\xi)\, d\xi,$$

where $K(x, \xi)$ is real and symmetric, is given approximately by

$$y(x) \approx F(x) + \frac{\lambda}{\lambda_m - \lambda} \left(\int_a^b F\phi_m\, dx \right) \phi_m(x),$$

where $\phi_m(x)$ is the corresponding normalized characteristic function.

49. Assume that a real symmetric kernel $K(x, \xi)$ can itself be expanded in a series of its orthogonalized and normalized characteristic functions,

$$K(x, \xi) = \sum_{n=1}^{\infty} a_n(\xi)\phi_n(x) \qquad (a \leqq x \leqq b, a \leqq \xi \leqq b),$$

where K is considered as a function of x for fixed values of ξ.

(a) Assuming also that term-by-term integration is permissible, show that the coefficient function $a_n(\xi)$ must be of the form

$$a_n(\xi) = \frac{1}{\lambda_n} \phi_n(\xi),$$

where λ_n is the nth characteristic number, and hence obtain the so-called *bilinear expansion* of a real symmetric kernel:

$$K(x, \xi) = \sum_{n=1}^{\infty} \frac{\phi_n(x)\phi_n(\xi)}{\lambda_n} \qquad (a \leqq x \leqq b, a \leqq \xi \leqq b).$$

[*Mercer's theorem* states that this series converges (absolutely and uniformly) to $K(x, \xi)$ for all x and ξ in (a, b) if $K(x, \xi)$ is continuous, real, and symmetric, and if all except a finite number of its characteristic numbers are of the same sign. When K is continuous and *Hermitian* (see Problem 46), the same conclusion follows if $\phi_n(\xi)$ is replaced by $\overline{\phi_n(\xi)}$ in the expansion. In most physically motivated problems, it is known in advance that all characteristic numbers are positive, so that the expansion is then valid. The series defines such a kernel completely, in terms of its characteristic functions and numbers, and is of basic importance in many considerations. See also Problem 83.]

(b) Deduce that a real and continuous symmetric kernel which possesses only a finite number of linearly independent characteristic functions must be a *separable* kernel (in the sense of Section 3.6).

(c) Verify the result of part (a) in the case of the kernel involved in equation (103).

Section 3.9.

50. Consider the integral equation

$$y(x) = \lambda \int_0^1 x\xi y(\xi)\, d\xi + 1.$$

(a) Make use of equation (182) to show in advance that the iterative procedure of Section 3.9 will converge when $|\lambda| < 3$.

(b) Show that the iterative procedure leads formally to the expression

$$y(x) = 1 + x\left(\frac{\lambda}{2} + \frac{\lambda^2}{6} + \frac{\lambda^3}{18} + \cdots\right).$$

(c) Use the method of Section 3.6 to obtain the exact solution of the problem in the form

$$y(x) = 1 + \frac{3\lambda x}{2(3 - \lambda)} \qquad (\lambda \neq 3).$$

[Notice that the leading terms in the series expansion in powers of λ are given correctly by part (b), and that the estimate of part (a) happens to provide the actual convergence limit on λ.]

51. Deal with the integral equation

$$y(x) = \lambda \int_0^1 (x + \xi) y(\xi)\, d\xi + 1$$

as in Problem 50. Show also that the estimate afforded by (182) is slightly conservative in this case.

52. Deal with the integral equation

$$y(x) = \sin x + \lambda \int_0^{2\pi} \cos(x + \xi)\, y(\xi)\, d\xi$$

as in Problem 50. [Compare with the result of Problem 41(c).]

53. Show that, for sufficiently small values of $|\epsilon|$, an approximate solution of the equation

$$y(x) = \epsilon \int_0^a e^{-|x-\xi|}\, y(\xi)\, d\xi + 1$$

is afforded by the expression

$$y(x) \approx 1 + \epsilon[2 - e^{-x} - e^{-(a-x)}].$$

54. (a) Apply the method of Section 3.9 to the equation

$$y(x) = \int_0^x (x + \xi) y(\xi)\, d\xi + 1,$$

taking $y^{(0)}(x) = 1$, and obtaining the results of three successive substitutions.

(b) Show that the problem considered is equivalent to that specified by the differential equation $y'' - 2xy' - 3y = 0$ and the initial conditions $y(0) = 1$, $y'(0) = 0$.

55. (a) If $K(x, \xi)$ is real and symmetric, $\phi_n(x)$ is the normalized characteristic function corresponding to λ_n, and

$$(f, g) \equiv \int_a^b f(x) g(x)\, dx,$$

show that the expansion theorem of Section 3.8 implies the validity of the expansion

$$\mathscr{K} u(x) = \sum_{n=1}^{\infty} \frac{(u, \phi_n)}{\lambda_n} \phi_n(x) \qquad (a \leqq x \leqq b),$$

with the notation of (168), when $u(x)$ is continuous in (a, b).

(b) From the result of part (a), deduce *Hilbert's formula*:

$$(\mathscr{K} u, v) = \sum_{n=1}^{\infty} \frac{(u, \phi_n)(v, \phi_n)}{\lambda_n}.$$

Section 3.10.

56. Determine the resolvent kernel associated with $K(x, \xi) = x\xi$ in the interval $(0, 1)$ in the form of a power series in λ, obtaining the first three terms.

57. Proceed as in Problem 56, for the kernel $K(x, \xi) = x + \xi$.

58. Determine the resolvent kernel associated with $K(x, \xi) = \cos(x + \xi)$ in $(0, 2\pi)$, in the form of a power series in λ. [Compare with the result of Problem 41(b).]

59. Determine the iterated kernel $K_2(x, \xi)$ associated with $K(x, \xi) = |x - \xi|$ in $(0, 1)$.

60. Determine the coefficient of λ in the expansion of the resolvent kernel associated with $K(x, \xi) = e^{-|x-\xi|}$ in the interval $(0, a)$.

61. Show that the equation (147) can be written in the form

$$y(x) = F(x) + \lambda \int_a^b \left[\sum_{n=1}^{\infty} \frac{\phi_n(x)\phi_n(\xi)}{\lambda_n - \lambda} \right] F(\xi)\, d\xi$$

and deduce the validity of the expansion

$$\Gamma(x, \xi, \lambda) = \sum_{n=1}^{\infty} \frac{\phi_n(x)\phi_n(\xi)}{\lambda_n - \lambda} \qquad (a \leqq x \leqq b, a \leqq \xi \leqq b)$$

when $K(x, \xi)$ is continuous, real, and symmetric and λ is not characteristic.

62. Suppose that the resolvent kernel in (208) is assumed, as the *ratio* of two power series in λ, in the form

$$\Gamma(x, \xi; \lambda) = \frac{\sum_{n=0}^{\infty} D_n(x, \xi)\lambda^n}{\sum_{n=0}^{\infty} C_n \lambda^n}.$$

(a) By replacing $y(x)$ by $F(x) + \lambda \int_a^b \Gamma(x, \xi; \lambda)F(\xi)\, d\xi$ in the basic equation (205), and clearing fractions, deduce formally that the relation

$$\sum_{n=0}^{\infty} \lambda^n \int_a^b D_n(x, \xi)F(\xi)\, d\xi - \left(\sum_{n=0}^{\infty} C_n \lambda^n \right) \int_a^b K(x, \xi)F(\xi)\, d\xi$$
$$- \sum_{n=0}^{\infty} \lambda^{n+1} \int_a^b \int_a^b K(x, \xi_1)D_n(\xi_1,\xi)F(\xi)\, d\xi_1\, d\xi = 0$$

must be satisfied identically in λ and F. [Assume that the order of integration and summation can be interchanged.]

(b) Show that this requirement implies the relation

$$D_n(x, \xi) = C_n K(x, \xi) + \int_a^b K(x, \xi_1)D_{n-1}(\xi_1, \xi)\, d\xi_1 \qquad (n = 1, 2, \cdots),$$

together with the initial condition

$$D_0(x, \xi) = C_0 K(x, \xi).$$

(c) Verify that the preceding determination leads to the resolvent kernel defined by equation (207) if we take $C_0 = 1$ and $C_n = 0$ when $n \geqq 1$. [Whereas this series converges only for $|\lambda| < |\lambda_1|$, the constants C_n can be chosen that the numerator and denominator series both converge *for all values of* λ. This result, which is discussed in Section 3.11, is of basic importance in the theory of linear integral equations.]

Section 3.11.

63. Obtain the resolvent kernel associated with $K(x, \xi) = x\xi$ in (0, 1), by the method of Section 3.11.

64. Proceed as in Problem 63 for the kernel $K(x, \xi) = x + \xi$ in (0, 1).

65. (a) Obtain the solution of the equation

$$y(x) = \lambda \int_a^b K(x, \xi)y(\xi)\, d\xi + F(x),$$

where

$$K(x, \xi) = u(x)v(\xi),$$

in the form

$$y(x) = F(x) + \lambda \frac{\displaystyle\int_a^b K(x, \xi)F(\xi)\, d\xi}{1 - \lambda \displaystyle\int_a^b K(x, x)\, dx},$$

by the methods of Section 3.6. [Write $y(x) = \lambda c u(x) + F(x)$, where

$$c = \int_a^b v(\xi)y(\xi)\, d\xi,$$

determine c, and identify the result with the form given.]

(b) Obtain the same result by the methods of Section 3.11.

[Notice that when $\int_a^b K(x, x)\, dx = 0$ there is *no* characteristic number, whereas otherwise the single characteristic number is $\lambda_1 = 1 \Big/ \int_a^b K(x, x)\, dx$.]

66. Show that $\Delta(\lambda)$, in equation (211), is not changed when $K(x, \xi)$ is replaced by $K(\xi, x)$ and deduce that *$K(x, \xi)$ and $K(\xi, x)$ have the same characteristic numbers.*

67. If $K(x, \xi)$ is not real, show that the condition (220) may be interpreted *either* as requiring that F be orthogonal to characteristic functions of the transposed kernel $K(\xi, x)$ *or*, equivalently, as requiring that F be orthogonal *in the Hermitian sense* (see Problem 46) to characteristic functions of the *conjugate* transposed kernel $\overline{K(\xi, x)}$.

Section 3.12.

68. If $F(x)$ is the Fourier sine transform of $f(x)$, show that

$$y_1(x) = \sqrt{\pi/2}\, f(x) + F(x)$$

is a characteristic function of (227) corresponding to $\lambda = \sqrt{2/\pi}$ and that

$$y_2(x) = \sqrt{\pi/2}\, f(x) - F(x)$$

is a characteristic function corresponding to $\lambda = -\sqrt{2/\pi}$. [Thus the functions defined by (231) and (232) are by no means unique!]

69. Show formally that the Green's function $G(x, \xi)$ associated with the expression $\dfrac{d^2y}{dx^2} - y$ over the *infinite* interval $(-\infty, \infty)$, subject to the requirement that y be *bounded* as $x \to \pm\infty$, is of the form

$$G(x, \xi) = \tfrac{1}{2}e^{-|x-\xi|}.$$

70. (a) If $I(x) = \int_{-\infty}^{\infty} e^{-|x-\xi|}\,\Phi(\xi)\,d\xi$, verify that $I''(x) = I(x) - 2\Phi(x)$ for any continuous function $\Phi(x)$ which is dominated by $e^{|x|}$ as $x \to \pm\infty$.

(b) Use this result to show that any continuous solution of the integral equation

$$y(x) = \lambda \int_{-\infty}^{\infty} e^{-|x-\xi|}\,y(\xi)\,d\xi + F(x)$$

must also satisfy the differential equation

$$y''(x) - (1 - 2\lambda)y(x) = F''(x) - F(x).$$

71. Suppose that $F(x) \equiv 0$ in Problem 70.

(a) In the case when $1 - 2\lambda$ is positive, and hence we may write $1 - 2\lambda = \alpha^2$ or $\lambda = (1 - \alpha^2)/2$, show that the general solution,

$$y(x) = c_1 e^{\alpha x} + c_2 e^{-\alpha x},$$

of the homogeneous differential equation, also satisfies the homogeneous integral equation when and only when $|\alpha| < 1$. Deduce that any *positive* value of λ of the form $\lambda = (1 - \alpha^2)/2$ (and hence any λ between 0 and $\frac{1}{2}$) is a characteristic number of multiplicity two, with the characteristic functions $e^{\pm \alpha x}$.

(b) In the case when $\lambda = \frac{1}{2}$, verify that the solution $y(x) = c_1 + c_2 x$ satisfies the integral equation, so that $\lambda = \frac{1}{2}$ is also a double characteristic number, with characteristic functions 1 and x.

(c) In the case when $1 - 2\lambda$ is negative, and hence we may then write $\lambda = (1 + \beta^2)/2$, show similarly that any such value of λ (and hence any $\lambda > \frac{1}{2}$) is a double characteristic number, corresponding to $y = \cos \beta x$ and $y = \sin \beta x$. Thus deduce that all positive values of λ, and only those values, are characteristic numbers.

(d) If *only functions which are bounded as* $x \to \pm\infty$ *are accepted as characteristic functions*, deduce that only those values of λ for which $\lambda \geqq \frac{1}{2}$ are then characteristic numbers, that the number $\lambda = \frac{1}{2}$ is then of multiplicity one, and that all values of λ greater than $\frac{1}{2}$ are of multiplicity two.

72. Suppose that $F(x) = \sin \mu x$ in Problem 70.

(a) Show that the differential equation admits a *bounded* solution if and only if $\lambda \neq (1 + \mu^2)/2$, where $\mu > 0$, and that this solution must be of the *unique* form

$$y(x) = \frac{1 + \mu^2}{1 + \mu^2 - 2\lambda} \sin \mu x$$

when $\lambda < \frac{1}{2}$. Verify that this expression also satisfies the integral equation for all values of λ such that $\lambda \neq (1 + \mu^2)/2$.

(b) Deduce that the integral equation

$$y(x) = \lambda \int_{-\infty}^{\infty} e^{-|x-\xi|}\,y(\xi)\,d\xi + \sin \mu x$$

possesses a unique bounded continuous solution

$$y(x) = \frac{1 + \mu^2}{1 + \mu^2 - 2\lambda} \sin \mu x$$

when $\lambda < \frac{1}{2}$, a single infinity of such solutions,

$$y(x) = \frac{1 + \mu^2}{\mu^2} \sin \mu x + c_1,$$

when $\lambda = \frac{1}{2}$, and a double infinity of such solutions,

$$y(x) = \frac{1 + \mu^2}{1 + \mu^2 - 2\lambda} \sin \mu x + c_1 \cos (\sqrt{2\lambda - 1}\, x) + c_2 \sin (\sqrt{2\lambda - 1}\, x),$$

when $\lambda > \frac{1}{2}$ but $\lambda \neq (1 + \mu^2)/2$, and that it possesses *no* continuous solution (bounded or otherwise) when $\lambda = (1 + \mu^2)/2$. [Notice that, whereas all values of $\lambda \geqq \frac{1}{2}$ are characteristic numbers, so that the *homogeneous* equation is then solvable, the *nonhomogeneous* equation here is *also* solvable unless λ coincides with a *particular* characteristic number which depends upon the right-hand member, rather than on the kernel itself. This is in contrast with the situation relevant to nonsingular equations, where (for a symmetric kernel) solvability of the *homogeneous* equation precludes solvability of the *nonhomogeneous* equation unless $F(x)$ is orthogonal to the corresponding homogeneous solutions.]

73. (a) Show that the equation

$$y(x) = \lambda \int_{-\infty}^{\infty} K(|x - \xi|) y(\xi)\, d\xi,$$

in which the kernel is a function only of $|x - \xi|$, can be written in the form

$$y(x) = \lambda \int_{0}^{\infty} K(u)[y(x + u) + y(x - u)]\, du.$$

(b) Show that this equation is satisfied by

$$y(x) = c_1 \cos \beta x + c_2 \sin \beta x.$$

if λ is of the form

$$\lambda = \frac{1}{2 \displaystyle\int_0^{\infty} K(u) \cos \beta u\, du}.$$

Deduce that all values of β for which the integral involved exists give rise to characteristic values of λ, for which solutions exist which are bounded as $x \to \pm\infty$.

(c) In the special case $K(|x - \xi|) = e^{-|x-\xi|}$, considered in Problems 70 to 72, show that the preceding result is in accordance with the result of Problem 71(d).

Section 3.13.

74. Obtain the solution of the generalized Abel integral equation

$$F(x) = \int_0^x \frac{y(\xi)}{(x - \xi)^\alpha}\, d\xi \qquad (0 < \alpha < 1)$$

in the form
$$y(x) = \frac{\sin \alpha\pi}{\pi} \frac{d}{dx} \int_0^x \frac{F(\xi)\, d\xi}{(x - \xi)^{1-\alpha}},$$

by dividing both sides of the given equation by $(s - x)^{1-\alpha}$ and proceeding as in Section 3.13. [Notice that

$$\int_0^1 t^{-\alpha}(1 - t)^{\alpha-1}\, dt = \Gamma(1 - \alpha)\Gamma(\alpha) = \frac{\pi}{\sin \alpha\pi}$$

when $0 < \alpha < 1$.]

75. Let the Laplace transform of a function $f(x)$ be considered as a function of a new independent variable s, so that

$$\mathscr{L}f(x) \equiv F(s) = \int_0^\infty e^{-sx} f(x)\, dx.$$

(a) By making the substitution $sx = u$ in the integral defining the transform, show that

$$\mathscr{L}x^{-\alpha} = \Gamma(1 - \alpha)s^{\alpha-1}, \qquad s^{-\alpha} = \frac{1}{\Gamma(\alpha)} \mathscr{L}x^{\alpha-1},$$

when $0 < \alpha < 1$.

(b) Establish the property

$$\frac{1}{s} \mathscr{L}f(x) = \mathscr{L} \int_0^x f(\xi)\, d\xi.$$

76. By making use of the results of Problem 75, and of the convolution property

$$\mathscr{L} \int_0^x \phi_1(x - \xi)\phi_2(\xi)\, d\xi = \mathscr{L}\phi_1(x)\, \mathscr{L}\phi_2(x),$$

verify that the solution of Abel's integral equation,

$$F(x) = \int_0^x (x - \xi)^{-1/2} y(\xi)\, d\xi,$$

can be obtained by the following steps:

$$\mathscr{L}F(x) = \mathscr{L}x^{-1/2}\, \mathscr{L}y(x); \qquad \mathscr{L}y(x) = \frac{1}{\Gamma(\frac{1}{2})} s^{1/2}\, \mathscr{L}F(x);$$

$$\frac{1}{s} \mathscr{L}y(x) = \frac{1}{[\Gamma(\frac{1}{2})]^2} \mathscr{L}x^{-1/2}\, \mathscr{L}F(x);$$

$$\int_0^x y(\xi)\, d\xi = \frac{1}{\pi} \int_0^x (x - \xi)^{-1/2} F(\xi)\, d\xi;$$

$$y(x) = \frac{1}{\pi} \frac{d}{dx} \int_0^x \frac{F(\xi)}{\sqrt{x - \xi}}\, d\xi.$$

77. Obtain the solution of Problem 74 by the method of Problem 76.

78. (a) If y satisfies the equation

$$F(x) = \int_0^x H(x - \xi)y(\xi)\, d\xi,$$

show that y also satisfies the equation

$$H(0)y(x) = F'(x) - \int_0^x H'(x - \xi)y(\xi)\, d\xi,$$

and that $Y(x) \equiv \int_0^x y(\xi)\, d\xi$ satisfies the equation

$$H(0)\,Y(x) = F(x) - \int_0^x H'(x - \xi)\,Y(\xi)\, d\xi.$$

(b) By formally integrating the equal members of the first equation of part (a) over $(0, \infty)$ with respect to x, and formally interchanging the order of integration in the resultant right-hand member, obtain the relation

$$\int_0^\infty F(x)\, dx = \int_0^\infty H(u)\, du \int_0^\infty y(\xi)\, d\xi.$$

79. If $H(0) = 0$ in Problem 78, show that $y(x)$ also satisfies the equation

$$H'(0)y(x) = F''(x) - \int_0^x H''(x - \xi)y(\xi)\, d\xi$$

and that $Z(x) \equiv \int_0^x (x - \xi)y(\xi)\, d\xi$ satisfies the equation

$$H'(0)Z(x) = F(x) - \int_0^x H''(x - \xi)Z(\xi)\, d\xi.$$

80. Suppose that $y(x)$ satisfies the equation

$$y(x) = F(x) + \lambda \int_a^b K(x, \xi)y(\xi)\, d\xi,$$

where $K(x, \xi)$ is real and continuous, but not necessarily symmetric.

(a) By multiplying the equal members of this equation by $K(x, t)$, and integrating the results over (a, b) with respect to x, deduce that there follows also

$$0 = \int_a^b K(\xi, x)y(\xi)\, d\xi - \int_a^b K(t, x)F(t)\, dt - \lambda \int_a^b \int_a^b K(t, x)K(t, \xi)y(\xi)\, dt\, d\xi,$$

after an appropriate relettering of variables.

(b) By multiplying the equal members of this equation by λ, and adding the results to the corresponding equal members of the original equation, show that any solution of that equation also satisfies the equation

$$y(x) = f(x) + \lambda \int_a^b K_s(x, \xi)y(\xi)\, d\xi,$$

where

$$f(x) = F(x) - \lambda \int_a^b K(t, x)F(t)\, dt,$$

and where $K_s(x, \xi)$ is the *symmetric* kernel

$$K_s(x, \xi) = K(x, \xi) + K(\xi, x) - \lambda \int_a^b K(t, x)K(t, \xi)\, dt.$$

81. Suppose that $y(x)$ satisfies the equation

$$F(x) = \int_a^b K(x, \xi)y(\xi)\, d\xi,$$

where $K(x, \xi)$ is real, but not necessarily symmetric. By multiplying the equal members by $K(x, t)$, and integrating the results over (a, b) with respect to x, show that $y(x)$ also satisfies the equation

$$\tilde{F}(x) = \int_a^b \tilde{K}(x, \xi)y(\xi)\, d\xi,$$

where

$$\tilde{F}(x) = \int_a^b K(\xi, x)F(\xi)\, d\xi$$

and where $\tilde{K}(x, \xi)$ is the *symmetric* kernel

$$\tilde{K}(x, \xi) = \int_a^b K(t, x)K(t, \xi)\, dt.$$

[Notice that the methods of Section 3.8 are applicable to the new equation.]

82. Let γ_n and $u_n(x)$ denote corresponding characteristic quantities for the real symmetric kernel

$$K_{\mathrm{I}}(x, \xi) = \int_a^b K(x, t)K(\xi, t)\, dt$$

over (a, b), so that

$$u_n(x) = \gamma_n \int_a^b K_{\mathrm{I}}(x, \xi)u_n(\xi)\, d\xi.$$

(a) By replacing K_{I} by its definition, and interchanging the order of integration in the resultant double integral, show that there follows

$$u_n(x) = \gamma_n \int_a^b K(x, t)V_n(t)\, dt,$$

where

$$V_n(x) = \int_a^b K(\xi, x)u_n(\xi)\, d\xi.$$

(b) By multiplying both members of the integral equation by $u_n(x)$, and integrating the results over (a, b), show that

$$\int_a^b [u_n(x)]^2\, dx = \gamma_n \int_a^b [V_n(x)]^2\, dx,$$

and hence deduce that *the characteristic numbers γ_n of $K_{\mathrm{I}}(x, \xi)$ are positive.*

(c) By writing $v_n(x) = \sqrt{\gamma_n}\, V_n(x)$, show that the equations of part (a) can be expressed in the symmetrical form

$$u_n(x) = \sqrt{\gamma_n} \int_a^b K(x, \xi) v_n(\xi)\, d\xi,$$

$$v_n(x) = \sqrt{\gamma_n} \int_a^b K(\xi, x) u_n(\xi)\, d\xi,$$

and that there then follows also

$$\int_a^b u_n^2\, dx = \int_a^b v_n^2\, dx.$$

(d) Show that the result of introducing the first relation of part (c) into the right-hand member of the second one is of the form

$$v_n(x) = \gamma_n \int_a^b K_{II}(x, \xi) v_n(\xi)\, d\xi,$$

where
$$K_{II}(x, \xi) = \int_a^b K(t, x) K(t, \xi)\, dt.$$

Hence deduce that *the auxiliary kernels K_I and K_{II}, associated with a real nonsymmetric kernel $K(x, \xi)$, possess the same characteristic numbers*, that *these numbers are all positive*, and that *the respective characteristic functions $u_n(x)$ and $v_n(x)$ corresponding to γ_n satisfy the simultaneous equations of part* (c). [Notice that the two sets of characteristic functions thus associated with a real nonsymmetric kernel are *orthogonal* sets, and that, by putting the equations of part (c) in a symmetrical form, we have ensured the fact that one of the sets is *normalized* when the other set has this property. Notice that the characteristic numbers γ_n are *not*, in general, characteristic numbers of the kernel $K(x, \xi)$ itself.]

83. With the notation of Problem 82, show formally that the assumption that $K(x, \xi)$ can be expanded in the form

$$K(x, \xi) = \sum_{n=1}^{\infty} u_n(x) a_n(\xi)$$

leads to the requirement that $a_n(\xi) = v_n(\xi)/\sqrt{\gamma_n}$ when the functions $u_n(x)$ and $v_n(x)$ are normalized,

$$\int_a^b u_n^2\, dx = \int_a^b v_n^2\, dx = 1,$$

so that there must follow

$$K(x, \xi) = \sum_{n=1}^{\infty} \frac{u_n(x) v_n(\xi)}{\sqrt{\gamma_n}} \qquad (a \leq x \leq b,\ a \leq \xi \leq b).$$

[This result is known to be valid for *any* continuous real kernel $K(x, \xi)$, the series converging *uniformly*, and it plays a role for such general kernels which is somewhat analogous to that played by the expansion of Problem 49 for real symmetric (or Hermitian) kernels (see Reference 8).]

84. The *Cauchy principal value* of an integral, $\oint_a^b f(x)\,dx$, in which $f(x)$ becomes infinite at a point $x = c$ inside the range of integration, is defined as the limit

$$\oint_a^b f(x)\,dx = \lim_{\epsilon\to 0}\left[\int_a^{c-\epsilon} f(x)\,dx + \int_{c+\epsilon}^b f(x)\,dx\right],$$

when that limit exists. When the *separate* limits on the right exist, the integral is convergent in the strict sense and the symbol $\oint$ may be replaced by the usual symbol $\int$.

(a) Verify that

$$\oint_{-1}^{2} \frac{dx}{x} = \log 2,$$

but that $\int_{-1}^{2} \frac{dx}{x}$ does not exist. [Recall that $\int \frac{dx}{x} = \log|x| + C$.]

(b) Verify that neither $\int_{-1}^{1} \frac{dx}{x^2}$ nor $\oint_{-1}^{1} \frac{dx}{x^2}$ exists.

85. The Hilbert integral representation of a suitably regular function $g(\theta)$, over the interval $0 < \theta < \pi$, is of the form

$$g(\theta) = \frac{1}{\pi^2}\oint_0^\pi \oint_0^\pi K(\theta, \phi_1)K(\phi, \phi_1)g(\phi)\,d\phi_1\,d\phi + \frac{1}{\pi}\int_0^\pi g(\phi)\,d\phi$$

where
$$K(\theta, \phi) = \frac{\sin\phi}{\cos\phi - \cos\theta}.$$

[The Cauchy principal values are necessary because of the strong singularity of $K(\theta, \phi)$ when $\phi = \theta$.] Show that, if we write

$$f(\phi_1) = \frac{1}{\pi}\oint_0^\pi K(\phi, \phi_1)g(\phi)\,d\phi,$$

there follows

$$g(\theta) = \frac{1}{\pi}\oint_0^\pi K(\theta, \phi_1)f(\phi_1)\,d\phi_1 + \frac{1}{\pi}\int_0^\pi g(\phi)\,d\phi,$$

and hence deduce the validity of the simultaneous equations

$$f(\theta) = \frac{1}{\pi}\oint_0^\pi g(\phi)\frac{\sin\theta}{\cos\theta - \cos\phi}\,d\phi,$$

$$g(\theta) = \frac{1}{\pi}\oint_0^\pi f(\phi)\frac{\sin\phi}{\cos\phi - \cos\theta}\,d\phi + \frac{1}{\pi}\int_0^\pi g(\phi)\,d\phi.$$

[The function f is often called the *Hilbert transform* of g over $(0, \pi)$.]

86. (a) With the terminology of Problem 85, let a function $g(\theta)$ be defined by the Fourier cosine series

$$g(\theta) = a_0 + \sum_{n=1}^{\infty} a_n \cos n\theta \qquad (0 < \theta < \pi).$$

By making use of the known relation

$$\oint_0^\pi \frac{\cos n\phi}{\cos\theta - \cos\phi}\, d\phi = -\pi \frac{\sin n\theta}{\sin\theta} \qquad (n = 0, 1, 2, \cdots),$$

and assuming the validity of a certain interchange of order of summation and integration, deduce that the Hilbert transform of $g(\theta)$ is of the form

$$f(\theta) = -\sum_{n=1}^{\infty} a_n \sin n\theta \qquad (0 < \theta < \pi).$$

[In particular, notice that the Hilbert transform of a constant is zero.]

(b) If $f(\theta)$ is defined by the preceding equation, verify that $g(\theta)$ is given correctly by the last equation of Problem 85. [Express the product of two sines as the difference between two cosines.]

87. (a) By making the change in variables

$$\cos\theta = x, \qquad \cos\phi = \xi,$$

and writing

$$\frac{f(\cos^{-1} x)}{\sqrt{1 - x^2}} = F(x), \qquad \frac{g(\cos^{-1} x)}{\sqrt{1 - x^2}} = y(x),$$

in the relations of Problem 85, show that the solution of the singular integral equation

$$F(x) = \frac{1}{\pi} \oint_{-1}^{1} y(\xi) \frac{d\xi}{\xi - x} \qquad (-1 < x < 1)$$

is defined by the relation

$$\sqrt{1 - x^2}\, y(x) = \frac{1}{\pi} \oint_{-1}^{1} \sqrt{1 - \xi^2}\, F(\xi) \frac{d\xi}{x - \xi} + \frac{1}{\pi} \int_{-1}^{1} y(\xi)\, d\xi.$$

[Notice that the solution is not unique, since the value of the integral $\int_{-1}^{1} y(\xi)\, d\xi$ may be prescribed.]

(b) If it is required that $y(-1)$ be finite, show that there must follow

$$\int_{-1}^{1} y(\xi)\, d\xi = \int_{-1}^{1} \sqrt{1 - \xi^2}\, F(\xi) \frac{d\xi}{1 + \xi},$$

and verify that the solution of part (a) can then be written in the form

$$y(x) = \frac{1}{\pi} \sqrt{\frac{1 + x}{1 - x}} \oint_{-1}^{1} \sqrt{\frac{1 - \xi}{1 + \xi}}\, F(\xi) \frac{d\xi}{x - \xi}.$$

[These results are of importance, for example, in the aerodynamic theory of airfoils.]

Section 3.14.

88. Obtain two successive approximations to the smallest characteristic number and the corresponding characteristic function of the problem

$$y(x) = \lambda \int_0^1 K(x, \xi) y(\xi)\, d\xi \quad \textit{where} \quad K(x, \xi) = \begin{cases} x, & x < \xi, \\ \xi, & x > \xi, \end{cases}$$

starting with the initial approximation $y^{(1)} = 1$. [Show that $f^{(1)} = \frac{1}{2}(2x - x^2)$, and that equations (261a,b,c) give the estimates $\lambda_1^{(1)} = 3$, 3, and 2.5, respectively. With $y^{(2)} = 2x - x^2$, show that there follows $f^{(2)} = \frac{1}{12}(8x - 4x^3 + x^4)$, and that (261a,b) give the respective estimates $\lambda_1^{(2)} = 2.5$ and 2.471. (Equation (261c) gives the estimate 2.4677.) By plotting the functions $y^{(2)}(x)$ and $\lambda_1^{(2)}f^{(2)}(x)$, show also that the difference between the input and output in the second cycle is less than about 3 per cent of the maximum value.]

89. Show that the integral equation of Problem 88 is equivalent to the problem $y'' + \lambda y = 0, y(0) = y'(1) = 0$, and deduce that the exact characteristic numbers are of the form $\frac{1}{4}(2n - 1)^2\pi^2$, with corresponding characteristic functions proportional to $\sin \frac{1}{2}(2n - 1)\pi x$. [Notice that $\lambda_1 = \pi^2/4 = 2.4674$.]

90. Obtain an approximation to the second characteristic number and corresponding characteristic function of Problem 88, assuming (for simplicity) that the fundamental characteristic function is given with sufficient accuracy by $y_1(x) = 2x - x^2$, and taking $F(x) = x$ in equation (262). [Show that, neglecting an arbitrary multiplicative factor in $y^{(1)}$, we may take $y^{(1)} = 25x^2 - 18x$, and so obtain $f^{(1)} = \frac{1}{12}(-8x + 36x^3 - 25x^4)$. Show also that equation (261a) here fails to estimate $\lambda_2^{(1)}$, whereas (261b) gives $\lambda_2^{(1)} = 23.3$. (Equation (261c) would give a more nearly accurate result.) Notice that the true value is $\lambda_2 = 9\pi^2/4 = 22.2$, in accordance with the result of Problem 89.]

91. Let λ_n denote a characteristic number associated with a real symmetric kernel $K(x, \xi)$ over (a, b), and denote by $\phi_n(x)$ a corresponding *normalized* characteristic function, so that

$$\phi_n(x) = \lambda_n \int_a^b K(x, \xi)\phi_n(\xi)\, d\xi$$

and

$$\int_a^b [\phi_n(x)]^2\, dx = 1.$$

(a) By multiplying both members of the first relation by $\phi_n(x)$, integrating over (a, b), and using the normalizing condition, deduce that

$$\lambda_n = \frac{1}{\int_a^b \int_a^b K(x, \xi)\phi_n(x)\phi_n(\xi)\, dx\, d\xi}.$$

(b) By making use of the Schwarz inequality (Problem 127 of Chapter 1), deduce that

$$\frac{1}{|\lambda_n|} \leq \sqrt{\int_a^b \int_a^b [K(x, \xi)]^2\, dx\, d\xi} \equiv A,$$

so that $|\lambda_n| \geqq 1/A$. [This result establishes equation (182).]

92. If $K(x, \xi)$ is a real symmetric kernel and $\mathscr{K}$ is the integral operator such that

$$\mathscr{K}f(x) = \int_a^b K(x, \xi)f(\xi)\, d\xi,$$

the form

$$J(y) \equiv \int_a^b y(x)\mathscr{K} y(x)\,dx \equiv \int_a^b\int_a^b K(x, \xi)y(x)y(\xi)\,dx\,d\xi$$

is known as the *quadratic integral form* associated with the operator $\mathscr{K}$. [Notice that this definition is in complete analogy with the definition of the quadratic form associated with a real symmetric *matrix* **A**, as the scalar product of a vector **x** and its transform **A x** (see Section 1.14).]

(a) Verify that the variation of the functional $J(y)$ can be expressed in the form

$$\delta J = 2\int_a^b\left[\int_a^b K(x, \xi)y(\xi)\,d\xi\right]\delta y(x)\,dx,$$

with the notation of Section 2.5.

(b) Deduce that the requirement that the expression

$$I(y) \equiv \lambda\int_a^b\int_a^b K(x, \xi)y(x)y(\xi)\,dx\,d\xi - \int_a^b [y(x)]^2\,dx + 2\int_a^b y(x)F(x)\,dx$$

be stationary, for arbitrary small continuous variations in the function $y(x)$, leads to the requirement that y satisfy the integral equation

$$\begin{aligned} y(x) &= \lambda\int_a^b K(x, \xi)y(\xi)\,d\xi + F(x) \\ &\equiv \lambda\mathscr{K}y(x) + F(x). \end{aligned}$$

93. Deduce from Problem 92 that when $K(x, \xi)$ is real and symmetric, the characteristic numbers of the operator $\mathscr{K}$ are stationary values of the ratio

$$\lambda = \frac{\int_a^b [y(x)]^2\,dx}{\int_a^b\int_a^b K(x, \xi)y(x)y(\xi)\,dx\,d\xi} \equiv \frac{\int_a^b y^2\,dx}{J(y)}$$

or of the ratio

$$\lambda = \frac{1}{\int_a^b\int_a^b K(x, \xi)\phi(x)\phi(\xi)dx\,d\xi} \equiv \frac{1}{J(\phi)}$$

where ϕ is subject to the constraint (normalizing condition)

$$\int_a^b \phi^2\,dx = 1,$$

and that the stationary functions are the corresponding characteristic functions. [Compare equation (94) of Chapter 2, and Problems 51 and 94 of that chapter. The *smallest* stationary value can be shown to be a *minimum* (see Problem 94). Hence, of *all* continuous functions ϕ for which $\int_a^b \phi^2\,dx = 1$, that function ϕ_1 which *maximizes* $J(\phi)$ is the characteristic function corresponding to λ_1, where here $1/\lambda_1 \geqq 1/\lambda_2 \geqq \cdots$, and $\lambda_1 = 1/J(\phi_1)$. Of all the functions ϕ which are normalized and which are *also* orthogonal to ϕ_1, that function ϕ_2 which maximizes $J(\phi)$ is the *second* characteristic function, corresponding to $\lambda_2 = 1/J(\phi_2)$, and so forth. (Compare Problem 127 of Chapter 1, noticing that λ here is analogous to $1/\lambda$ in Chapter 1.)]

94. (a) Use *Hilbert*'s *formula* (Problem 54) to deduce the expansion

$$J(y) = \sum_{n=1}^{\infty} \frac{c_n^2}{\lambda_n}, \quad c_n = \int_a^b y(x)\phi_n(x)\,dx,$$

when y is continuous in (a, b).

(b) If $1/\lambda_1 \geqq 1/\lambda_2 \geqq \cdots$, show that

$$\frac{1}{\lambda_1} - \frac{J(y)}{\sum_{n=1}^{\infty} c_n^2} = \frac{\left(\frac{1}{\lambda_1} - \frac{1}{\lambda_2}\right) c_2^2 + \cdots}{\sum_{n=1}^{\infty} c_n^2} \geqq 0,$$

and hence that

$$J(y) \leqq \frac{1}{\lambda_1} \sum_{n=1}^{\infty} c_n^2.$$

(c) Use Bessel's inequality [equation (309) of Chapter 1] to deduce that

$$\frac{J(y)}{\int_a^b y^2\,dx} \leqq \frac{1}{\lambda_1},$$

and show that equality is obtained if and only if y is a characteristic function corresponding to λ_1. [Compare Problem 116 of Chapter 1.]

95. (a) Noticing that $f^{(n)}(x) = \int_a^b K(x, \xi) y^{(n)}(\xi)\,d\xi$, verify that equation (261b) can be obtained by replacing $y(x)$ in the ratio of Problem 93 by an approximation $y^{(n)}(x)$ to a characteristic function.

(b) Recalling also that a constant multiple of the output $f^{(n)}(x)$ is taken as the input $y^{(n+1)}(x)$ in the following cycle, verify that equation (261c) can be written in the form

$$\lambda_1 \approx \frac{\int_a^b y^{(n+1)}(x) y^{(n)}(x)\,dx}{\int_a^b \int_a^b K(x, \xi) y^{(n+1)}(x) y^{(n)}(\xi)\,dx\,d\xi},$$

and so verify that the estimate afforded by (261c) corresponds to that obtained by replacing y^2 by $y^{(n)}y^{(n+1)}$ in the ratio of Problem 93. [Notice that the estimates of (261b,c) are obtained without the necessity of explicitly evaluating a double integral.

A still better approximation (but one requiring such an evaluation) would be obtained by replacing y by $f^{(n)}$ in the ratio of Problem 93, to give

$$\lambda_1 \approx \frac{1}{J(f_n)} \int_a^b \{f^{(n)}(x)\}^2\,dx.]$$

96. An operator $\mathscr{K}$, associated with a real symmetric kernel $K(x, \xi)$, whose quadratic integral form

$$J(y) \equiv \int_a^b y(x)\mathscr{K}y(x)\,dx \equiv \int_a^b\int_a^b K(x, \xi)y(x)y(\xi)\,dx\,d\xi$$

is *nonnegative* for any real function y, is called a *positive* integral operator. If, in addition, $J(y)$ is zero only when $y(x)$ is the zero function, then $\mathscr{K}$ is said to be *positive definite.*

(a) Use the result of Problem 93 (or of Problem 91) to show that *the characteristic numbers corresponding to a positive integral operator are positive.*

(b) Show that the operators involving the kernels

$$K_{\text{I}}(x, \xi) = \int_a^b K(x, t)K(\xi, t)\,dt, \qquad K_{\text{II}}(x, \xi) = \int_a^b K(t, x)K(t, \xi)\,dt,$$

associated with a nonsymmetric kernel $K(x, \xi)$, are positive operators. [In the case of K_{I}, show that $J(y) = \int_a^b \left\{\int_a^b K(x, t)y(x)\,dx\right\}^2 dt.$]

97. A kernel which has the property

$$K(\xi, x) = -K(x, \xi)$$

is called a *skew-symmetric* kernel.

(a) If $K(x, \xi)$ is real and skew symmetric, show that the equation

$$y(x) = \lambda \int_a^b K(x, \xi)y(\xi)\,d\xi$$

implies the equation

$$y(x) = -\lambda^2 \int_a^b K_{\text{II}}(x, \xi)y(\xi)\,d\xi,$$

with the notation of Problem 96(b).

(b) Use the result of Problem 96(b) to deduce that *a real skew-symmetric kernel possesses no real characteristic numbers.* [Notice that a nonhomogeneous equation with a real continuous skew-symmetric kernel (over a finite interval) therefore possesses a solution for any continuous prescribed function $F(x)$.]

(c) Verify this conclusion in the special case of the kernel $K(x, \xi) = x - \xi$, associated with the interval $(0, 1)$.

Section 3.15.

98. Obtain approximate values of the solution of the equation

$$\frac{1}{2}(x - x^2) = \int_0^1 K(x, \xi)y(\xi)\, d\xi,$$

where $K(x, \xi)$ is defined by equation (278), at the points $x = 0, 0.25, 0.5, 0.75$, and 1, by the methods of Section 3.15. Use the weighting coefficients of the trapezoidal rule.

99. With the notation of Section 3.15, show that approximate characteristic numbers relevant to the problem

$$y(x) = \lambda \int_a^b K(x, \xi)y(\xi)\, d\xi$$

are afforded by *reciprocals* of the characteristic numbers relevant to the corresponding matrix $\mathbf{K}\,\mathbf{D}$.

100. Determine approximately the smallest characteristic value of λ for the problem

$$y(x) = \lambda \int_0^1 e^{-x\xi}\, y(\xi)\, d\xi,$$

by the method of Problem 99, using the ordinates at the points $x = 0$, 0.5, and 1, and the weighting coefficients of Simpson's rule. Determine also the approximate ratios of the ordinates of the corresponding characteristic function at those points. [Write $\kappa = 6/\lambda$, and use the iterative methods of Chapter 1, retaining three significant figures.]

Section 3.16.

101. (a) If $K(x, \xi)$ is of the form

$$K(x, \xi) = \begin{cases} x & \textit{when} \quad x < \xi, \\ \xi & \textit{when} \quad x > \xi, \end{cases}$$

show that the assumption $y(x) \approx c_1 + c_2 x + c_3 x^2$ reduces the equation

$$y(x) = F(x) + \lambda \int_0^1 K(x, \xi)y(\xi)\, d\xi \qquad (0 < x < 1)$$

to the requirement

$$c_1[1 - \lambda(x - \tfrac{1}{2}x^2)] + c_2[x - \tfrac{1}{2}\lambda(x - \tfrac{1}{3}x^3)] \\ + c_3[x^2 - \tfrac{1}{3}\lambda(x - \tfrac{1}{4}x^4)] \approx F(x) \qquad (0 < x < 1).$$

(b) Show that the problem of part (a) can be reduced to the problem

$$y''(x) + \lambda y(x) = F''(x), \qquad y(0) = F(0), \quad y'(1) = F'(1).$$

102. (a) Show that the assumption

$$y(\theta) \approx \sum_k c_k \sin k\theta \qquad (0 < \theta < \pi)$$

reduces the *integro-differential* equation

$$y(\theta) = F(\theta) + \lambda f(\theta) \oint_0^{\pi} \frac{dy(\phi)}{d\phi} \frac{d\phi}{\cos\theta - \cos\phi} \qquad (0 < \theta < \pi)$$

to the requirement

$$\sum_k c_k \left[1 + \lambda\pi k \frac{f(\theta)}{\sin\theta}\right] \sin k\theta \approx F(\theta) \qquad (0 < \theta < \pi).$$

[The symbol $\oint$ denotes the Cauchy principal value. See Problems 84 and 86(a). With a suitable interpretation of the symbols, this equation is one form of the basic equation of the *Prandtl lifting-line* theory of aerodynamics.]

(b) In the special case when $f(\theta) = \sin\theta$, show that the formal solution of the equation of part (a) is given by

$$y(\theta) = \sum_{k=1}^{\infty} \frac{a_k}{1 + \lambda\pi k} \sin k\theta \qquad (0 < \theta < \pi),$$

where a_k is the kth coefficient in the Fourier sine-series development of $F(\theta)$ over $(0, \pi)$. [Notice also that the numbers $-1/\pi$, $-1/2\pi$, $\cdots$ are hence characteristic values of λ when $f(\theta) = \sin\theta$.]

Section 3.17.

103. (a) Obtain an approximate solution of the special case of Problem 101(a) in which $F(x) = x$ and $\lambda = 1$, using the method of collocation at the points $x = 0$, 0.5, and 1.

(b) Compare the approximate solution with the exact solution, obtained from Problem 101(b), at the points $x = 0, 0.25, 0.5, 0.75$, and 1.

104. (a) Obtain an approximate solution of the equation

$$\sin\frac{\pi x}{2} = \int_0^1 K(x, \xi) y(\xi)\, d\xi,$$

where $K(x, \xi)$ is defined in Problem 101(a), by assuming $y(x) \approx c_1 + c_2 x + c_3 x^2$, and using the method of collocation at the points $x = 0, 0.5$, and 1.

(b) Compare the result with the exact solution $\frac{1}{4}\pi^2 \sin\frac{1}{2}\pi x$ at the points $x = 0, 0.25, 0.5, 0.75$, and 1.

105. Obtain an approximate solution of the special case of Problem 102(a) in which $f(\theta) = F(\theta) = 1$ and $\lambda = 1/\pi$, assuming a three-term approximation of the form

$$y(\theta) \approx c_1 \sin\theta + c_3 \sin 3\theta + c_5 \sin 5\theta,$$

and collocating at $\theta = 0$, $\pi/4$, and $\pi/2$. [The exact solution is not known. Notice that, if $f(\theta)$ and $F(\theta)$ are symmetric with respect to $\theta = \pi/2$, only harmonics of odd order need be considered, and recall that

$$\lim_{\theta\to 0} \frac{\sin k\theta}{\sin\theta} = k.]$$

106. (a) Suppose that the quantities

$$\Phi_j(x_i) \equiv \int_a^b K(x_i, \xi)\phi_j(\xi)\, d\xi \qquad (i = 1, 2, \cdots, n)$$

cannot be conveniently evaluated by direct integration, but are to be determined approximately, as weighted sums of the ordinates at a set of N points $\xi_1, \xi_2, \cdots, \xi_N$. Show that the matrix $\mathbf{\Phi}$, for which

$$\Phi_{ij} = \Phi_j(x_i) \qquad (i = 1, 2, \cdots, n; j = 1, 2, \cdots, n)$$

can be obtained by matrix multiplication, in the form

$$\mathbf{\Phi} = \mathbf{K}\,\mathbf{D}\,\boldsymbol{\phi}_\xi,$$

where $\mathbf{K}$ is the $n \times N$ matrix

$$\mathbf{K} = \begin{bmatrix} K(x_1, \xi_1) & \cdots & K(x_1, \xi_N) \\ \cdots & \cdots & \cdots \\ K(x_n, \xi_1) & \cdots & K(x_n, \xi_N) \end{bmatrix}$$

and $\boldsymbol{\phi}_\xi$ is the $N \times n$ matrix

$$\boldsymbol{\phi}_\xi = \begin{bmatrix} \phi_1(\xi_1) & \cdots & \phi_n(\xi_1) \\ \cdots & \cdots & \cdots \\ \phi_1(\xi_N) & \cdots & \phi_n(\xi_N) \end{bmatrix},$$

and where $\mathbf{D}$ is the diagonal $N \times N$ matrix such that the diagonal element $D_j \equiv D_{jj}$ is the weighting coefficient associated with the point ξ_j in the numerical integrations. [Notice that the matrix $\mathbf{K}\,\mathbf{D}$ is hence obtained by multiplying all elements of the jth column of $\mathbf{K}$ by D_j.]

(b) Deduce that the matrix of coefficients of the c's in equation (293) can be expressed in the form

$$\mathbf{S} = \boldsymbol{\phi}_x - \lambda\,\mathbf{K}\,\mathbf{D}\,\boldsymbol{\phi}_\xi,$$

where, in addition to the matrices defined in part (a), the matrix $\boldsymbol{\phi}_x$ is the square $n \times n$ matrix

$$\boldsymbol{\phi}_x = \begin{bmatrix} \phi_1(x_1) & \cdots & \phi_n(x_1) \\ \cdots & \cdots & \cdots \\ \phi_1(x_n) & \cdots & \phi_n(x_n) \end{bmatrix}.$$

107. (a) Verify that the application of the procedure of Problem 106 to the approximate solution of the equation

$$y(x) = x + \frac{1}{4}\int_0^1 \frac{\sin^2(x - \xi)}{(x - \xi)^2}\, y(\xi)\, d\xi,$$

with the assumption $y(x) \approx c_1 + c_2 x$, with approximate integration involving the three ordinates $\xi = 0, 0.5$, and 1 according to Simpson's rule, and with collocation at $x = 0$ and $x = 1$, leads to the equations specified by the matrix relation $\mathbf{S}\,\mathbf{c} = \mathbf{f}$, where $\mathbf{f} = \{0, 1\}$, and where $\mathbf{S}$ is evaluated as follows:

$$\mathbf{S} = \begin{bmatrix} 1 & 0 \\ 1 & 1 \end{bmatrix} - \frac{1}{24}\begin{bmatrix} 1 & 0.919 & 0.708 \\ 0.708 & 0.919 & 1 \end{bmatrix}\begin{bmatrix} 1 & 0 & 0 \\ 0 & 4 & 0 \\ 0 & 0 & 1 \end{bmatrix}\begin{bmatrix} 1 & 0 \\ 1 & \frac{1}{2} \\ 1 & 1 \end{bmatrix}$$

$$= \begin{bmatrix} 0.817 & -0.106 \\ 0.817 & 0.882 \end{bmatrix}.$$

(b) Obtain the corresponding approximate solution.

Section 3.18.

108. (a) Obtain an approximate solution of the integral equation considered in Problem 103(a), determining the parameters by use of the weighting functions 1, x, and x^2.

(b) Compare the results so obtained with the data of Problem 103(b).

109. (a) Apply the method of weighting functions to the equation treated in Problem 105, first multiplying both sides of the approximate relation by $\sin\theta$, and then using the weighting functions $\sin\theta$, $\sin 3\theta$, and $\sin 5\theta$.

(b) Compare the results with those of Problem 105.

Section 3.19.

110. (a) Apply the modified method of least squares to the integral equation treated in Problem 103(a), using the points $x = 0, 0.25, 0.5, 0.75$, and 1 as the points x_i, and using weighting coefficients corresponding to Simpson's rule.

(b) Compare the results with those of Problems 103(b) and 108.

111. (a) Proceed as in Problem 110 in dealing with Problem 105, introducing the two additional points $\theta = \pi/8$ and $3\pi/8$.

(b) Compare the results with those of Problems 105 and 109.

112. (a) Apply the modified method of least squares to the treatment of Problem 107, introducing the one additional point $x = 0.5$. [First obtain the three equations corresponding to (319) by the method of Problem 106, taking $n = N = 3$.]

(b) Compare the results with those of Problem 107.

Section 3.20.

113. (a) If $K(x, \xi) = x$ when $x < \xi$ and $K(x, \xi) = \xi$ when $x > \xi$, determine the coefficients in the approximation

$$K(x, \xi) \approx c_1 + c_2 x + c_3 x^2 \qquad (0 \leq x \leq 1),$$

as functions of ξ, in such a way that the two members are equal at $x = 0$, and at

$x = 1$, and that they possess equal integrals over (0, 1). [Thus obtain the approximation $K(x, \xi) \approx 4x\xi - 3x\xi^2 - 3x^2\xi + 3x^2\xi^2$.]

(b) Use this approximation to obtain an approximate solution of the problem $y(x) = x + \int_0^1 K(x, \xi)y(\xi)\,d\xi$, and compare the result with the exact solution. [See Problem 101(b).]

114. By replacing $K(x, \xi) = \dfrac{\sin^2(x - \xi)}{(x - \xi)^2}$ by the first two terms of its expansion in powers of $(x - \xi)$,

$$K(x, \xi) = 1 - \tfrac{1}{3}(x - \xi)^2 + \tfrac{2}{45}(x - \xi)^4 + \cdots,$$

obtain an approximate solution of the problem

$$y(x) = x + \frac{1}{4}\int_0^1 K(x, \xi)y(\xi)\,d\xi,$$

and compare the result with those of Problems 107 and 112.

115. (a) Determine the constants c_1 and c_2 in such a way that the integral of the squared difference between the two members of the relation

$$\frac{\sin^2 u}{u^2} \approx c_1 + c_2 u^2 \qquad (0 \leqq u \leqq 1)$$

is as small as possible. [Evaluate $\int_0^1 \dfrac{\sin^2 u}{u^2}\,du$ numerically, by use of series or otherwise.]

(b) Treat Problem 114 by replacing $K(x, \xi)$ by the corresponding approximation $c_1 + c_2(x - \xi)^2$, and compare the result with the results of Problem 114.

116. Obtain an estimate of the smallest characteristic value of λ for the problem $y(x) = \lambda\int_0^1 K(x, \xi)y(\xi)\,d\xi$, where $K(x, \xi)$ is defined in Problem 114.

117. Suppose that $y(x)$ is the required solution of the integral equation

$$y(x) = F(x) + \lambda\int_a^b K(x, \xi)y(\xi)\,d\xi,$$

and that $\bar{y}(x)$ is the solution of the equation obtained by approximating $K(x, \xi)$ by a *separable* kernel $S(x, \xi)$. If the error $y(x) - \bar{y}(x)$ in the solution is denoted by $\varepsilon(x)$, and the error $K(x, \xi) - S(x, \xi)$ in the approximation to the kernel is denoted by $E(x, \xi)$, show that the true solution is of the form

$$y(x) = \bar{y}(x) + \varepsilon(x),$$

where $\varepsilon(x)$ satisfies the equation

$$\varepsilon(x) = \Phi(x) + \lambda\int_a^b K(x, \xi)\varepsilon(\xi)\,d\xi,$$

in which

$$\Phi(x) = \lambda \int_a^b E(x, \xi)\bar{y}(\xi)\, d\xi.$$

[Notice that this equation is of the same form as the original equation, with the prescribed fun:tion $F(x)$ replaced by the calculable function $\Phi(x)$. Thus, if $K(x, \xi)$ is again approximated by $S(x, \xi)$ in the integral equation for the correction $\varepsilon(x)$, the linear algebraic equations which are then to be solved differ only in the right-hand members from those already determined in obtaining the first approximation. The function $\Phi(x)$, and the relevant integrals involving that function, may be evaluated by numerical methods, extreme accuracy usually being unnecessary if the kernel error $E(x, \xi)$ is small over (a, b).]

118. Suppose that the kernel $K(x, \xi)$ in the equation

$$y(x) = F(x) + \lambda \int_a^b K(x, \xi)y(\xi)\, d\xi \tag{A}$$

is expressed as the sum of a *separable* kernel $S(x, \xi)$ and a remainder $E(x, \xi)$, so that

$$K(x, \xi) = S(x, \xi) + E(x, \xi),$$

in such a way that

$$\max_{\substack{a \leq x \leq b \\ a \leq \xi \leq b}} |E(x, \xi)| < \frac{1}{(b - a)\,|\lambda|}.$$

(a) With the temporary abbreviation

$$\phi(x) \equiv F(x) + \lambda \int_a^b S(x, \xi)y(\xi)\, d\xi,$$

show that

$$y(x) = \phi(x) + \lambda \int_a^b E(x, \xi)y(\xi)\, d\xi.$$

(b) If $\gamma(x, \xi; \lambda)$ is the resolvent kernel of $E(x, \xi)$ over (a, b), show that there follows

$$y(x) = \phi(x) + \lambda \int_a^b \gamma(x, \xi; \lambda)\phi(\xi)\, d\xi,$$

and that this relation can be written in the form

$$y(x) = f(x) + \lambda \int_a^b k(x, \xi; \lambda)y(\xi)\, d\xi, \tag{B}$$

where

$$f(x) \equiv F(x) + \lambda \int_a^b \gamma(x, \xi; \lambda)F(\xi)\, d\xi$$

and where

$$k(x, \xi; \lambda) \equiv S(x, \xi) + \lambda \int_a^b \gamma(x, t; \lambda)S(t, \xi)\, dt.$$

(c) If $S(x, \xi)$ is of the form

$$S(x, \xi) = \sum_{n=1}^{N} a_n(x) b_n(\xi),$$

show that

$$k(x, \xi; \lambda) = \sum_{n=1}^{N} A_n(x, \lambda) b_n(\xi),$$

where

$$A_n(x, \lambda) = a_n(x) + \lambda \int_a^b \gamma(x, t; \lambda) a_n(t)\, dt,$$

so that $k(x, \xi; \lambda)$ also is a *separable* kernel.

[This method can be used when $|\lambda K(x, \xi)| \nless 1/(b - a)$ in (a, b), so that the iterative method of Section 3.9 may not be applicable to the original equation (A). Under the assumption $|\lambda E(x, \xi)| < 1/(b - a)$ in (a, b), the Neumann series of $\gamma(x, \xi; \lambda)$ is convergent, and its first few terms may be used to supply an *approximation* to γ, after which corresponding approximations to $f(x)$ and the separable kernel $k(x, \xi; \lambda)$ are determined and the new integral equation (B) can be dealt with by the algebraic methods of Section 3.6.]

APPENDIX

The Crout Method for Solving Sets of Linear Algebraic Equations

A. The procedure. The method of Crout, for solving a set of n linear algebraic equations in n unknowns, is basically equivalent to the method of Gauss (see footnote to page 4). However, the calculations are systematized in such a way that they are conveniently carried out (for example) on a desk calculator, with a minimum number of separate machine operations. Furthermore, a very considerable saving of time and labor results from the fact that the recording of auxiliary data is minimized and compactly arranged.

Only a description and illustration of the method is given here; an analytic justification is included in the original paper.*

The calculation proceeds from the *augmented matrix* of the system,

$$\mathbf{M} \equiv \left[\begin{array}{cccc|c} a_{11} & a_{12} & \cdots & a_{1n} & c_1 \\ a_{21} & a_{22} & \cdots & a_{2n} & c_2 \\ \cdots & \cdots & \cdots & \cdots & \cdot\cdot \\ a_{n1} & a_{n2} & \cdots & a_{nn} & c_n \end{array}\right] \equiv [\mathbf{A} \mid \mathbf{c}], \tag{1}$$

which may be considered as partitioned into the coefficient matrix $\mathbf{A}$ and the column vector $\mathbf{c}$, to an *auxiliary matrix*

$$\mathbf{M}' \equiv \left[\begin{array}{cccc|c} a'_{11} & a'_{12} & \cdots & a'_{1n} & c'_1 \\ a'_{21} & a'_{22} & \cdots & a'_{2n} & c'_2 \\ \cdots & \cdots & \cdots & \cdots & \cdot\cdot \\ a'_{n1} & a'_{n2} & \cdots & a'_{nn} & c'_n \end{array}\right] \equiv [\mathbf{A}' \mid \mathbf{c}'], \tag{2}$$

* See Reference 3 to Chapter 1.

of the same dimensions, and thence to the required *solution vector*

$$\mathbf{x} \equiv \begin{Bmatrix} x_1 \\ x_2 \\ \cdot \\ \cdot \\ \cdot \\ x_n \end{Bmatrix}. \tag{3}$$

It is convenient to define the *diagonal element* of any element to the *right* of the principal diagonal of a matrix as that element of the principal diagonal which lies in the same *row* as the given element. The diagonal element of any element *below* the principal diagonal is defined as that element of the principal diagonal which lies in the same *column* as the given element.

With this definition, the procedure for obtaining the elements of $\mathbf{M}'$ from those of the given matrix $\mathbf{M}$ may be described by the four rules which follow:

1. The elements of $\mathbf{M}'$ are determined in the following order: elements of the first column, then elements of the first row to the right of the first column; elements of the second column below the first row, then elements of the second row to the right of the second column; and so on, until all elements are determined.

2. The first column of $\mathbf{M}'$ is identical with the first column of $\mathbf{M}$. Each element of the first row of $\mathbf{M}'$ except the first is obtained by dividing the corresponding element of $\mathbf{M}$ by the leading element a_{11}.

3. Each element a'_{ij} on or below the principal diagonal of $\mathbf{M}'$ is obtained by subtracting from the corresponding element a_{ij} of $\mathbf{M}$ the sum of the products of elements in the ith row and corresponding elements in the jth column of $\mathbf{M}'$, all uncalculated elements being imagined to be zeros. In symbols, we thus have

$$a'_{ij} = a_{ij} - \sum_{k=1}^{j-1} a'_{ik}a'_{kj} \qquad (i \geqq j). \tag{4}$$

4. Each element a'_{ij} to the right of the principal diagonal is calculated by the procedure of Rule 3 *followed by a division by the diagonal element* a'_{ii} in $\mathbf{M}'$. Thus there follows

$$a'_{ij} = \frac{a_{ij} - \sum_{k=1}^{i-1} a'_{ik}a'_{kj}}{a'_{ii}} \qquad (i < j). \tag{5}$$

In the important cases when the coefficient matrix $\mathbf{A}$ is *symmetric* ($a_{ji} = a_{ij}$), it can be shown that *any element* a'_{ij} *to the right of the principal diagonal is equal to the result of dividing the symmetrically placed element* a'_{ji} (below the diagonal) *by its diagonal element* a'_{ii}. This fact reduces the labor involved in the formation of $\mathbf{M}'$, in such a case, by a factor of nearly two, since each element *below* the diagonal may be recorded as a by-product of the calculation of the symmetrically placed element, before the final division is effected.

The procedure for obtaining the final solution vector $\mathbf{x}$ from the matrix $\mathbf{A}'$ and the vector $\mathbf{c}'$, into which $\mathbf{M}'$ is partitioned, may be described by the three rules which follow:

1. The elements of $\mathbf{x}$ are determined in the reverse order $x_n, x_{n-1}, x_{n-2}, \cdots, x_1$, from the last element to the first.

2. The last element x_n is identical with the last element c'_n of $\mathbf{c}'$.

3. Each remaining element x_i of $\mathbf{x}$ is obtained by subtracting from the corresponding element c'_i of $\mathbf{c}'$ the sum of the products of elements in the ith row of $\mathbf{A}'$ by corresponding elements of the column $\mathbf{x}$, all uncalculated elements of $\mathbf{x}$ being imagined to be zeros. Thus there follows

$$x_i = c'_i - \sum_{k=i+1}^{n} a'_{ik} x_k. \tag{6}$$

The solution may, of course, be checked completely by substitution into the n basic linear equations. However, if desired, a *check column* may be carried along in the calculation to provide a *continuous* check on the work. Each element of the initial check column, corresponding to the augmented matrix $\mathbf{M}$, is the sum of the elements of the corresponding row of $\mathbf{M}$. If this column is recorded to the right of the augmented matrix, *and is treated in the same manner as the column* $\mathbf{c}$, corresponding check columns are thus obtained for the auxiliary matrix $\mathbf{M}'$, and for the solution vector $\mathbf{x}$. Continuous checks on the calculation are then afforded by the two rules which follow:

1. In the auxiliary matrix, any element of the check column should *exceed by unity* the sum of the other elements in its row which lie *to the right of the principal diagonal.*

2. Each element of the check column associated with the solution vector should *exceed by unity* the corresponding element of the solution vector.

The preceding checks serve, not only to display numerical errors, but also to give an estimate of the effect of round-off errors resulting from the retention of insufficiently many significant figures.*

The simplicity and efficiency of this procedure can be appreciated only when it is applied to specific sets of equations and compared with other procedures.

It is of some interest to notice that (as is shown in the reference cited) the set of equations which would be obtained by the direct use of the Gauss reduction would possess the augmented matrix

$$\widetilde{\mathbf{M}} \equiv \left[\begin{array}{cccc|c} 1 & a'_{12} & \cdots & a'_{1n} & c'_1 \\ 0 & 1 & \cdots & a'_{2n} & c'_2 \\ \cdots & \cdots & \cdots & \cdots & \cdots \\ 0 & 0 & \cdots & 1 & c'_n \end{array}\right] \equiv [\widetilde{\mathbf{A}} \mid \mathbf{c}'], \tag{7}$$

which differs from (2) only in the substitution of ones in the principal diagonal and zeros below it. The transition from the set of equations represented by (7) to the solution (6) is seen to correspond to the "back solution" of the Gauss procedure. *The compactness of the Crout procedure is achieved by recording auxiliary data in the spaces of* $\widetilde{\mathbf{M}}$ *which would otherwise be occupied by ones and zeros.*

* Appreciable loss of accuracy may be encountered if a *small diagonal element* appears in $\mathbf{M}'$ at an early stage of the calculation. Such a situation may often be remedied by reordering the equations and/or unknowns.

The ith diagonal element of $\mathbf{A}'$ happens to be the coefficient by which the ith equation is divided, before that equation is used to eliminate the ith unknown from succeeding equations in the Gauss reduction. Since all other steps in the reduction do not affect the *determinant* of the coefficient matrix, it follows that *the determinant of* $\mathbf{A}$ *is equal to the product of the diagonal elements of* $\mathbf{A}'$,

$$|\mathbf{A}| = a'_{11}a'_{22}\cdots a'_{nn}. \tag{8}$$

Thus the Crout procedure is useful also in evaluating determinants, the columns $\mathbf{c}$ and $\mathbf{c}'$, as well as the final vector $\mathbf{x}$, then being omitted.

Furthermore, it is found that the product $a'_{11}a'_{22}\cdots a'_{rr}$ is the determinant Δ_r of the $r \times r$ submatrix

$$\mathbf{D}_r = \begin{bmatrix} a_{11} & \cdots & a_{1r} \\ \cdots & \cdots & \cdots \\ a_{r1} & \cdots & a_{rr} \end{bmatrix}$$

of $\mathbf{A}$. The results of Section 1.20 thus state that *when the coefficient matrix* $\mathbf{A}$ *is real and symmetric, it is also positive definite if and only if all the diagonal elements* $a'_{11}, \cdots, a'_{nn}$ *of* $\mathbf{A}'$ *are positive.*

In the reference cited, it is shown that the method can be extended to the convenient treatment of equations with *complex* coefficients, and to the general case of m equations in n unknowns.

B. A numerical example. In order to illustrate the procedure numerically, we apply it to the system

$$\begin{aligned} 554.11\,x_1 - 281.91\,x_2 - 34.240x_3 &= 273.02, \\ -281.91\,x_1 + 226.81\,x_2 + 38.100x_3 &= -63.965, \\ -34.240x_1 + 38.100x_2 + 80.221x_3 &= 34.717, \end{aligned}$$

with the *augmented matrix* (and associated check column)

$$\mathbf{M} = \left[\begin{array}{ccc|c} \underline{554.11} & -281.91 & -34.240 & 273.02 \\ -281.91 & \underline{226.81} & 38.100 & -63.965 \\ -34.240 & 38.100 & \underline{80.221} & 34.717 \end{array}\right] \quad \begin{array}{c} \textit{Check} \\ 510.98 \\ -80.965\,. \\ 118.80 \end{array}$$

The *auxiliary matrix* (and associated check column) are obtained in the form

$$\mathbf{M}' = \left[\begin{array}{ccc|c} \underline{554.11} & -0.50876 & -0.061793 & 0.49272 \\ -281.91 & \underline{83.385} & 0.24801 & 0.89870 \\ -34.240 & 20.680 & \underline{72.976} & 0.45224 \end{array}\right] \quad \begin{array}{c} \textit{Check} \\ 0.92216 \\ 2.14668\,, \\ 1.45228 \end{array}$$

and the *solution vector* (and final check column) are found to be

$$\mathbf{x} = \begin{Bmatrix} 0.92083 \\ 0.78654 \\ 0.45224 \end{Bmatrix} \quad \begin{array}{c} \textit{Check} \\ 1.92080 \\ 1.78650\,, \\ 1.45228 \end{array}$$

if all calculated values are rounded off to five significant figures throughout the calculation. Thus there follows

$$x_1 = 0.92083, \qquad x_2 = 0.78654, \qquad x_3 = 0.45224,$$

where reference to the final check column suggests the possible presence of round-off errors of the order of four units in the last place retained. Such errors would be decreased by retaining additional significant figures in the formation of $\mathbf{M}'$ and $\mathbf{x}$.

The elements of the first column of $\mathbf{M}'$ are identical with the corresponding elements of $\mathbf{M}$; the elements of the first row of $\mathbf{M}'$ following the first element are obtained by dividing the corresponding elements of $\mathbf{M}$ by 554.11. The remaining elements of $\mathbf{M}'$ are determined as follows:

$$a'_{22} = 226.81 - (-281.91)(-0.50876) = 83.385.$$

$$a'_{32} = 38.100 - (-0.50876)(-34.240) = 20.680.$$

$$a'_{23} = \frac{38.100 - (-0.061793)(-281.91)}{83.385} = \frac{20.680}{83.385} = 0.24801.$$

$$c'_2 = \frac{-63.965 - (0.49272)(-281.91)}{83.385} = 0.89870.$$

$$a'_{33} = 80.221 - (-0.061793)(-34.240) - (0.24801)(20.680) = 72.976.$$

$$c'_3 = \frac{34.717 - (0.49272)(-34.240) - (0.89870)(20.680)}{72.976} = 0.45224.$$

The last element of $\mathbf{x}$ is identical with c'_3. The remaining elements of $\mathbf{x}$ are determined as follows:

$$x_2 = 0.89870 - (0.24801)(0.45224) = 0.78654.$$

$$x_1 = 0.49272 - (-0.50876)(0.78654) - (-0.061793)(0.45224) = 0.92083.$$

It may be noticed that, because of the symmetry of the coefficient matrix, it is not actually necessary to calculate a'_{23} and a'_{32} independently. If a'_{23} is calculated, the numerator (20.680) may be recorded as a'_{32} before the final division is effected.

The advantages of the procedure (and the additional simplifications introduced by symmetry of the matrix $\mathbf{A}$) increase with the number of equations involved.

It is particularly important to notice that the calculation of each element of either $\mathbf{M}'$ or $\mathbf{x}$ involves only a *single continuous machine operation* (a sum of products, with or without a final division), without the necessity of intermediate tabulation or transfer of auxiliary data.

If the *determinant* of the coefficient matrix is required, it can be obtained as the product of the diagonal elements of $\mathbf{A}'$:

$$|\mathbf{A}| = (554.11)(83.385)(72.976) = 3.3718 \times 10^6.$$

Since the diagonal elements of $\mathbf{A}'$ are all *positive*, it follows also that $\mathbf{A}$ is *positive definite*.

C. Application to tridiagonal systems. Systems of equations of the special form

$$\left.\begin{aligned}
&d_1x_1 + f_1x_2 &&= c_1,\\
&e_2x_1 + d_2x_2 + f_2x_3 &&= c_2,\\
&\qquad e_3x_2 + d_3x_3 + f_3x_4 &&= c_3,\\
&\cdots\cdots\cdots\cdots\cdots\cdots\cdots\cdots\cdots\cdots\cdots\cdots\\
&\qquad\qquad e_{n-1}x_{n-2} + d_{n-1}x_{n-1} + f_{n-1}x_n &&= c_{n-1},\\
&\qquad\qquad\qquad e_nx_{n-1} + d_nx_n &&= c_n
\end{aligned}\right\} \tag{9}$$

are of frequent occurrence in practice and are often said to be *tridiagonal,* since only the diagonal elements d_i and the adjacent elements e_i and f_i may differ from zero in the coefficient matrix **A** associated with any such system. It can be seen that the Crout reduction preserves this property, in the sense that the submatrix **A**′ in the auxiliary matrix **M**′ of equation (2) then also is of tridiagonal form. Furthermore, by virtue of the fact that the e's have no nonzero left neighbors and the f's no nonzero upper neighbors, it is easily verified that here the relationships between the elements of **M**′ and the corresponding elements of **M** take the simplified forms

$$e_i' = e_i \qquad (i = 2, 3, \cdots, n), \tag{10a}$$

$$d_1' = d_1, \qquad d_i' = d_i - e_i'f_{i-1}' \qquad (i = 2, 3, \cdots, n), \tag{10b}$$

$$f_i' = \frac{f_i}{d_i'} \qquad (i = 1, 2, \cdots, n-1), \tag{10c}$$

and

$$c_1' = \frac{c_1}{d_1'}, \qquad c_i' = \frac{c_i - e_i'c_{i-1}'}{d_i'} \qquad (i = 2, 3, \cdots, n). \tag{10d}$$

Finally, the relations of equation (6) reduce to the forms

$$x_n = c_n', \qquad x_i = c_i' - f_i'x_{i+1} \qquad (i = n-1, n-2, \cdots, 1). \tag{11}$$

For the purpose of compactness, it is convenient to record the coefficients and right-hand members of (9) in the array

$$\mathbf{P} = \left[\begin{array}{ccc|c}
 & d_1 & f_1 & c_1\\
e_2 & d_2 & f_2 & c_2\\
\cdots & \cdots & \cdots & \cdots\\
e_{n-1} & d_{n-1} & f_{n-1} & c_{n-1}\\
e_n & d_n & & c_n
\end{array}\right], \tag{12}$$

so that the diagonal elements of **A** are in the second column, the subdiagonal elements in the first column, and the superdiagonal elements in the third column. If a corresponding array **P**′, with primed elements, is defined in correspondence with the Crout auxiliary matrix, its elements may be determined by use of equations (10a–d), after which the elements of the solution vector **x** are obtained by use of (11).

For this purpose, we notice first that the *first column* of $\mathbf{P}'$ is identical with the first column of $\mathbf{P}$. To complete the *first row* of $\mathbf{P}'$, we next record the quantities

$$d_1' = d_1, \qquad f_1' = \frac{f_1}{d_1'}, \qquad c_1' = \frac{c_1}{d_1'}.$$

Then, to complete the *second row* of $\mathbf{P}'$, we determine successively

$$d_2' = d_2 - e_2' f_1', \qquad f_2' = \frac{f_2}{d_2'}, \qquad c_2' = \frac{c_2 - e_2' c_1'}{d_2'}.$$

The *third row* is completed by use of the same formulas with all subscripts advanced by unity and the remaining rows of $\mathbf{P}'$ are completed, in order, in the same way (except that f_n' is not needed in the *last* row).

The elements of **x** then are determined successively, from the last two columns of $\mathbf{P}'$, in the reverse order

$$x_n = c_n', \qquad x_{n-1} = c_{n-1}' - f_{n-1}' x_n, \qquad \cdots, \qquad x_1 = c_1' - f_1' x_2.$$

It may be noticed that the complete process requires a maximum of $3n - 3$ multiplications, $2n - 1$ divisions, and $3n - 3$ additions, when the system comprises n equations, and it compares favorably with other procedures in this respect.

As a simple illustration, the system

$$\begin{aligned} 2x_1 - x_2 \qquad\qquad\qquad &= 6, \\ -x_1 + 3x_2 - 2x_3 \qquad\quad &= 1, \\ -2x_2 + 4x_3 - 3x_4 &= -2, \\ -3x_3 + 5x_4 &= 1 \end{aligned}$$

corresponds to the array

$$\mathbf{P} = \left[\begin{array}{cccc|c} & 2 & -1 & & 6 \\ -1 & 3 & -2 & & 1 \\ -2 & 4 & -3 & & -2 \\ -3 & 5 & & & 1 \end{array}\right],$$

and it may be verified that the auxiliary array $\mathbf{P}'$ and the solution vector **x** are obtained in the forms

$$\mathbf{P}' = \left[\begin{array}{ccc|c} & 2 & -\frac{1}{2} & 3 \\ -1 & \frac{5}{2} & -\frac{4}{5} & \frac{8}{5} \\ -2 & \frac{12}{5} & -\frac{5}{4} & \frac{1}{2} \\ -3 & \frac{5}{4} & & 2 \end{array}\right], \qquad \mathbf{x} = \begin{Bmatrix} 5 \\ 4 \\ 3 \\ 2 \end{Bmatrix}.$$

The check column can be used exactly as before, if so desired, when account is taken of the fact that the *second column* of **P** (or of **P**′) comprises the elements in the *principal diagonal* of **M** (or **M**′).

It follows from equation (8) that the determinant of the coefficient matrix associated with (9) is the product of the elements in the second column of **P**′. In the above example, this determinant thus has the value 15.

When the coefficient matrix associated with (9) is symmetric, it is also positive definite if and only if all the elements in the second column of **P**′ are positive, as is true in the case of the example.

Answers to Problems

Chapter 1

1. (a) $x_1 = 1, x_2 = -1, x_3 = 1.$

   (b) $x_1 = 2 - \frac{1}{2}C,\ x_2 = -\frac{5}{2} + \frac{3}{4}C,\ x_3 = C.$

   (c) $x_1 = 5, x_2 = 4, x_3 = 3, x_4 = 2.$

2. (a) $\begin{bmatrix} 3 & -2 & 3 \\ 0 & 1 & 0 \end{bmatrix}.$ (b) $\begin{bmatrix} 0 & 0 \\ 0 & 0 \end{bmatrix}.$ (c) $[a_1b_1 + a_2b_2 + \cdots + a_nb_n].$

   (d) $\begin{bmatrix} a_1b_1 & a_2b_1 & \cdots & a_nb_1 \\ a_1b_2 & a_2b_2 & \cdots & a_nb_2 \\ \cdots & \cdots & \cdots & \cdots \\ a_1b_n & a_2b_n & \cdots & a_nb_n \end{bmatrix}.$ (e) $\begin{bmatrix} c_1a_{11} & c_1a_{12} \\ c_2a_{21} & c_2a_{22} \end{bmatrix}.$

   (f) $\begin{bmatrix} c_1a_{11} & c_2a_{12} \\ c_1a_{21} & c_2a_{22} \end{bmatrix}.$

6. $\mathbf{A} = \begin{bmatrix} \alpha\beta & -\alpha^2 \\ \beta^2 & -\alpha\beta \end{bmatrix}$ where α and β are arbitrary complex numbers.

9. 0.319, 0.363, 0.462, 0.587, 0.724.

11. $\lambda = 1: \mathbf{x} = C\{1, -1, 2\};\ \lambda = -9: \mathbf{x} = C\{3, 9, -2\}.$

13. $\begin{vmatrix} x_1^2 & x_1y_1 & y_1^2 & 1 \\ x_2^2 & x_2y_2 & y_2^2 & 1 \\ x_3^2 & x_3y_3 & y_3^2 & 1 \\ x_4^2 & x_4y_4 & y_4^2 & 1 \end{vmatrix} = 0.$

18. (a) δ_{ij}. (b) $d_i\delta_{ij}$. (c) $d_i\delta_{ij}$.

20. $|\mathbf{A}| \neq 0$.

21. $$\mathbf{A}^T = \begin{bmatrix} 1 & 2 & -1 \\ 2 & 1 & 0 \\ 1 & 0 & 1 \end{bmatrix}, \quad \text{Adj}\,\mathbf{A} = \begin{bmatrix} 1 & -2 & -1 \\ -2 & 2 & 2 \\ 1 & -2 & -3 \end{bmatrix},$$

$$\mathbf{A}^{-1} = \begin{bmatrix} -\frac{1}{2} & 1 & \frac{1}{2} \\ 1 & -1 & -1 \\ -\frac{1}{2} & 1 & \frac{3}{2} \end{bmatrix}.$$

22. $$\mathbf{M} = \begin{bmatrix} 0 & -2 \\ 0 & 4 \\ 0 & 1 \end{bmatrix}.$$

35. (b) $\mathbf{x} = \{\frac{6}{5}, \frac{2}{5}, 0\} + C\{\frac{1}{5}, -\frac{3}{5}, 1\} = \frac{1}{5}\{6 + C, 2 - 3C, 5C\}$.

36. (b) $\lambda = 1: \mathbf{x} = C_1\{2, 1, 0\} + C_2\{-1, 0, 1\}$; $\lambda = -3: \mathbf{x} = C\{1, 2, -1\}$.

37. (a) $a \neq 0, b \neq 1$. (b) $a \neq 0, b = 1$.
(c) $a = 0, b = 1$. (d) $a = 0, b \neq 1$.

38. (a) $r = 3$. (b) $r = 3$. (c) $r = 2$.

39. Yes.

40. (a) 45°. (b) 45°.

47. (b) The set $\mathbf{A}\,\mathbf{x} = \mathbf{c}$ possesses a solution for any $\mathbf{c}$.

50. (a) $\lambda_1 = 1: \mathbf{e}_1 = \pm\{1/\sqrt{5}, -2/\sqrt{5}\}$; $\lambda_2 = 6: \mathbf{e}_2 = \pm\{2/\sqrt{5}, 1/\sqrt{5}\}$.
(c) With upper signs in answer to part (a): $\alpha_1 = -1/\sqrt{5}$, $\alpha_2 = 3/\sqrt{5}$.
(d) $\mathbf{x} = \{2/(6 - \lambda), 1/(6 - \lambda)\}$ if $\lambda \neq 1$ or 6. If $\lambda = 6$, there is no solution. If $\lambda = 1$, $\mathbf{x} = \{\frac{2}{5} + C, \frac{1}{5} - 2C\}$, where C is arbitrary.

54. If $\mathbf{e}_1$ is taken as a multiple of the first vector, and $\mathbf{e}_2$ a combination of the first two, then

$$\mathbf{e}_1 = \pm\frac{1}{3}\{1, 0, 2, 2\}, \quad \mathbf{e}_2 = \pm\frac{\sqrt{2}}{6}\{2, 3, -2, 1\}, \quad \mathbf{e}_3 = \pm\frac{\sqrt{2}}{6}\{2, 1, 2, -3\}.$$

55. With upper signs in answer to Problem 54: $\mathbf{v} = 2\mathbf{e}_1 + (\sqrt{2}/2)\mathbf{e}_2 + (3\sqrt{2}/2)\mathbf{e}_3$.

57. $$\mathbf{Q} = \begin{bmatrix} 1 & 0 & 0 \\ 0 & \frac{1}{2}\sqrt{2} & \frac{1}{2}\sqrt{2} \\ 0 & \frac{1}{2}\sqrt{2} & -\frac{1}{2}\sqrt{2} \end{bmatrix}, \quad \mathbf{Q}^T\mathbf{A}\,\mathbf{Q} = \begin{bmatrix} 1 & 0 & 0 \\ 0 & 2 & 0 \\ 0 & 0 & 4 \end{bmatrix}.$$

(Columns may be interchanged. Also, the signs of all elements in any column of $\mathbf{Q}$ may be changed.)

58. With $\mathbf{Q}$ as in answer to Problem 57: $\mathbf{A} = x_1'^2 + 2x_2'^2 + 4x_3'^2$.

61. $\mathbf{P} = \begin{bmatrix} 0 & 1 \\ 1 & 0 \end{bmatrix}, \quad \mathbf{Q} = \begin{bmatrix} 1 & 0 & 2 \\ 0 & 1 & 0 \\ 0 & 0 & 1 \end{bmatrix}$.

62. (a) $\lambda_1 = 1: \mathbf{e}_1 = \pm \dfrac{\sqrt{2}}{6}\{1 + i, -4\}$; $\lambda_2 = 10: \mathbf{e}_2 = \pm \dfrac{1}{3}\{2 + 2i, 1\}$.

(c) With upper signs in answer to part (a): $\alpha_1 = -\sqrt{2}/3, \ \alpha_2 = 5/3$.

72. A is not positive definite.

73. With $x_1 = \frac{1}{2}(\alpha_1 + \alpha_2)$, $x_2 = \frac{1}{2}(\alpha_1 - \alpha_2)$, there follows

$$A = \alpha_1^2 + 2\alpha_2^2, \qquad B = \alpha_1^2 + \alpha_2^2.$$

(Other sign combinations are possible in the definitions of x_1 and x_2.)

74. 7; 10.

75. $\mathbf{A}$ is not positive definite.

76. (a) All characteristic numbers must be negative.
(b) Discriminants with odd subscripts must be negative; those with even subscripts must be positive.

77. $-1 < c < \frac{1}{2}$.

80. $\mathbf{x}' = \{\sqrt{2}, 0, 1\}$.

81. $\mathbf{y}' = \begin{bmatrix} \frac{1}{2}\sqrt{2} & \frac{1}{2}\sqrt{2} & 0 \\ \frac{1}{2}\sqrt{2} & -\frac{1}{2}\sqrt{2} & 0 \\ 0 & 0 & 1 \end{bmatrix} \begin{Bmatrix} 3 \\ 2 \\ 1 \end{Bmatrix} = \begin{bmatrix} \frac{3}{2} & -\frac{1}{2} & \sqrt{2} \\ \frac{1}{2} & \frac{1}{2} & 0 \\ 0 & 0 & 1 \end{bmatrix} \begin{Bmatrix} \sqrt{2} \\ 0 \\ 1 \end{Bmatrix} = \begin{Bmatrix} \frac{5}{2}\sqrt{2} \\ \frac{1}{2}\sqrt{2} \\ 1 \end{Bmatrix}$.

85. (a) Characteristic numbers: -1 and 5; corresponding characteristic vectors: multiples of $\{1, -1\}$ and $\{1, 1\}$, respectively.

(b) $\mathbf{B}$ is not positive definite.

(c) $\begin{bmatrix} \dfrac{3^{100} + 1}{2} & \dfrac{3^{100} - 1}{2} \\ \dfrac{3^{100} - 1}{2} & \dfrac{3^{100} + 1}{2} \end{bmatrix}$.

88. (b) $\mathbf{x} = \dfrac{1}{2}\begin{bmatrix} e^{3t} + e^{-t} & e^{3t} - e^{-t} \\ e^{3t} - e^{-t} & e^{3t} + e^{-t} \end{bmatrix}\mathbf{c}$.

90. $\lambda_3 = 8.12$: multiple of $\{0.229, 0.631, 1.000\}$.

91. The zero vector is inevitably obtained after two iterations. The only characteristic number is $\lambda = 0$.

92. Successive approximations oscillate. The characteristic numbers are complex.

93. $\lambda_4 = 8.290$: multiple of $\{0.347, 0.653, 0.879, 1.000\}$;
$\lambda_3 = 1$: multiple of $\{1, 1, 0, -1\}$.

94. $\lambda_1 = 0$: multiple of $\{1, 1, 1, 1\}$;
$\lambda_2 = 0.586$: multiple of $\{-1, -0.414, 0.414, 1\}$;
$\lambda_3 = 2$: multiple of $\{1, -1, -1, 1\}$;
$\lambda_4 = 3.414$: multiple of $\{-1, 2.414, -2.414, 1\}$.

95. $\lambda_1 = 0.308$.

99. $\lambda_1 = 1$: multiple of $\{1, -2\}$; $\lambda_2 = \frac{11}{2}$: multiple of $\{4, 1\}$.

100. $\mathbf{M} = \begin{bmatrix} \frac{1}{3} & \frac{2}{3}\sqrt{2} \\ -\frac{2}{3} & \frac{1}{6}\sqrt{2} \end{bmatrix}.$

(Columns may be interchanged and the signs of all elements in any column may be changed.)

101. With $\mathbf{M}$ as in answer to Problem 100, $\mathbf{x} = \mathbf{M}\,\boldsymbol{\alpha}$ leads to desired forms with $\lambda_1 = 1$ and $\lambda_2 = \frac{11}{2}$.

104. $\lambda_3 = 0.398$.

105. $\lambda_1 = 0.0513$.

106. $\lambda_1 = 0$: $\mathbf{u}_1 = C_1\{1, -2\}$, $\mathbf{u}_1' = C_1'\{1, 1\}$;
$\lambda_2 = 1$: $\mathbf{u}_2 = C_2\{1, -1\}$, $\mathbf{u}_2' = C_2'\{2, 1\}$.

110. (b) $\mathbf{P} = k\begin{bmatrix} 1 & \alpha \\ -1 & \beta \end{bmatrix}$ where $\alpha + \beta = 1$, $k \neq 0$.

112. $\omega_1 = 0.518\sqrt{k/m}$: $\mathbf{x} = C_1\{0.268, 0.732, 1\}$;
$\omega_2 = 1.41\sqrt{k/m}$: $\mathbf{x} = C_2\{1, 1, -1\}$;
$\omega_3 = 1.93\sqrt{k/m}$: $\mathbf{x} = C_3\{3.73, -2.73, 1\}$.

113. $\omega_1 = 0.404\sqrt{k/m}$: $\mathbf{x} = C_1\{0.238, 0.674, 1\}$;
$\omega_2 = 1.30\sqrt{k/m}$: $\mathbf{x} = C_2\{1.79, 2.36, -1\}$;
$\omega_3 = 1.91\sqrt{k/m}$: $\mathbf{x} = C_3\{9.53, -6.33, 1\}$.

114. $\omega_1 = 0$: $\mathbf{x} = C_1\{1, 1, 1\}$;
$\omega_2 = 1.00\sqrt{k/m}$: $\mathbf{x} = C_2\{1, 0, -1\}$;
$\omega_3 = 1.73\sqrt{k/m}$: $\mathbf{x} = C_3\{1, -2, 1\}$.

115. $\omega_1 = 0$: $\mathbf{x} = C_1\{1, 1, 1\}$;
$\omega_2 = 0.85\sqrt{k/m}$: $\mathbf{x} = C_2\{1.56, 0.44, -1\}$;
$\omega_3 = 1.67\sqrt{k/m}$: $\mathbf{x} = C_3\{2.56, -4.56, 1\}$.

136. $A_0 = \dfrac{3}{2}$, $A_k = \dfrac{2}{\mu_k(1 + \cos \mu_k)}$. (This expression can also be written in various other forms.)

137. (b) $y(x) = \dfrac{3x}{2\lambda} + \displaystyle\sum_{k=1}^{\infty} \frac{2 \sin \mu_k x}{\mu_k(\lambda - \mu_k^2)(1 + \cos \mu_k)}$ $(0 < x < 1)$.

Chapter 2

2. Lengths $a/\sqrt{3}$ in the x direction, $b/\sqrt{3}$ in the y direction, and $c/\sqrt{3}$ in the z direction.

3. Squares of semiaxes are $2[(A + C) \pm \sqrt{(A - C)^2 + 4B^2}]^{-1}$.

4. $y = \frac{1}{2}(5x^2 - 3x)$.

8. (b) $I(\epsilon) = 2 + \frac{1}{3}\epsilon^2$.

9. (a) $y'' - y = 0$. (b) $2(xy')' = 1$.
(c) $2y'' + k^2 \sin y = 0$. (d) $(ay')' + by = 0$.

11. $Az + B\theta = C$.

12. $Ar = B \sec (\theta \sin \alpha + C)$.

13. $A \cot \phi = B \cos \theta + C \sin \theta$.

14. $A\theta = B \displaystyle\int \frac{\sqrt{1 + f'^2}}{r\sqrt{r^2 - B^2}}\, dr + C$.

16. (a) $y = x$. (b) $y = \beta x/(1 - \alpha)$ $(\alpha \neq 1)$.
(c) $y = [1 - \beta + (\alpha + \beta)x]/(1 + \alpha)$ $(\alpha \neq -1)$.
(d) $y = -\beta/\alpha$ $(\alpha \neq 0)$, $y = \beta x +$ constant $(\alpha = 0)$.

17. (a) $\Big[(2y' + y)\,\eta\Big]_0^1 = 0$. (b) $\Big[(2xy' - y)\,\eta\Big]_0^1 = 0$.
(c) $\Big[y'\,\eta\Big]_0^1 = 0$. (d) $\Big[ay'\,\eta\Big]_0^1 = 0$.

18. $y = \begin{cases} -2x & (0 \leq x \leq \frac{1}{4}), \\ 2x - 1 & (\frac{1}{4} \leq x \leq 1). \end{cases}$

19. $I(x) = \sqrt{2}$; $I(\cosh x) = \sinh 1$.

20. (a) 2ϵ. (b) 3ϵ.

21. $\Delta I = \frac{3}{2}\epsilon + \frac{17}{15}\epsilon^2$; $\delta I = \frac{3}{2}\epsilon$.

23. $y = 1 - \frac{1}{2}x$.

24. $\dfrac{d}{dx}\left(\dfrac{\partial F}{\partial y'}\right) - \dfrac{\partial F}{\partial y} = 0$; $\left[\dfrac{\partial F}{\partial y'}\right]_{x=b} = \beta$, $\left[\dfrac{\partial F}{\partial y'}\right]_{x=a} = \alpha$.

25. $x = c_1t + c_2,\ y = d_1t + d_2;\ ax + by = c.$

27. Euler equation: $(ay'')'' + (by')' + cy = 0.$

Natural boundary conditions: $\Big[\{(ay'')' + by'\}\,\delta y\Big]_{x_1}^{x_2} = \Big[ay''\,\delta y'\Big]_{x_1}^{x_2} = 0.$

30. Euler equation: $(au_x)_x + (bu_y)_y + cu = 0.$
Natural boundary condition:

$$\left(a\frac{\partial u}{\partial x}\cos\nu + b\frac{\partial u}{\partial y}\sin\nu\right)\delta u = 0 \quad on\ \mathscr{C}.$$

32. Euler equation: $u_{xxxx} + 2u_{xxyy} + u_{yyyy} \equiv \nabla^4 u = 0.$
Natural boundary conditions:

$$\left[\left\{\frac{\partial}{\partial x}\nabla^2 u + (1-\alpha)u_{xyy}\right\}\delta u\right]_{x_1}^{x_2} = \Big[(u_{xx} + \alpha u_{yy})\,\delta u_x\Big]_{x_1}^{x_2} = 0,$$

$$\left[\left\{\frac{\partial}{\partial y}\nabla^2 u + (1-\alpha)u_{xxy}\right\}\delta u\right]_{y_1}^{y_2} = \Big[(u_{yy} + \alpha u_{xx})\,\delta u_y\Big]_{y_1}^{y_2} = 0.$$

38. $(x - \frac{1}{2})^2 + (y + k)^2 = k^2 + \frac{1}{4}$, where k satisfies the equation

$$(4k^2 + 1)\cot^{-1} 2k = 4A + 2k.$$

50. $3\sqrt{2}/8$, along line $y = \frac{7}{4} - x.$

54. $$\omega_n \approx \frac{n^2\pi^2}{l^2}\sqrt{\frac{\int_0^l EI\sin^2\frac{n\pi x}{l}\,dx}{\int_0^l \rho\sin^2\frac{n\pi x}{l}\,dx}}.$$

60. $m_1\ddot{x}_1 = k_2(x_2 - x_1) - k_1x_1,\ m_2\ddot{x}_2 = k_3(x_3 - x_2) - k_2(x_2 - x_1),$
$m_3\ddot{x}_3 = -k_3(x_3 - x_2).$

61. Equation of motion is $\ddot{q}_1 = \frac{1}{23}g.$

70. $H = \frac{1}{2m}\left(p_r^2 + \frac{1}{r^2}p_\theta^2\right) + V(r).\quad \dot{r} = \frac{p_r}{m},\ \dot{\theta} = \frac{p_\theta}{mr^2};$
$\dot{p}_r = p_\theta^2/mr^3 - V'(r),\ \dot{p}_\theta = 0.$

75. (b) $m\ddot{x} + \sum_i k_i\hat{l}_i(\hat{l}_ix + \hat{m}_iy + \hat{n}_iz) = 0,$

$m\ddot{y} + \sum_i k_i\hat{m}_i(\hat{l}_ix + \hat{m}_iy + \hat{n}_iz) = 0,$

$m\ddot{z} + \sum_i k_i\hat{n}_i(\hat{l}_ix + \hat{m}_iy + \hat{n}_iz) = 0.$

80. $\omega^4 - g(A + C)\omega^2 + g^2(AC - B^2) = 0.$

83. (a) $V = \dfrac{Q^2}{2C} - EQ,\ T = \dfrac{1}{2} L\dot{Q}^2,\ F = \dfrac{1}{2} R\dot{Q}^2;$

$$L\ddot{Q} + R\dot{Q} + \frac{Q}{C} - E = 0;\ \omega = \frac{1}{\sqrt{LC}}.$$

(b) $V = \dfrac{Q_1^2}{2C_1} + \dfrac{(Q_1 - Q_2)^2}{2C_{12}} - E_1Q_1,\ T = \frac{1}{2}(L_1\dot{Q}_1^2 + L_2\dot{Q}_2^2),$

$$F = \tfrac{1}{2}R_{12}(\dot{Q}_1 - \dot{Q}_2)^2;$$

$$L_1\ddot{Q}_1 + \frac{1}{C_1} Q_1 + \frac{1}{C_{12}} (Q_1 - Q_2) + R_{12}(\dot{Q}_1 - \dot{Q}_2) - E_1 = 0,$$

$$L_2\ddot{Q}_2 + \frac{1}{C_{12}} (Q_2 - Q_1) + R_{12}(\dot{Q}_2 - \dot{Q}_1) = 0;$$

$$\begin{vmatrix} \left(L_1\omega^2 - \dfrac{1}{C_1} - \dfrac{1}{C_{12}}\right) & \dfrac{1}{C_{12}} \\ \dfrac{1}{C_{12}} & \left(L_2\omega^2 - \dfrac{1}{C_{12}}\right) \end{vmatrix} = 0.$$

88. $\delta \displaystyle\int_a^b \left[\frac{1}{2}(x^2y_1'^2 + x^4y_2'^2 - x^2y_1^2 + 2x^3y_1y_2 - x^4y_2^2) + x\phi_1y_1 + x^3\phi_2y_2\right] dx = 0.$

95. $c_1 = \frac{85}{26} \doteq 3.27,\ c_2 = -\frac{35}{13} \doteq -2.69.$

96. $\lambda_1^{(1)} = 15;\ \lambda_1^{(2)} \doteq 14.42,\ \lambda_2^{(1)} \doteq 63.6.$

Chapter 3

7. $F(x) = f(x) - y_0'A(x) - [y_0 + y_0'(x - a)]B(x).$

10. (a) $y(x) = -\displaystyle\int_0^1 G(x, \xi)f(\xi)\, d\xi + \int_0^1 G(x, \xi)[y(\xi) + \epsilon y^2(\xi)]\, d\xi$

where $G(x, \xi) = \begin{cases} x(1 - \xi) & (x < \xi), \\ \xi(1 - x) & (x > \xi). \end{cases}$

(b) $y(x) = -\displaystyle\int_0^1 G(x, \xi)f(\xi)\, d\xi + \epsilon \int_0^1 G(x, \xi)y^2(\xi)\, d\xi$

where $G(x, \xi) = \begin{cases} \dfrac{\sin x \sin (1 - \xi)}{\sin 1} & (x < \xi), \\ \dfrac{\sin (1 - x) \sin \xi}{\sin 1} & (x > \xi). \end{cases}$

11. $y(x) = \dfrac{1}{6}(x^3 - 3x) + \displaystyle\int_0^1 G(x, \xi)y(\xi)\, d\xi$ where $G(x, \xi) = \begin{cases} x & (x < \xi), \\ \xi & (x > \xi). \end{cases}$

12. $y(x) = \dfrac{1}{6}(x^3 - 3x + 6) + \displaystyle\int_0^1 G(x, \xi)y(\xi)\, d\xi$, where G is as in Problem 11.

14. (b) $y(x) = \lambda \int_0^1 G(x, \xi)\, \xi y(\xi)\, d\xi$, where G is defined in part (a).

39. $\Gamma(x, \xi; \lambda) = \dfrac{4}{4 - \lambda^2}\left[1 + \lambda - \dfrac{3}{2}\lambda(x + \xi) - 3(1 - \lambda)x\xi\right] \quad (\lambda \neq \pm 2).$

40. (b) $F(x) = x: y = \dfrac{2\pi^2\lambda^2}{\pi^2\lambda^2 - 1}\sin x + \dfrac{2\pi\lambda}{\pi^2\lambda^2 - 1}\cos x + x.$

$F(x) = 1: y = 1.$

(c) $F(x) = 1: y = 1 + c(\cos x + \sin x).$

(d) $F(x) = A\cos x + B\sin x.$

41. (a) $\lambda_1 = \dfrac{1}{\pi}, y_1(x) = \cos x;\ \lambda_2 = -\dfrac{1}{\pi}, y_2(x) = \sin x.$

(b) $\Gamma(x, \xi; \lambda) = \dfrac{\cos(x + \xi) + \pi\lambda\cos(x - \xi)}{1 - \pi^2\lambda^2} \quad \left(\lambda \neq \pm\dfrac{1}{\pi}\right).$

(c) $y = \dfrac{\sin x}{1 + \pi\lambda}$ if $\lambda \neq \pm\dfrac{1}{\pi}$; $y = \dfrac{1}{2}\sin x + A\cos x$ (A arbitrary) if $\lambda = \dfrac{1}{\pi}$;

no solution if $\lambda = -\dfrac{1}{\pi}$.

42. $y = 1 + \dfrac{2\lambda\sin\pi\omega}{(1 - 2\pi\lambda)\omega}e^{i\omega x} \quad \left(\lambda \neq \dfrac{1}{2\pi}, \omega \neq 0\right);$

$y = 1 + \dfrac{2\pi\lambda}{1 - 2\pi\lambda}e^{i\omega x} \quad \left(\lambda \neq \dfrac{1}{2\pi}, \omega = 0\right)$; no solution if $\lambda = \dfrac{1}{2\pi}$.

43. $y \approx 0.363x + x^2 - 0.039x^3.$

49. (c) $1 - 3x\xi = \dfrac{\sqrt{3}(1 - x)\sqrt{3}(1 - \xi)}{2} + \dfrac{(1 - 3x)(1 - 3\xi)}{-2}.$

51. $y = 1 + \dfrac{12\lambda x + 6\lambda + \lambda^2}{12 - 12\lambda - \lambda^2} \quad (\lambda \neq -6 \pm 4\sqrt{3}).$

Estimated convergence limit: $|\lambda| < \sqrt{42}/7 \doteq 0.926.$
True convergence limit: $|\lambda| < 4\sqrt{3} - 6 \doteq 0.928.$

52. $y = \dfrac{\sin x}{1 + \pi\lambda} \quad \left(\lambda \neq \pm\dfrac{1}{\pi}\right).$

Estimated convergence limit: $|\lambda| < \dfrac{1}{\sqrt{2}\,\pi}.$

True convergence limit: $|\lambda| < \dfrac{1}{\pi}.$

54. (a) $y^{(3)}(x) = 1 + \frac{3}{2}x^2 + \frac{7}{8}x^4 + \frac{77}{240}x^6.$

56. $\Gamma(x, \xi; \lambda) = x\xi(1 + \frac{1}{3}\lambda + \frac{1}{9}\lambda^2 + \cdots).$

57. $\Gamma(x, \xi; \lambda) = (x + \xi) + \lambda[\frac{1}{3} + \frac{1}{2}(x + \xi) + x\xi] + \lambda^2[\frac{1}{3} + \frac{7}{12}(x + \xi) + x\xi] + \cdots.$

58. $\Gamma(x, \xi; \lambda) = [1 + \pi^2\lambda^2 + \pi^4\lambda^4 + \cdots] \cos(x + \xi) + \pi\lambda[1 + \pi^2\lambda^2 + \pi^4\lambda^4 + \cdots] \cos(x - \xi).$

59. When $x < \xi$:

$$K_2(x, \xi) = \int_0^x (x - \xi_1)(\xi - \xi_1)\, d\xi_1 + \int_x^\xi (\xi_1 - x)(\xi - \xi_1)\, d\xi_1 + \int_\xi^1 (\xi_1 - x)(\xi_1 - \xi)\, d\xi_1$$

$$= \frac{1}{3} - \frac{1}{2}(x + \xi) + x\xi + \frac{1}{3}(\xi - x)^3.$$

60. $$K_2(x, \xi) = \begin{cases} (1 + \xi - x)e^{x-\xi} - \frac{1}{2}[e^{-(x+\xi)} + e^{x+\xi-2a}] & (x < \xi), \\ (1 + x - \xi)e^{\xi-x} - \frac{1}{2}[e^{-(x+\xi)} + e^{x+\xi-2a}] & (x > \xi). \end{cases}$$

63. $\Gamma(x, \xi; \lambda) = \dfrac{3x\xi}{3 - \lambda}.$

64. $\Gamma(x, \xi; \lambda) = \dfrac{12(x + \xi) + \lambda[4 - 6(x + \xi) + 12x\xi]}{12 - 12\lambda - \lambda^2}.$

98. Exact solution is $y(x) = 1$.

100. $\lambda_1 \approx 1.24$; $y(\frac{1}{2})/y(0) \approx 0.801$, $y(1)/y(0) \approx 0.656$.

105. $y(\theta) \approx 0.541 \sin\theta + 0.031 \sin 3\theta + 0.007 \sin 5\theta$.

107. (b) $y(x) \approx 0.131 + 1.012x$.

116. $\lambda_1 \approx 1.06$.

Index*

* **Boldface** figures in parentheses refer to problem numbers.

P

Q

R

S

A CATALOG OF SELECTED
DOVER BOOKS
IN SCIENCE AND MATHEMATICS

A CATALOG OF SELECTED

DOVER BOOKS

IN SCIENCE AND MATHEMATICS

Astronomy

BURNHAM'S CELESTIAL HANDBOOK, Robert Burnham, Jr. Thorough guide to the stars beyond our solar system. Exhaustive treatment. Alphabetical by constellation: Andromeda to Cetus in Vol. 1; Chamaeleon to Orion in Vol. 2; and Pavo to Vulpecula in Vol. 3. Hundreds of illustrations. Index in Vol. 3. 2,000pp. 6⅛ x 9¼.
23567-X, 23568-8, 23673-0 Three-vol. set

THE EXTRATERRESTRIAL LIFE DEBATE, 1750–1900, Michael J. Crowe. First detailed, scholarly study in English of the many ideas that developed from 1750 to 1900 regarding the existence of intelligent extraterrestrial life. Examines ideas of Kant, Herschel, Voltaire, Percival Lowell, many other scientists and thinkers. 16 illustrations. 704pp. 5⅜ x 8½. 40675-X

A HISTORY OF ASTRONOMY, A. Pannekoek. Well-balanced, carefully reasoned study covers such topics as Ptolemaic theory, work of Copernicus, Kepler, Newton, Eddington's work on stars, much more. Illustrated. References. 521pp. 5⅜ x 8½.
65994-1

AMATEUR ASTRONOMER'S HANDBOOK, J. B. Sidgwick. Timeless, comprehensive coverage of telescopes, mirrors, lenses, mountings, telescope drives, micrometers, spectroscopes, more. 189 illustrations. 576pp. 5⅝ x 8¼. (Available in U.S. only.)
24034-7

STARS AND RELATIVITY, Ya. B. Zel'dovich and I. D. Novikov. Vol. 1 of *Relativistic Astrophysics* by famed Russian scientists. General relativity, properties of matter under astrophysical conditions, stars, and stellar systems. Deep physical insights, clear presentation. 1971 edition. References. 544pp. 5⅝ x 8¼. 69424-0

Chemistry

CHEMICAL MAGIC, Leonard A. Ford. Second Edition, Revised by E. Winston Grundmeier. Over 100 unusual stunts demonstrating cold fire, dust explosions, much more. Text explains scientific principles and stresses safety precautions. 128pp. 5⅜ x 8½. 67628-5

THE DEVELOPMENT OF MODERN CHEMISTRY, Aaron J. Ihde. Authoritative history of chemistry from ancient Greek theory to 20th-century innovation. Covers major chemists and their discoveries. 209 illustrations. 14 tables. Bibliographies. Indices. Appendices. 851pp. 5⅜ x 8½. 64235-6

CATALYSIS IN CHEMISTRY AND ENZYMOLOGY, William P. Jencks. Exceptionally clear coverage of mechanisms for catalysis, forces in aqueous solution, carbonyl- and acyl-group reactions, practical kinetics, more. 864pp. 5⅜ x 8½.
65460-5

THE HISTORICAL BACKGROUND OF CHEMISTRY, Henry M. Leicester. Evolution of ideas, not individual biography. Concentrates on formulation of a coherent set of chemical laws. 260pp. 5⅜ x 8½. 61053-5

A SHORT HISTORY OF CHEMISTRY, J. R. Partington. Classic exposition explores origins of chemistry, alchemy, early medical chemistry, nature of atmosphere, theory of valency, laws and structure of atomic theory, much more. 428pp. 5⅜ x 8½. (Available in U.S. only.) 65977-1

GENERAL CHEMISTRY, Linus Pauling. Revised 3rd edition of classic first-year text by Nobel laureate. Atomic and molecular structure, quantum mechanics, statistical mechanics, thermodynamics correlated with descriptive chemistry. Problems. 992pp. 5⅜ x 8½. 65622-5

Engineering

DE RE METALLICA, Georgius Agricola. The famous Hoover translation of greatest treatise on technological chemistry, engineering, geology, mining of early modern times (1556). All 289 original woodcuts. 638pp. 6¾ x 11. 60006-8

FUNDAMENTALS OF ASTRODYNAMICS, Roger Bate et al. Modern approach developed by U.S. Air Force Academy. Designed as a first course. Problems, exercises. Numerous illustrations. 455pp. 5⅜ x 8½. 60061-0

DYNAMICS OF FLUIDS IN POROUS MEDIA, Jacob Bear. For advanced students of ground water hydrology, soil mechanics and physics, drainage and irrigation engineering and more. 335 illustrations. Exercises, with answers. 784pp. 6⅛ x 9¼. 65675-6

ANALYTICAL MECHANICS OF GEARS, Earle Buckingham. Indispensable reference for modern gear manufacture covers conjugate gear-tooth action, gear-tooth profiles of various gears, many other topics. 263 figures. 102 tables. 546pp. 5⅜ x 8½. 65712-4

MECHANICS, J. P. Den Hartog. A classic introductory text or refresher. Hundreds of applications and design problems illuminate fundamentals of trusses, loaded beams and cables, etc. 334 answered problems. 462pp. 5⅜ x 8½. 60754-2

MECHANICAL VIBRATIONS, J. P. Den Hartog. Classic textbook offers lucid explanations and illustrative models, applying theories of vibrations to a variety of practical industrial engineering problems. Numerous figures. 233 problems, solutions. Appendix. Index. Preface. 436pp. 5⅜ x 8½. 64785-4

STRENGTH OF MATERIALS, J. P. Den Hartog. Full, clear treatment of basic material (tension, torsion, bending, etc.) plus advanced material on engineering methods, applications. 350 answered problems. 323pp. 5⅜ x 8½. 60755-0

A HISTORY OF MECHANICS, René Dugas. Monumental study of mechanical principles from antiquity to quantum mechanics. Contributions of ancient Greeks, Galileo, Leonardo, Kepler, Lagrange, many others. 671pp. 5⅜ x 8½. 65632-2

Math–Geometry and Topology

ELEMENTARY CONCEPTS OF TOPOLOGY, Paul Alexandroff. Elegant, intuitive approach to topology from set-theoretic topology to Betti groups; how concepts of topology are useful in math and physics. 25 figures. 57pp. 5⅜ x 8½. 60747-X

COMBINATORIAL TOPOLOGY, P. S. Alexandrov. Clearly written, well-organized, three-part text begins by dealing with certain classic problems without using the formal techniques of homology theory and advances to the central concept, the Betti groups. Numerous detailed examples. 654pp. 5⅜ x 8½. 40179-0

EXPERIMENTS IN TOPOLOGY, Stephen Barr. Classic, lively explanation of one of the byways of mathematics. Klein bottles, Moebius strips, projective planes, map coloring, problem of the Koenigsberg bridges, much more, described with clarity and wit. 43 figures. 210pp. 5⅜ x 8½. 25933-1

CONFORMAL MAPPING ON RIEMANN SURFACES, Harvey Cohn. Lucid, insightful book presents ideal coverage of subject. 334 exercises make book perfect for self-study. 55 figures. 352pp. 5⅝ x 8¼. 64025-6

THE GEOMETRY OF RENÉ DESCARTES, René Descartes. The great work founded analytical geometry. Original French text, Descartes's own diagrams, together with definitive Smith-Latham translation. 244pp. 5⅜ x 8½. 60068-8

THE THIRTEEN BOOKS OF EUCLID'S ELEMENTS, translated with introduction and commentary by Sir Thomas L. Heath. Definitive edition. Textual and linguistic notes, mathematical analysis. 2,500 years of critical commentary. Unabridged. 1,414pp. 5⅜ x 8½. Three-vol. set.
Vol. I: 60088-2 Vol. II: 60089-0 Vol. III: 60090-4

GEOMETRY OF COMPLEX NUMBERS, Hans Schwerdtfeger. Illuminating, widely praised book on analytic geometry of circles, the Moebius transformation, and two-dimensional non-Euclidean geometries. 200pp. 5⅝ x 8¼. 63830-8

DIFFERENTIAL GEOMETRY, Heinrich W. Guggenheimer. Local differential geometry as an application of advanced calculus and linear algebra. Curvature, transformation groups, surfaces, more. Exercises. 62 figures. 378pp. 5⅜ x 8½. 63433-7

CURVATURE AND HOMOLOGY: Enlarged Edition, Samuel I. Goldberg. Revised edition examines topology of differentiable manifolds; curvature, homology of Riemannian manifolds; compact Lie groups; complex manifolds; curvature, homology of Kaehler manifolds. New Preface. Four new appendixes. 416pp. 5⅜ x 8½. 40207-X

TOPOLOGY, John G. Hocking and Gail S. Young. Superb one-year course in classical topology. Topological spaces and functions, point-set topology, much more. Examples and problems. Bibliography. Index. 384pp. 5⅝ x 8¼. 65676-4

Physics

OPTICAL RESONANCE AND TWO-LEVEL ATOMS, L. Allen and J. H. Eberly. Clear, comprehensive introduction to basic principles behind all quantum optical resonance phenomena. 53 illustrations. Preface. Index. 256pp. 5⅜ x 8½. 65533-4

ULTRASONIC ABSORPTION: An Introduction to the Theory of Sound Absorption and Dispersion in Gases, Liquids and Solids, A. B. Bhatia. Standard reference in the field provides a clear, systematically organized introductory review of fundamental concepts for advanced graduate students, research workers. Numerous diagrams. Bibliography. 440pp. 5⅜ x 8½. 64917-2

QUANTUM THEORY, David Bohm. This advanced undergraduate-level text presents the quantum theory in terms of qualitative and imaginative concepts, followed by specific applications worked out in mathematical detail. Preface. Index. 655pp. 5⅜ x 8½. 65969-0

ATOMIC PHYSICS (8th edition), Max Born. Nobel laureate's lucid treatment of kinetic theory of gases, elementary particles, nuclear atom, wave-corpuscles, atomic structure and spectral lines, much more. Over 40 appendices, bibliography. 495pp. 5⅜ x 8½. 65984-4

AN INTRODUCTION TO HAMILTONIAN OPTICS, H. A. Buchdahl. Detailed account of the Hamiltonian treatment of aberration theory in geometrical optics. Many classes of optical systems defined in terms of the symmetries they possess. Problems with detailed solutions. 1970 edition. xv + 360pp. 5⅜ x 8½. 67597-1

THIRTY YEARS THAT SHOOK PHYSICS: The Story of Quantum Theory, George Gamow. Lucid, accessible introduction to influential theory of energy and matter. Careful explanations of Dirac's anti-particles, Bohr's model of the atom, much more. 12 plates. Numerous drawings. 240pp. 5⅜ x 8½. 24895-X

ELECTRONIC STRUCTURE AND THE PROPERTIES OF SOLIDS: The Physics of the Chemical Bond, Walter A. Harrison. Innovative text offers basic understanding of the electronic structure of covalent and ionic solids, simple metals, transition metals and their compounds. Problems. 1980 edition. 582pp. 6⅛ x 9¼. 66021-4

HYDRODYNAMIC AND HYDROMAGNETIC STABILITY, S. Chandrasekhar. Lucid examination of the Rayleigh-Benard problem; clear coverage of the theory of instabilities causing convection. 704pp. 5⅜ x 8¼. 64071-X

INVESTIGATIONS ON THE THEORY OF THE BROWNIAN MOVEMENT, Albert Einstein. Five papers (1905–8) investigating dynamics of Brownian motion and evolving elementary theory. Notes by R. Fürth. 122pp. 5⅜ x 8½. 60304-0

THE PHYSICS OF WAVES, William C. Elmore and Mark A. Heald. Unique overview of classical wave theory. Acoustics, optics, electromagnetic radiation, more. Ideal as classroom text or for self-study. Problems. 477pp. 5⅜ x 8½. 64926-1

CATALOG OF DOVER BOOKS

METHODS OF THERMODYNAMICS, Howard Reiss. Outstanding text focuses on physical technique of thermodynamics, typical problem areas of understanding, and significance and use of thermodynamic potential. 1965 edition. 238pp. 5⅜ x 8½. 69445-3

TENSOR ANALYSIS FOR PHYSICISTS, J. A. Schouten. Concise exposition of the mathematical basis of tensor analysis, integrated with well-chosen physical examples of the theory. Exercises. Index. Bibliography. 289pp. 5⅜ x 8½. 65582-2

RELATIVITY IN ILLUSTRATIONS, Jacob T. Schwartz. Clear nontechnical treatment makes relativity more accessible than ever before. Over 60 drawings illustrate concepts more clearly than text alone. Only high school geometry needed. Bibliography. 128pp. 6⅛ x 9¼. 25965-X

THE ELECTROMAGNETIC FIELD, Albert Shadowitz. Comprehensive undergraduate text covers basics of electric and magnetic fields, builds up to electromagnetic theory. Also related topics, including relativity. Over 900 problems. 768pp. 5⅝ x 8¼. 65660-8

GREAT EXPERIMENTS IN PHYSICS: Firsthand Accounts from Galileo to Einstein, edited by Morris H. Shamos. 25 crucial discoveries: Newton's laws of motion, Chadwick's study of the neutron, Hertz on electromagnetic waves, more. Original accounts clearly annotated. 370pp. 5⅜ x 8½. 25346-5

RELATIVITY, THERMODYNAMICS AND COSMOLOGY, Richard C. Tolman. Landmark study extends thermodynamics to special, general relativity; also applications of relativistic mechanics, thermodynamics to cosmological models. 501pp. 5⅜ x 8½. 65383-8

LIGHT SCATTERING BY SMALL PARTICLES, H. C. van de Hulst. Comprehensive treatment including full range of useful approximation methods for researchers in chemistry, meteorology and astronomy. 44 illustrations. 470pp. 5⅜ x 8½. 64228-3

STATISTICAL PHYSICS, Gregory H. Wannier. Classic text combines thermodynamics, statistical mechanics and kinetic theory in one unified presentation of thermal physics. Problems with solutions. Bibliography. 532pp. 5⅜ x 8½. 65401-X